Classical Mechanics

A professor–student collaboration

Classical Mechanics

A professor-student collaboration

Edited by
Mario Campanelli
Department of Physics and Astronomy, University College of London, London, UK

With contributions from:
Antonio d'Alfonso del Sordo
Camilla Tacconis
Enrico Caprioglio
Lodovico Scarpa
Muhammad Tayyab Shabbir
Sheila María Pérez García

IOP Publishing, Bristol, UK

ISBN 978-0-7503-2690-2 (ebook)
ISBN 978-0-7503-2688-9 (print)
ISBN 978-0-7503-2691-9 (myPrint)
ISBN 978-0-7503-2689-6 (mobi)

DOI 10.1088/978-0-7503-2690-2

Version: 20200801

IOP ebooks

British Library Cataloguing-in-Publication Data: A catalogue record for this book is available from the British Library.

Published by IOP Publishing, wholly owned by The Institute of Physics, London

IOP Publishing, Temple Circus, Temple Way, Bristol, BS1 6HG, UK

US Office: IOP Publishing, Inc., 190 North Independence Mall West, Suite 601, Philadelphia, PA 19106, USA

Physics is hard, we do what we can.

—Reddit user

It's trivial, no?

—Professor Mario Campanelli

Contents

Preface

Why study it?

As children, many amongst the readers of this book will have gone for a ride on a rocking horse or, perhaps, will have had fun on a merry-go-round. There is no doubt that every child looks at a roller coaster with some sort of excitement. Many of you will have wondered at some point: how is this all possible? We hope this book can help you find an answer to this question and many more. It is, therefore, our aim, pleasure and duty to guide you through this process and reignite that juvenile sense of fascination for such phenomena as those mentioned above, which are all successfully described by the laws of physics.

The importance of classical mechanics is undeniable, as it is the foundation for most of physics. Remarkably, with just a few equations, it enables us to describe the motion of a variety of physical systems, such as a ball flying through the air, or planets and galaxies.

Despite being the oldest branch of physics, the term *classical mechanics* is relatively new. It was born only in the last century to distinguish the theory developed prior to 1900 to describe objects of everyday sizes and speeds from the new theories of special and general relativity and quantum mechanics.

The theory, based on **Newton's laws of motion** (and therefore often called *Newtonian mechanics*), provides a very good approximation of the motion of macroscopic bodies under most of the conditions we encounter in our everyday life.

The following cases fall outside the limits of applicability of classical mechanics:

- *microscopic systems*, e.g. atoms, molecules, nuclei, (i.e. at distances smaller than $\hbar/mv$) described by the language of **quantum mechanics**.
- *particles travelling at speeds close to the speed of light*, i.e. $c = 299\ 792\ 458$ m s^{-1}—for which we need to use **relativistic mechanics**.

However, for sizes larger than atoms and smaller than the solar system, and ordinary velocities, classical mechanics is an incredibly successful approximation!

Classical mechanics is an excellent example of how physics and mathematics are tied together—Newton developed calculus in order to solve problems in mechanics. So it is the perfect setting to learn how to apply maths to physical problems, and to learn problem-solving techniques.

Like most areas of physics, the laws of classical mechanics are expressed in terms of differential equations, often very similar to those found in very different areas, so mathematical techniques developed to solve mechanics problems have found their applications many other areas of physics.

Have a look at Newton's second law written as a second order differential equation (you will become more confident with it later on in the course[1]):

[1] To be completely honest, we wrote it in a slightly different form from the one you will find in this book. This form has been chosen to emphasise the similarity with the second equation!

$$\frac{d^2x^i}{dt^2} = \frac{1}{m}F^i(x, v, t)$$

where (x^i, F^i) for $i = 1, 2, 3$ are the components of the position and force vector. It has an uncanny similarity with the geodesic equation of general relativity[2],

$$\frac{d^2x^\alpha}{d\tau^2} = -\Gamma^\alpha_{\mu\nu}u^\mu u^\nu,$$

where $\alpha = 0, 1, 2, 3$ and the right-hand side depends on $\Gamma^\alpha_{\mu\nu}(x^\beta)$ and the components of the velocity vector, u^μ.

The subject is usually divided into:

1. **statics**—systems at rest and in *equilibrium*.
2. **kinematics**—description of systems in motion, which can also be quite complex.
3. **dynamics**—study of the actions (forces) responsible for the motion.

Problem solving

Classical mechanics, like all of physics, is also a great way to teach students how to solve real-life problems using mathematical tools, and it is usually the first occasion a student is confronted with such a challenge. The best way to learn how to solve problems is of course working on a lot of them, but also reading them and their solution will help; this textbook has several problems with worked out solutions. There are several strategies to deal with a new problem. The most common are:

- draw a diagram;
- write down what you know, and where you are aiming at;
- solve it symbolically first;
- always check units and dimensions;
- check limits and special cases;
- if computing numerical answers, check orders of magnitude.

Often solving a problem requires writing the relation between physical expressions like the equation of motion, the speed and the acceleration as a function of time, and solving a differential equation. The general solution of this equation will depend on parameters that have to be fixed using initial or boundary conditions.

Modelling assumption

In order to describe the complexity of Nature in mathematical form, many explicit and implicit assumptions are needed. For instance, space is assumed to be infinite and flat, and time to flow in a constant, identical way. The first part of this book will deal with the motion of bodies, just considering their position and not their internal structure or movement. For this, a very useful modelling assumption to simplify a problem in mechanics is that of a **particle**.

[2] Don't panic, this is not going to be in the course!

Definition: Particle
A **particle** is an object whose dimensions may be neglected when describing its motion. It is often referred to as a *point-particle* or *point-mass* as its mass can be considered to be concentrated at a single point.

Other assumptions will be described over the course of the book.

Acknowledgments

This book is based on the lectures given by Professor Mario Campanelli at University College London during the first term of the academic year 2017–18. The contributors are among the students who attended the course and they contacted the lecturer with the idea of the book, for which they deserve the whole credit.

The book contains elements from the old notes and exams of this course. We acknowledge the previous lecturers, in particular Professor Andrew Fisher and Professor Mark Lancaster, for part of the material contained in this book.

The contributors would also like to thank Dr Mark Fuller, as he helped them to organise an introductory course in classical mechanics for which a simplified version of the material covered in the book was employed. The course aimed at introducing year 12 and 13 students to university physics and was taught by the contributors. The four-week course covered the first four chapters of the book and the last lecture introduced some concepts of quantum mechanics, at the students' request. This course provided the contributors with some useful guidance on how to convey the ideas of classical mechanics in a simple yet inspiring way.

Last but not least, the contributors would like to thank Professor Mario Campanelli for embracing the idea without hesitation, and for his support and availability throughout the composition of the book.

Editor biography

Professor Mario Campanelli

Mario Campanelli is an experimental particle physicist with decades of experience on the world's highest energy colliders. He graduated in Rome in 1995 working on the development of a calorimeter for the L3 experiment at CERN. Then he earned his PhD in ETH Zurich on measurement of the properties of W bosons, always with L3. He was researcher at ETHZ on the ICARUS experiment, and in Geneva University on the CDF experiment at Fermilab, near Chicago, Illinois. In 2007 he joined the faculty of University College London to work on the ATLAS experiment at CERN, where he mainly studies the properties of strong interactions. In 2018 he was nominated professor at UCL. He has 25 years of experience in teaching graduate- and undergraduate-level courses.

Short biographies of contributors

Camilla Tacconis

Camilla will soon graduate from University College London with a BSc in physics. She has always nurtured a passion for physics which, combined with the desire to share her knowledge and educate younger people, naturally drove her to launch this project. Coupling her background in both the Italian and English academic systems she has put together the merits of both to make her contributions to the book as helpful as possible for future students. She aspires to become a researcher in solid state physics, and one day use this textbook to lecture on classical mechanics at University.

Lodovico Scarpa

Lodovico is a theoretical physics student who dreams of becoming a researcher. His interests lie in the fields of (loop) quantum gravity and quantum information. He was fascinated by classical mechanics when he first learned it and wanted to transmit the ideas behind the equations to other students. He thinks gyroscopes are really cool! Having moved from the mesmerising city of Venice to study at University College London helped him understand how to make this textbook accessible to students from different educational backgrounds than the UK's.

Antonio d'Alfonso del Sordo

Antonio d'Alfonso del Sordo is a Mathematics and Physics student in his third year at UCL. Always appreciative of the beauty of mathematics describing physical phenomena, he decided to share his passion by contributing to this publication. Antonio is also a CELTA qualified English language teacher, with three-year-long experience in language summer schools across the UK, and recently took a position as Assistant Director of Studies. He has also volunteered as a mathematics and physics tutor in London high schools through UCL, and worked as a student ambassador for his department, as well as mentor for year 1 students. These experiences have made him conscious of the needs of learners from a wide range of backgrounds.

Enrico Caprioglio

Enrico Caprioglio is a third year theoretical physics student, graduating in 2020. He has great experience in science communication, having worked at the London International Youth Science Forum for three years. When not studying physics he is either making coffee or talking to someone about physics, as any Italian physicist would do.

Sheila María Pérez García

Being a Joint Honours Mathematics and Physics student currently in her third year of study, Sheila provides a fresh perspective on a very well-known topic in physics. Her focus was to make the content as accessible as possible, no matter which education system people come from. With a strong mathematical background, she contributed to establish a strong foundation in mathematics, which then propels the book forward to tackle more complex physical concepts.

Muhammad Tayyab Shabbir

Tayyab is a third year undergraduate student at University College London who is passionate about physics and making it accessible. Tayyab has a background in the British and Pakistani education system and based on this experience makes his contributions to the book to make it accessible to students from all educational backgrounds. Tayyab believes that many of life's big questions can be answered using physics and that most if not everything around us is constrained by the laws of physics if you look deep enough.

IOP Publishing

Classical Mechanics
A professor-student collaboration
Mario Campanelli

Chapter 1

Mathematical preliminaries

1.0 Introduction

Galileo once said, 'And, believe me, if I were again beginning my studies, I should follow the advice of Plato and start with mathematics.'

In this chapter, we will review some of the main mathematical tools which will be employed throughout the course. You may be covering these topics in more depth in a course dedicated to mathematical methods. The purpose of this chapter is merely to give some grounding on which you can build your knowledge of classical mechanics.

Some of you may find this section redundant, some of you may find it helpful to refresh some previous knowledge forgotten over the summer, some of you may find it essential.

This chapter cannot replace the many mathematical methods textbooks on the market; yet we made a great effort to condense some essential knowledge before diving into classical mechanics.

1.1 Vectors

In mechanics, numerical quantities mainly appear in two different forms:

Definition 1.1.1: Scalar

A **scalar** quantity is only defined by a single number indicating its magnitude, e.g. mass, energy, speed. Its value is, in fact, a *scalar*, i.e. a real number.

doi:10.1088/978-0-7503-2690-2ch1

Definition 1.1.2: Vector
A **vector** quantity has both a magnitude and a direction, e.g. position, displacement, velocity, acceleration, force acted upon or by a particle.

Vectors can be visualised as **directed line segments**, where:

- the length of the line represents the magnitude of the vector;
- the direction of the line shows the direction of the vector;
- the point of application of the vector (its starting point) does not usually have a specific meaning; if it is relevant for the description of the physics, a reference frame must be defined, and another vector is needed to join the origin with this starting point.

Remark 1.1. Throughout this course, vectors will be denoted by boldface characters, e.g. $\boldsymbol{a}$; in hand-written text, we underline the character to indicate that it is in boldface.

Let A and B be any two points in space. Then, we denote the vector from A to B by $\overrightarrow{AB} \equiv \boldsymbol{a}$, as shown in figure 1.1.

To describe the position of a vector, we need a **reference frame**, namely a set of axes, with an origin, usually defined as point O.

Given any point P in space, we call the vector $\overrightarrow{OP}$, from the origin O to P, the **position vector** of P, $\boldsymbol{r} \equiv \overrightarrow{OP}$. Since any vector can be thought of as a position vector (by sliding it so that its starting point, or tail, is at the origin), there is a one-to-one correspondence between points and vectors.

A Cartesian coordinate system has perpendicular axes, that can be defined by unit vectors (vectors whose magnitude equals unity) $\hat{\boldsymbol{i}}$, $\hat{\boldsymbol{j}}$ and $\hat{\boldsymbol{k}}$ which point respectively along the positive x-, y- and z-axis.

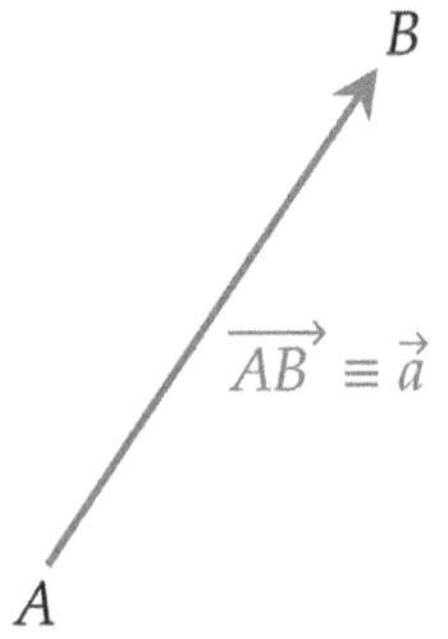

Figure 1.1. Vector $\overrightarrow{AB}$

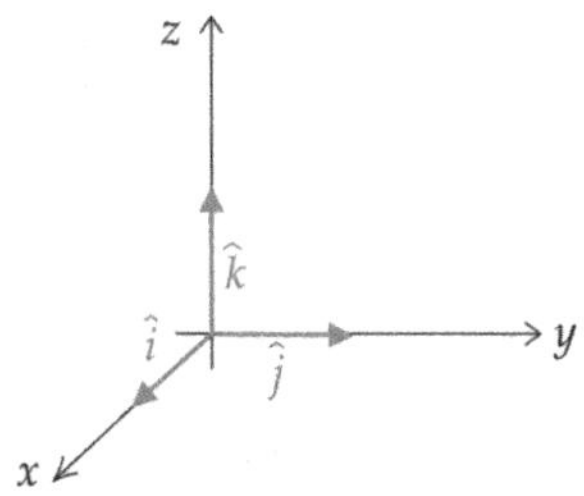

Figure 1.2. Cartesian unit vectors.

A vector $\boldsymbol{r}$ may then be written as the sum of three vectors, each parallel to one of the unit vectors:

$$\boldsymbol{r} = x\hat{\boldsymbol{i}} + y\hat{\boldsymbol{j}} + z\hat{\boldsymbol{k}}. \tag{1.1}$$

Alternatively, considering the unit vectors as known, we can also represent the same vector in the form of a column:

$$\boldsymbol{r} = \begin{pmatrix} x \\ y \\ z \end{pmatrix}. \tag{1.2}$$

Following the same convention, unit vectors may thus be represented as:

$$\hat{\boldsymbol{i}} = \begin{pmatrix} 1 \\ 0 \\ 0 \end{pmatrix}; \quad \hat{\boldsymbol{j}} = \begin{pmatrix} 0 \\ 1 \\ 0 \end{pmatrix}; \quad \hat{\boldsymbol{k}} = \begin{pmatrix} 0 \\ 0 \\ 1 \end{pmatrix}. \tag{1.3}$$

These are called the **Cartesian unit vectors**, which can be represented as shown in figure 1.2.

> **Remark 1.2: Cartesian unit vectors**
> **Cartesian unit vectors** can also be referred to as the *versors* of the coordinate system, and they form a set of mutually orthogonal unit vectors (i.e. a *standard basis* in linear algebra).
> Notations such as ($\hat{x}$, $\hat{y}$, $\hat{z}$), ($\hat{x}_1$, $\hat{x}_2$, $\hat{x}_3$), ($\hat{e}_x$, $\hat{e}_y$, $\hat{e}_z$), or ($\hat{e}_1$, $\hat{e}_2$, $\hat{e}_3$), with or without '*hat*' may also be used, in particular when the ($\hat{\boldsymbol{i}}$, $\hat{\boldsymbol{j}}$, $\hat{\boldsymbol{k}}$)-notation may lead to a confusion with other quantities.

Example 1.1. Consider Newton's second law (details later in the course):

$$\boldsymbol{F} = m\boldsymbol{a}$$

This is a vector equation: not only are the magnitudes on both sides of the equation equal, but also their directions. Equivalently, every component has to be equal on both sides of the equation.

$$F = F_x\hat{i} + F_y\hat{j} + F_z\hat{k}$$
$$a = a_x\hat{i} + a_y\hat{j} + a_z\hat{k}.$$

Thus,

$$F_x = ma_x$$
$$F_y = ma_y$$
$$F_z = ma_z$$

◆

1.1.1 Linear operations

Vector addition $\boldsymbol{a} + \boldsymbol{b}$

Definition 1.1.3: Addition of vectors

Vector addition is the operation of adding two or more vectors together into a **vector sum** (figure 1.3).

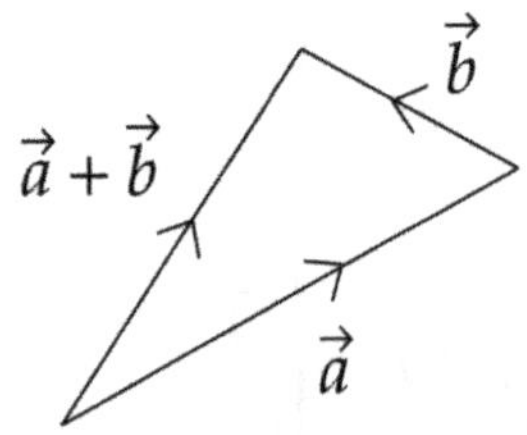

Figure 1.3. Addition of vectors.

The so-called **parallelogram law** gives the rule for summing two or more vectors. For any two vectors $\boldsymbol{a}$ and $\boldsymbol{b}$, the **vector sum** $\boldsymbol{a} + \boldsymbol{b}$ is obtained by placing them head to tail and drawing the vector from the free tail to the free head.

In Cartesian coordinates, vector addition can be performed simply by adding the corresponding components of the vectors. Thus, if $\boldsymbol{a} = (a_1, a_2, \ldots, a_n)$ and $\boldsymbol{b} = (b_1, b_2, \ldots, b_n)$,

$$\boldsymbol{a} + \boldsymbol{b} = \begin{pmatrix} a_1 + b_1 \\ a_2 + b_2 \\ \cdots \\ a_n + b_n \end{pmatrix}. \tag{1.4}$$

To sum more than two vectors, first take the sum of the first two, then add the third, the fourth, etc.

Law 1.1.1: Laws of algebra for the vector sum

(i) Commutative law: $\boldsymbol{b} + \boldsymbol{a} = \boldsymbol{a} + \boldsymbol{b}$.
(ii) Associative law: $\boldsymbol{a} + (\boldsymbol{b} + \boldsymbol{c}) = (\boldsymbol{a} + \boldsymbol{b}) + \boldsymbol{c}$.

Definition 1.1.4: Negative of a vector
Let $\boldsymbol{b}$ be any vector. Then, the vector with the same magnitude as $\boldsymbol{b}$ and the **opposite** direction is called the **negative** of $\boldsymbol{b}$ and is written as $-\boldsymbol{b}$.

Definition 1.1.5: Subtraction of two vectors
The subtraction of two vectors is very similar to their addition (figure 1.4):

$$\boldsymbol{a} - \boldsymbol{b} = \boldsymbol{a} + (-\boldsymbol{b}). \tag{1.5}$$

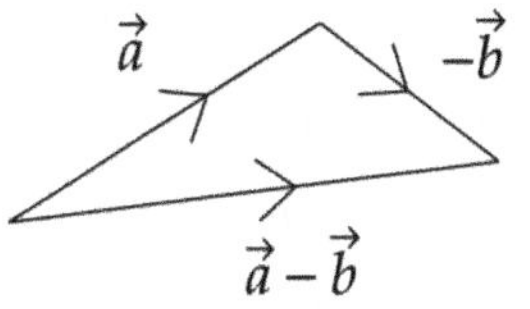

Figure 1.4. Subtraction of vectors.

Remark 1.3. The subtraction of two equal vectors yields the zero vector, 0, which has zero magnitude and no associated direction.

The scalar multiple $\lambda\boldsymbol{a}$

Definition 1.1.6: Scalar multiple
Let $\boldsymbol{a}$ be a vector and λ be a scalar (a real number). Then the **scalar multiple** $\lambda\boldsymbol{a}$ is the vector whose magnitude is $|\lambda||\boldsymbol{a}|$ and whose direction is
(i) the same as $\boldsymbol{a}$ if λ is positive,
(ii) undefined if λ is zero (the answer is a zero-size vector),
(iii) the same as $-\boldsymbol{a}$ if λ is negative.

It follows that $-(\lambda\boldsymbol{a}) = (-\lambda)\boldsymbol{a}$.

Law 1.1.2: Laws of algebra for the scalar multiple

(i) Associative law: $\lambda(\mu \boldsymbol{a}) = (\lambda\mu)\boldsymbol{a}$.
(ii) Distributive law: $\lambda(\boldsymbol{a} + \boldsymbol{b}) = \lambda\boldsymbol{a} + \lambda\boldsymbol{b}$ and $(\lambda + \mu)\boldsymbol{a} = \lambda\boldsymbol{a} + \mu\boldsymbol{a}$.

1.1.2 Scalar product

Definition 1.1.7: Scalar product
The **scalar** (or **dot**) **product** of two vectors yields a scalar which is equal to the product of the magnitudes of the two vectors times the cosine of the angle between them.

$$\boldsymbol{a} \cdot \boldsymbol{b} = |\boldsymbol{a}||\boldsymbol{b}|\cos\theta, \quad (0 \leqslant \theta \leqslant \pi) \tag{1.6}$$

where θ is the angle between $\boldsymbol{a}$ and $\boldsymbol{b}$.

You can think of it as *magnitude of* $\boldsymbol{a}$ *times the component of* $\boldsymbol{b}$ *in direction of* $\boldsymbol{a}$. It can be:
- positive for vectors going in the same direction;
- negative for vectors going into opposite directions;
- zero for *perpendicular* vectors.

Law 1.1.3: Laws of algebra for the scalar product

(i) Commutative law: $\boldsymbol{b} \cdot \boldsymbol{a} = \boldsymbol{a} \cdot \boldsymbol{b}$ [since *cosine* is an even function, i.e. $\cos(-\theta) = \cos\theta$].
(ii) Distributive law: $\boldsymbol{a} \cdot (\boldsymbol{b} + \boldsymbol{c}) = \boldsymbol{a} \cdot \boldsymbol{b} + \boldsymbol{a} \cdot \boldsymbol{c}$.
(iii) Associative with scalar multiplication: $(\lambda\boldsymbol{a}) \cdot \boldsymbol{b} = \lambda(\boldsymbol{a} \cdot \boldsymbol{b})$.

Applying scalar products to combinations of the Cartesian unit vectors, we obtain

$$\hat{\boldsymbol{i}} \cdot \hat{\boldsymbol{i}} = \hat{\boldsymbol{j}} \cdot \hat{\boldsymbol{j}} = \hat{\boldsymbol{k}} \cdot \hat{\boldsymbol{k}} = 1; \qquad \hat{\boldsymbol{i}} \cdot \hat{\boldsymbol{j}} = \hat{\boldsymbol{j}} \cdot \hat{\boldsymbol{k}} = \hat{\boldsymbol{k}} \cdot \hat{\boldsymbol{i}} = 0. \tag{1.7}$$

This set of properties can be summed up by stating that the set of Cartesian unit vectors is **orthonormal**, i.e. of unit length (or *normalised to 1*) and orthogonal to one another. Since the operation is also *distributive*, we can multiply out the brackets to express the dot product in terms of vectors' components:

$$\boldsymbol{a} \cdot \boldsymbol{b} = (a_1\hat{\boldsymbol{i}} + a_2\hat{\boldsymbol{j}} + a_3\hat{\boldsymbol{k}}) \cdot (b_1\hat{\boldsymbol{i}} + b_2\hat{\boldsymbol{j}} + b_3\hat{\boldsymbol{k}}) = a_1b_1 + a_2b_2 + a_3b_3. \tag{1.8}$$

Definition 1.1.8: Magnitude of a vector

The **magnitude** of a vector is given by the square root of the dot product of the vector with itself:

$$|\boldsymbol{a}|^2 = \boldsymbol{a} \cdot \boldsymbol{a} = a_1^2 + a_2^2 + a_3^2 \Longrightarrow |\boldsymbol{a}| = \sqrt{a_1^2 + a_2^2 + a_3^2}. \tag{1.9}$$

(This is obvious from taking $\boldsymbol{b} = \boldsymbol{a}$ in the original definition or writing out the dot product in terms of components and using Pythagoras' theorem.)

Example 1.2. The speed of a particle is the magnitude of its velocity vector:

$$v = |\boldsymbol{v}| = \sqrt{\boldsymbol{v} \cdot \boldsymbol{v}} = \sqrt{v_1^2 + v_2^2 + v_3^2}.$$

Remark 1.4. We shall use the simple, non-underlined letter as a shorthand for the magnitude of the corresponding vector: for example, the magnitude of the acceleration vector is

$$a = |\boldsymbol{a}|$$

and the speed (magnitude of the velocity) is

$$v = |\boldsymbol{v}|.$$

Definition 1.1.9: Unit vector

In general, a vector whose magnitude equals unity is called a **unit vector**. The Cartesian basis vectors are examples of unit vectors. In general, the unit vector in the direction of vector $\boldsymbol{a}$ is

$$\boldsymbol{a} = \frac{\boldsymbol{a}}{|\boldsymbol{a}|}. \tag{1.10}$$

1.1.3 Matrices

Definition 1.1.10: Matrix

An $m \times n$ **matrix A** is an array of real numbers $a_{ij} \in \mathbb{R}$, where $1 \leqslant i \leqslant m$ and $1 \leqslant j \leqslant n$, arranged in m rows and n columns as follows:

$$A = \begin{pmatrix} a_{11} & a_{12} & \cdots & a_{1n} \\ a_{21} & a_{22} & \cdots & a_{2n} \\ \cdots & \cdots & \cdots & \cdots \\ a_{m1} & a_{m2} & \cdots & a_{mn} \end{pmatrix}.$$

> **Remark 1.5.** A **vector** is therefore a simple example of a matrix with just one column.

The most common types of matrices are:

- **Square matrix**: a matrix where the numbers of rows and columns are the same, i.e. an $m \times m$ matrix.
- **Zero matrix**: a matrix all of whose elements are zero, i.e. $a_{ij} = 0$ for all $i = 1, \dots, m$ and $j = 1, \dots, n$. It is generally denoted by 0.
- **Diagonal matrix**: a matrix all of whose entries, except the main diagonal, are zero. That is, $a_{ij} = 0$ for $i \neq j$.
- A special case of a diagonal matrix is the **identity matrix**, where all entries of the main diagonal are 1 and the remaining elements are 0. Identity matrices are generally denoted by $\mathbf{I}_k$, where k describes the size. For example, the 3×3 identity matrix is:

$$\mathbf{I}_3 = \begin{pmatrix} 1 & 0 & 0 \\ 0 & 1 & 0 \\ 0 & 0 & 1 \end{pmatrix}.$$

Matrix addition and scalar multiplication

- To add or subtract matrices, one adds or subtracts the corresponding elements of each matrix. Only matrices of the same size (also called, *additively conformable*) can be added or subtracted.
- To multiply a matrix by a scalar, you multiply every element in the matrix by that scalar.

Example 1.3. Let

$$\mathbf{A} = \begin{pmatrix} 2 & 3 \\ 4 & 5 \end{pmatrix}$$

and

$$\mathbf{B} = \begin{pmatrix} 1 & 3 \\ 2 & 5 \end{pmatrix}.$$

We want to find $\mathbf{A} + 2\mathbf{B}$. Thus,

$$\mathbf{A} + 2\mathbf{B} = \begin{pmatrix} 2 + 2 \times 1 & 3 + 2 \times 3 \\ 4 + 2 \times 2 & 5 + 2 \times 5 \end{pmatrix} = \begin{pmatrix} 4 & 9 \\ 8 & 15 \end{pmatrix}.$$

◆

Matrix multiplication

Matrices can be multiplied together if the number of columns in the first matrix is equal to the number of rows in the second matrix.

If **AB** exists, then **A** is said to be *multiplicatively conformable* with matrix **B**.

> **Definition 1.1.11: Matrix product**
>
> Let $\mathbf{A} = (a_{ij})_{m \times n}$ and $\mathbf{B} = (b_{ij})_{r \times p}$. Then the **product AB** is defined if $n = r$ and is the $m \times p$ matrix with elements
>
> $$\sum_{k=1}^{n} a_{ik} b_{kj}, \quad i = 1, \ldots, m, \quad j = 1, \ldots, p.$$

Above $n = r$ means that the number of columns of **A** is the same as the number of rows of **B**, and the product is the matrix where the ijth entry is the sum of the products of corresponding entries from the ith column of **A** with the jth column of **B**. The product is not element-wise, and therefore the usual properties of real numbers are not automatically inherited. The *order* in which they are multiplied is important. In fact, there are genuinely new features:

- In general, $\mathbf{AB} \neq \mathbf{BA}$ (even if **A** and **B** are both squared matrices).
- If **AB** exists, **BA** does not necessarily exist.

> **Remark 1.6. TOP TIP**
>
> In order to find the product of two multiplicatively conformable matrices, one has to multiply the elements in each row in the left-hand matrix by the corresponding elements in each column in the right-hand matrix, then add the results together.

Example 1.4. Let

$$\mathbf{A} = \begin{pmatrix} 1 & 2 \\ 3 & 4 \end{pmatrix}$$

and

$$\mathbf{B} = \begin{pmatrix} -1 \\ -2 \end{pmatrix}.$$

We want to:

(i) find **AB**. First calculate the size of **AB**.

$$(2 \times 2) \times (2 \times 1) \quad \text{gives} \quad 2 \times 1$$

$$\mathbf{AB} = \begin{pmatrix} 1 & 2 \\ 3 & 4 \end{pmatrix} \begin{pmatrix} -1 \\ -2 \end{pmatrix} = \begin{pmatrix} p \\ q \end{pmatrix}$$

p is the total of the first row of **A** multiplied by the first column of **B**.

$$p = 1 \times (-1) + 2 \times (-2) = -5$$

q is the total of the second row of **A** multiplied by the first column of **B**.

$$q = 3 \times (-1) + 4 \times (-2) = -11.$$

Therefore,

$$\mathbf{AB} = \begin{pmatrix} -5 \\ -11 \end{pmatrix}.$$

(ii) explain why it is not possible to find **BA**.

BA cannot be found, since the number of columns in **B** is not the same as the number of rows in **A**.

◆

Determinants

One can calculate the **determinant** of a square matrix. The determinant is a scalar value associated with that matrix.

- For a 2×2 matrix $\mathbf{M} = \begin{pmatrix} a & b \\ c & d \end{pmatrix}$, the determinant of **M**, denoted by $\det\{\mathbf{M}\}$, is

$$\det\{\mathbf{M}\} = ad - bc.$$

- For a 3×3 matrix, finding the determinant is slightly more difficult. We have to reduce the 3×3 determinant to 2×2 determinants using the formula:

$$\begin{vmatrix} a & b & c \\ d & e & f \\ g & h & i \end{vmatrix} = a\begin{vmatrix} e & f \\ h & i \end{vmatrix} - b\begin{vmatrix} d & f \\ g & i \end{vmatrix} + c\begin{vmatrix} d & e \\ g & h \end{vmatrix}.$$

Exercise 1.1. Given $\mathbf{A} = \begin{pmatrix} 1 & 2 \\ 3 & 1 \end{pmatrix}$ and $\mathbf{I}_2$, which is the 2×2 identity matrix. Prove that

$$\mathbf{A}^2 = 2\mathbf{A} + 5\mathbf{I}_2.$$

Exercise 1.2. Let

$$\mathbf{A} = \begin{pmatrix} 2 & 1 & -1 \\ 1 & 0 & 4 \\ -4 & 2 & 1 \end{pmatrix}$$

and

$$\mathbf{B} = \begin{pmatrix} 3 & 1 & 2 \\ \alpha & 4 & 5 \\ 0 & 2 & 3 \end{pmatrix},$$

where α is a constant.

(i) Evaluate $\det\{\mathbf{A}\}$.
(ii) Given $\det\{\mathbf{B}\} = 2$, find the value of α.
(iii) For α found in the previous part, find **AB**.
(iv) Verify that $\det\{\mathbf{AB}\} = \det\{\mathbf{A}\}\det\{\mathbf{B}\}$.

◆

1.1.4 Vector product

Definition 1.1.12: Vector product

The **vector** (or **cross**) **product** is defined as follows:

$$\boldsymbol{a} \times \boldsymbol{b} = (|\boldsymbol{a}||\boldsymbol{b}|\sin\theta)\hat{\boldsymbol{n}} \tag{1.11}$$

where the magnitude is $|\boldsymbol{a} \times \boldsymbol{b}| = |\boldsymbol{a}||\boldsymbol{b}|\sin\theta$ and θ is the angle between $\boldsymbol{a}$ and $\boldsymbol{b}$ $(0 \leqslant \theta \leqslant \pi)$.

The unit vector $\hat{\boldsymbol{n}}$ is in a direction perpendicular to the plane spanned by $\boldsymbol{a}$ and $\boldsymbol{b}$. The direction of $\hat{\boldsymbol{n}}$ is given by the right-hand rule: if your index finger points in the direction of $\boldsymbol{a}$ and your middle finger in the direction of $\boldsymbol{b}$, then your thumb gives the direction of $\hat{\boldsymbol{n}}$.

Law 1.1.4: Laws of algebra for the vector product

(i) **Anti-**commutative law: $\boldsymbol{b} \times \boldsymbol{a} = -\boldsymbol{a} \times \boldsymbol{b}$, i.e. the cross product is *anti-symmetric.*
(ii) Distributive law: $\boldsymbol{a} \times (\boldsymbol{b} + \boldsymbol{c}) = \boldsymbol{a} \times \boldsymbol{b} + \boldsymbol{a} \times \boldsymbol{c}$.
(iii) Associative with scalar multiplication: $(\lambda\boldsymbol{a}) \times \boldsymbol{b} = \lambda(\boldsymbol{a} \times \boldsymbol{b})$.

Since the vector product is anti-commutative, the *order of the terms in vector products must be preserved.* The vector product is not associative.

The anti-commutative law follows from the definition of the right-hand rule for the direction, i.e. reversing the order of the vectors the resulting vector will be in the opposite direction. You can also think of it as being consistent with the fact that the *sine* function is odd, i.e. $\sin(-\theta) = -\sin\theta$. Hence, the cross product of any vector with itself is zero:

$$\boldsymbol{a} \times \boldsymbol{a} = -\boldsymbol{a} \times \boldsymbol{a} \Longrightarrow \boldsymbol{a} \times \boldsymbol{a} = 0. \tag{1.12}$$

This can also be seen by noticing that the *angle between the two vectors* is zero in this case, so $\sin\theta = 0$.

If $\{\hat{\boldsymbol{i}}, \hat{\boldsymbol{j}}, \hat{\boldsymbol{k}}\}$ is a standard basis, then

$$\begin{aligned} \hat{\boldsymbol{i}} \times \hat{\boldsymbol{i}} &= \hat{\boldsymbol{j}} \times \hat{\boldsymbol{j}} = \hat{\boldsymbol{k}} \times \hat{\boldsymbol{k}} = 0 \\ \hat{\boldsymbol{i}} \times \hat{\boldsymbol{j}} &= -\hat{\boldsymbol{j}} \times \hat{\boldsymbol{i}} = \hat{\boldsymbol{k}} \\ \hat{\boldsymbol{j}} \times \hat{\boldsymbol{k}} &= -\hat{\boldsymbol{k}} \times \hat{\boldsymbol{j}} = \hat{\boldsymbol{i}} \\ \hat{\boldsymbol{k}} \times \hat{\boldsymbol{i}} &= -\hat{\boldsymbol{i}} \times \hat{\boldsymbol{k}} = \hat{\boldsymbol{j}}. \end{aligned}$$

Therefore, for general vectors $\boldsymbol{a}$, $\boldsymbol{b}$ given in terms of their components with respect to the basis $\{\hat{\boldsymbol{i}}, \hat{\boldsymbol{j}}, \hat{\boldsymbol{k}}\}$, the vector product is calculated as

$$\begin{aligned}\boldsymbol{a} \times \boldsymbol{b} &= (a_1\hat{\boldsymbol{i}} + a_2\hat{\boldsymbol{j}} + a_3\hat{\boldsymbol{k}}) \times (b_1\hat{\boldsymbol{i}} + b_2\hat{\boldsymbol{j}} + b_3\hat{\boldsymbol{k}}) \\ &= a_1b_2(\hat{\boldsymbol{i}} \times \hat{\boldsymbol{j}}) + a_1b_3(\hat{\boldsymbol{i}} \times \hat{\boldsymbol{k}}) + a_2b_1(\hat{\boldsymbol{j}} \times \hat{\boldsymbol{i}}) \\ &\quad + a_2b_3(\hat{\boldsymbol{j}} \times \hat{\boldsymbol{k}}) + a_3b_1(\hat{\boldsymbol{k}} \times \hat{\boldsymbol{i}}) + a_3b_2(\hat{\boldsymbol{k}} \times \hat{\boldsymbol{j}}) \\ &= (a_2b_3 - a_3b_2)\hat{\boldsymbol{i}} + (a_3b_1 - a_1b_3)\hat{\boldsymbol{j}} + (a_1b_2 - a_2b_1)\hat{\boldsymbol{k}}.\end{aligned}$$

Using short-hand notation, this can be written compactly,

$$\boldsymbol{a} \times \boldsymbol{b} = \begin{pmatrix} a_2b_3 - a_3b_2 \\ a_3b_1 - a_1b_3 \\ a_1b_2 - a_2b_1 \end{pmatrix}. \tag{1.13}$$

You might be familiar with the following way of calculating a vector product using a matrix determinant:

$$\begin{aligned} a \times b &= \begin{vmatrix} \hat{i} & \hat{j} & \hat{k} \\ a_1 & a_2 & a_3 \\ b_1 & b_2 & b_3 \end{vmatrix} \\ &= (a_2b_3 - a_3b_2)\hat{\boldsymbol{i}} + (a_3b_1 - a_1b_3)\hat{\boldsymbol{j}} + (a_1b_2 - a_2b_1)\hat{\boldsymbol{k}}. \end{aligned} \tag{1.14}$$

Similarly to the dot product, the result of the cross product is independent of any particular basis set. The above also holds for any right-handed set of orthonormal basis vectors, which, in this case, means that the cross product of the first vector with the second gives the third.

Example 1.5. The angle between the unit vectors $\hat{\boldsymbol{i}}$ and $\hat{\boldsymbol{j}}$ is 90° (or $\pi/2$ radians). Therefore, the magnitude of their cross product is:

$$|\hat{\boldsymbol{i}} \times \hat{\boldsymbol{j}}| = 1 \times 1 \times \sin(\pi/2) = 1,$$

and its direction is along the positive z-axis (right-hand screw rule). Hence,

$$\hat{\boldsymbol{i}} \times \hat{\boldsymbol{j}} = \hat{\boldsymbol{k}}.$$

Example 1.6. Find an expression for the area of the parallelogram formed by the vectors $\boldsymbol{a}$ and $\boldsymbol{b}$ in terms of the cross product.

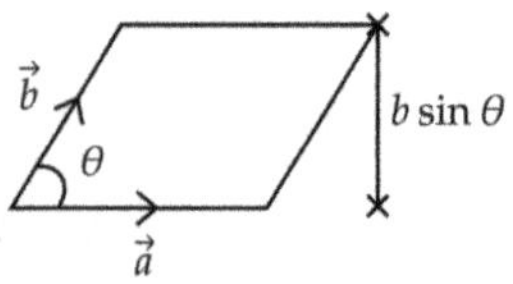

The area is the base of the parallelogram times its perpendicular height. Take the base to be along $\boldsymbol{a}$ so its length is $|\boldsymbol{a}|$; the height is then $|\boldsymbol{b}||\sin\theta|$. The area is therefore

$$|\boldsymbol{a}||\boldsymbol{b}||\sin\theta| = |\boldsymbol{a} \times \boldsymbol{b}|.$$

Example 1.7. Find the Cartesian components of the cross product of the vectors

$$\boldsymbol{a} = \hat{\imath} + 2\hat{\jmath} + 2\hat{k}; \quad \boldsymbol{b} = 3\hat{\imath} + \hat{\jmath} + 2\hat{k}.$$

Solution.

$$\boldsymbol{a} \times \boldsymbol{b} = (2 \times 2 - 2 \times 1)\hat{\imath} + (2 \times 3 - 1 \times 2)\hat{\jmath} + (1 \times 1 - 2 \times 3)\hat{k} = 2\hat{\imath} + 4\hat{\jmath} - 5\hat{k}.$$

Exercise 1.3. Consider the vectors

$$\boldsymbol{a} = \hat{\imath} + 2\hat{\jmath} + \hat{k}; \qquad \boldsymbol{b} = \hat{\imath} - 3\hat{\jmath} + 2\hat{k}.$$

Using the definitions of the unit-vector, scalar and vector products, compute

(i) $\boldsymbol{a}$.
(ii) $\boldsymbol{a} \cdot \boldsymbol{b}$.
(iii) $\boldsymbol{a} \times \boldsymbol{b}$.
(iv) the area of the parallelogram whose sides are formed by the vectors $\boldsymbol{a}$ and $\boldsymbol{b}$.

Exercise 1.4. Let $\boldsymbol{a} = \hat{\imath} + \hat{\jmath} + 2\hat{k}$ and $\boldsymbol{b} = \hat{\imath} + 7\hat{k}$. Find the following:

(i) $|\boldsymbol{a}|$, $|\boldsymbol{b}|$ and $|\boldsymbol{a} + 2\boldsymbol{b}|$.
(ii) $\boldsymbol{a} \cdot \boldsymbol{b}$.
(iii) the angle between $\boldsymbol{a}$ and $\boldsymbol{b}$.

Exercise 1.5. Let

$$\boldsymbol{v} = \hat{\imath} - 2\hat{\jmath} - 2\hat{k}$$

(i) Find the unit vector pointing in the opposite direction to $\boldsymbol{v}$.
(ii) Find a vector of length two perpendicular to $\boldsymbol{v}$. [There are infinitely many such vectors.]

Exercise 1.6. Let $\boldsymbol{a}$ and $\boldsymbol{b}$ be non-zero vectors such that $\boldsymbol{b}$ has twice the length of $\boldsymbol{a}$. If the vectors $(2\boldsymbol{a} - 5\boldsymbol{b})$ and $(6\boldsymbol{a} - \boldsymbol{b})$ are perpendicular, find the angle between $\boldsymbol{a}$ and $\boldsymbol{b}$.

Exercise 1.7. Let $\boldsymbol{c}$ be a non-zero vector and suppose that $\boldsymbol{a}$ and $\boldsymbol{b}$ satisfy

$$\boldsymbol{a} \cdot \boldsymbol{c} = \boldsymbol{b} \cdot \boldsymbol{c} \qquad (*)$$

Does it follow that $\boldsymbol{a} = \boldsymbol{b}$? Give a geometrical meaning to equation (*).

Exercise 1.8. Show that if three non-zero vectors $\boldsymbol{a}$, $\boldsymbol{b}$ and $\boldsymbol{c}$ satisfy

$$(\boldsymbol{a} \cdot \boldsymbol{c})\boldsymbol{b} = (\boldsymbol{b} \cdot \boldsymbol{c})\boldsymbol{a}$$

then either $\boldsymbol{a}$ and $\boldsymbol{b}$ are parallel (or anti-parallel) or $\boldsymbol{a}$ and $\boldsymbol{b}$ are both perpendicular to $\boldsymbol{c}$.

Exercise 1.9. Let $\boldsymbol{a}$, $\boldsymbol{b}$ and $\boldsymbol{c}$ be the position vectors of the non-collinear points A, B and C. Show that the magnitude of the vector

$$\boldsymbol{a} \times \boldsymbol{b} + \boldsymbol{b} \times \boldsymbol{c} + \boldsymbol{c} \times \boldsymbol{a}$$

is twice the area of the triangle with vertices A, B and C.

> **Remark 1.7: Appendix A**
> If you want to find out more about other methods for vector algebra, you may have a look at appendix A: index notation.

1.2 Complex numbers

> **Definition 1.2.1: Complex number**
> A complex number is a number of the form
>
> $$z = x + iy,$$
>
> where x and y are real numbers and i satisfies
>
> $$i^2 = -1.$$

The set of complex numbers is denoted by $\mathbb{C}$. The real part of z, $\mathrm{Re}\{z\}$, is the real number x, i.e. $\mathrm{Re}\{z\} = x$. The imaginary part of z, $\mathrm{Im}\{z\}$, is the real number y, i.e. $\mathrm{Im}\{z\} = y$. Furthermore, the set $\mathbb{C}$ is a field, i.e. addition and multiplication satisfy the usual rules. Complex numbers can be plotted in the **Argand diagram**, as shown in figure 1.5.

Example 1.8. Some examples of complex numbers include: the set of real numbers $\mathbb{R}$ (take $y = 0$ in the definition of complex number); $3 + 4i$; $\sqrt{2} - \frac{1}{\pi}i$.

◆

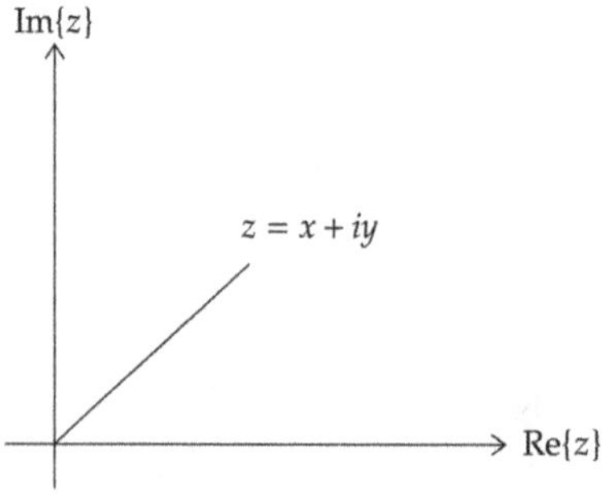

Figure 1.5. Argand diagram.

1.2.1 Arithmetic in $\mathbb{C}$

Addition and subtraction

Addition and subtraction in $\mathbb{C}$ is very straightforward: we simply add (or subtract, when subtracting complex numbers) the real and imaginary parts. Let $z_1 = x_1 + iy_1$ and $z_2 = x_2 + iy_2$, $z_1, z_2 \in \mathbb{C}$, then

$$w = z_1 + z_2 = (x_1 + iy_1) + (x_2 + iy_2) = (x_1 + x_2) + i(y_1 + y_2).$$

Multiplication

Multiplication exploits the fact that $i^2 = -1$. Let $z_1 = x_1 + iy_1$ and $z_2 = x_2 + iy_2$, $z_1, z_2 \in \mathbb{C}$, then

$$z_1 \cdot z_2 = (x_1 + iy_1) \cdot (x_2 + iy_2) = (x_1x_2 - y_1y_2) + i(x_1y_2 + y_1x_2).$$

Division

To perform division, one needs to first define the complex conjugate of z.

Definition 1.2.2: Complex conjugate, z^*

The complex conjugate z^* of z is defined by switching the sign of the imaginary part, that is (figure 1.6),

$$z = x + iy \iff z^* = x - iy.$$

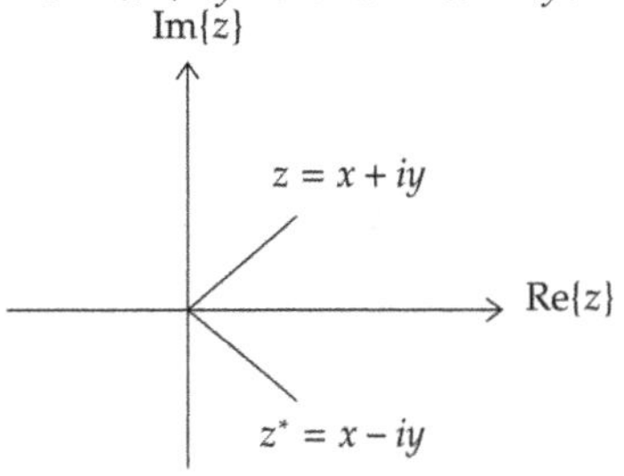

Figure 1.6. Complex conjugate.

Remark 1.8. It might be helpful to note that

$$zz^* = (x + iy) \cdot (x - iy) = x^2 + y^2.$$

One may now calculate

$$\frac{x_1 + iy_1}{x_2 + iy_2} = \frac{x_1 + iy_1}{x_2 + iy_2} \cdot \frac{x_2 - iy_2}{x_2 - iy_2} = \left(\frac{x_1x_2 + y_1y_2}{x_2^2 + y_2^2}\right) + i\left(\frac{y_1x_2 - x_1y_2}{x_2^2 + y_2^2}\right).$$

The general formula for the quotient of one complex number by another does not look all that pleasant, but in specific cases it is straightforward to calculate.

The *modulus* or *absolute value* of z is defined as

$$|z| = \sqrt{x^2 + y^2}. \tag{1.15}$$

Alternatively,

$$|z|^2 = zz^*.$$

1.2.2 Polar coordinates

One may now introduce polar coordinates (r, θ) in the plane

$$x = r\cos\theta, \;\; y = r\sin\theta, \;\; r \geqslant 0, \; -\pi < \theta \leqslant \pi.$$

Any complex number z can be written as

$$z = x + iy = r(\cos\theta + i\sin\theta)$$

where $r \equiv |z|$.

1.2.3 Complex exponential and Euler's formula

Euler's formula is

$$e^{i\theta} = \cos\theta + i\sin\theta.$$

This leads to the polar form (figure 1.7)

$$z = re^{i\theta}.$$

1.3 Calculus

Mechanics has to do with the rate of change of *things*. We therefore need calculus to describe it.

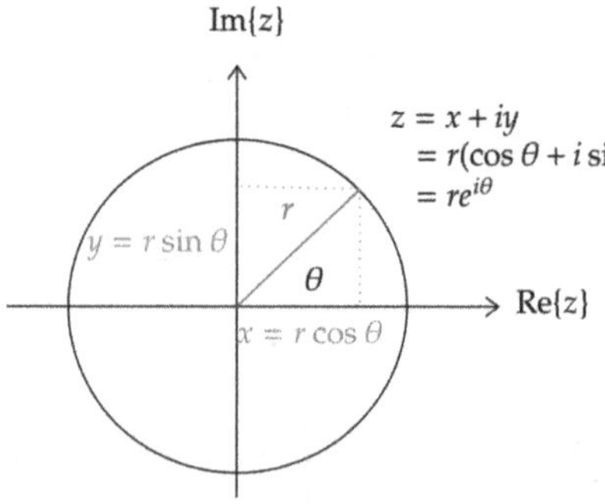

Figure 1.7. Complex exponential and polars coordinates.

1.3.1 Differentiation

Definition 1.3.1: Derivative

The **derivative** with respect to time of a function f is defined as follows:

$$\frac{\mathrm{d}f}{\mathrm{d}t} \equiv \lim_{\delta t \to 0} \frac{f(t+\delta t) - f(t)}{\delta t}. \tag{1.16}$$

It corresponds to the *rate of change* of the quantity described by f with time. Equivalently, it is the slope of the graph of f against time at time t, or the tangent of the angle that the function makes with the x-axis.

Suppose a particle is moving in space. We can describe its position vector $\boldsymbol{r}$ as a function of time t (relative to a suitable origin O), i.e. the *trajectory* (figure 1.8).

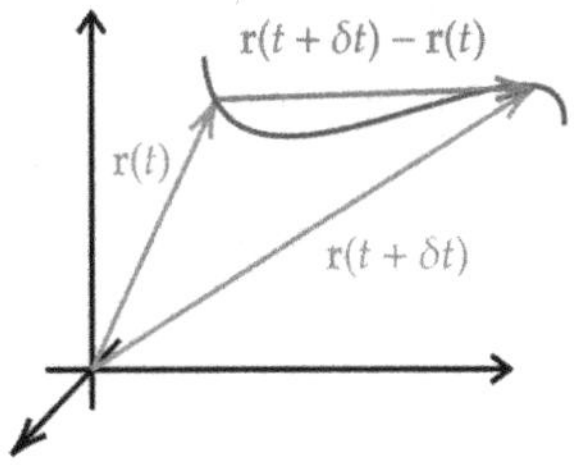

Figure 1.8. Position vectors.

In Cartesian coordinates,

$$\boldsymbol{r} = x\hat{\boldsymbol{i}} + y\hat{\boldsymbol{j}} + x\hat{\boldsymbol{k}}. \tag{1.17}$$

Then, we can define **velocity** as the derivative of the vector $\boldsymbol{r}$ with respect to time by

$$\boldsymbol{v} = \frac{\mathrm{d}\boldsymbol{r}}{\mathrm{d}t} = \dot{\boldsymbol{r}} = \lim_{\delta t \to 0} \frac{\boldsymbol{r}(t+\delta t) - \boldsymbol{r}(t)}{\delta t}. \tag{1.18}$$

In terms, of components, since the basis vectors are time-independent, we have

$$\boldsymbol{v} = \frac{\mathrm{d}x}{\mathrm{d}t}\hat{\boldsymbol{i}} + \frac{\mathrm{d}y}{\mathrm{d}t}\hat{\boldsymbol{j}} + \frac{\mathrm{d}z}{\mathrm{d}t}\hat{\boldsymbol{k}} = \dot{x}\hat{\boldsymbol{i}} + \dot{y}\hat{\boldsymbol{j}} + \dot{z}\hat{\boldsymbol{k}}. \tag{1.19}$$

The magnitude of $|\boldsymbol{v}|$ is called the *speed.* By definition, it is non-negative and it is determined by using Pythagoras' theorem:

$$|\boldsymbol{v}| = \sqrt{\dot{x}^2 + \dot{y}^2 + \dot{z}^2}. \tag{1.20}$$

In a similar way it is possible to define the acceleration as time derivative of the speed, or second derivative of the position:

$$\boldsymbol{a} = \frac{\mathrm{d}\boldsymbol{v}}{\mathrm{d}t} = \frac{\mathrm{d}^2\boldsymbol{r}}{\mathrm{d}t^2} = \ddot{\boldsymbol{r}}. \tag{1.21}$$

Example 1.9. A particle of mass m moves such that its position vector as a function of time is given by

$$\boldsymbol{r}(t) = 2\cos(t)\hat{\boldsymbol{i}} + 2\sin(t)\hat{\boldsymbol{j}} + t^2\hat{\boldsymbol{k}}$$

Calculate the following quantities as a function of time: (i) the particle's velocity; (ii) the acceleration.

(i) Velocity: $\boldsymbol{v}(t) = \dot{\boldsymbol{r}}(t) = -2\sin(t)\hat{\boldsymbol{i}} + 2\cos(t)\hat{\boldsymbol{j}} + 2t\hat{\boldsymbol{k}}$.

(ii) Acceleration: $\boldsymbol{a}(t) = \ddot{\boldsymbol{r}}(t) = -2\cos(t)\hat{\boldsymbol{i}} - 2\sin(t)\hat{\boldsymbol{j}} + 2\hat{\boldsymbol{k}}$.

Exercise 1.10. Various particles of mass m have position vectors $\boldsymbol{r}$ as a function of time t given by:

(i) $\boldsymbol{r}(t) = t^2\hat{\boldsymbol{i}} + 2t\hat{\boldsymbol{j}} + 3t^2\hat{\boldsymbol{k}}$.

(ii) $\boldsymbol{r}(t) = \cos(t)\hat{\boldsymbol{i}} + \sin(t\sqrt{3})\hat{\boldsymbol{j}} + te^{10t}\hat{\boldsymbol{k}}$.

(iii) $\boldsymbol{r}(t) = \cos(2t)\hat{\boldsymbol{i}} + 2t^2\hat{\boldsymbol{j}} + e^{t\sqrt{2}}\hat{\boldsymbol{k}}$.

(iv) $\boldsymbol{r}(t) = \sin(2t)\hat{\boldsymbol{i}} + \cos(t\sqrt{3})\hat{\boldsymbol{j}} + te^{2t}\hat{\boldsymbol{k}}$.

Find the particles' (a) velocity and (b) its acceleration at a time t.

Exercise 1.11. A particle of unit mass moves in the xy-plane so that its position vector at time t is

$$\boldsymbol{r}(t) = a\cos(\omega t)\hat{\boldsymbol{i}} + b\sin(\omega t)\hat{\boldsymbol{j}}$$

where a, b, $\omega > 0$ and $a > b$.

(i) Show that the particle moves in an ellipse. [*Hint*: recall that the equation of the ellipse can be taken as $\frac{x^2}{a^2} + \frac{y^2}{b^2} = 1$.]

(ii) Find the particle's velocity.

(iii) Find the particle's acceleration and show that it is indeed centripetal, i.e. directed towards the centre.

◆

Partial derivatives

Calculus concerns the rate of change of quantities. For a function f of more than one variable, we can talk about its rate of change in different directions. In two dimensions for instance, the rates of change of the function $f(x, y)$ in the x- and y-directions, respectively, are given by the partial derivatives

$$\frac{\partial f}{\partial x} = \lim_{h\to 0} \frac{f(x+h, y) - f(x, y)}{h}, \tag{1.22}$$

$$\frac{\partial f}{\partial y} = \lim_{h\to 0} \frac{f(x, y+h) - f(x, y)}{h}, \tag{1.23}$$

provided these limits exist. To be noticed that, in taking the derivative with respect to a variable, the other variable(s) is(are) taken to be constant.

Remark 1.9: Notation

We often say 'partial d' or 'curly d' for '∂'. Note that the symbol '∂' is NOT a delta (δ) and you should never write delta instead of '∂'. Another common notation for partial derivative is

$$f_x(x, y) \equiv \frac{\partial f(x, y)}{\partial x} \quad \text{and} \quad f_y(x, y) \equiv \frac{\partial f(x, y)}{\partial y}.$$

Example 1.10. Given a scalar field $f(x, y) = x^2 + 3xy + y^2$, calculate the partial derivative of the field with respect to the variable x.

$$\frac{\partial f}{\partial x} = 2x + 3y.$$

Notice how the variable y is treated as a constant.

Definition 1.3.2: The gradient

The Nabla operator is a vector of partial derivatives along the directions of the axes of a coordinate system.

$$\nabla \equiv \left(\frac{\partial}{\partial x}, \frac{\partial}{\partial y}, \frac{\partial}{\partial z}\right) = \frac{\partial}{\partial x}\hat{\boldsymbol{i}} + \frac{\partial}{\partial y}\hat{\boldsymbol{j}} + \frac{\partial}{\partial z}\hat{\boldsymbol{k}}. \tag{1.24}$$

When applied to a field, we say we are taking the gradient of that field. Writing ∇f is equivalent to 'grad f'.

Higher derivatives. We denote the second derivative of f with respect to x by $\frac{\partial^2 f}{\partial x^2}$ or f_{xx} and, similarly, with respect to y by $\frac{\partial^2 f}{\partial y^2}$ or f_{yy}.

There are also the mixed derivatives: the derivative with respect to y of the derivative of f with respect to x is denoted by

$$\frac{\partial}{\partial y}\left(\frac{\partial f}{\partial x}\right) = \frac{\partial^2 f}{\partial y \partial x} \quad \text{or} \quad (f_x)_y = f_{xy}.$$

Remark 1.10: Mixed derivatives

Notice that $f_{xy} = f_{yx}$ is true for any function f with continuous second derivatives. More generally, for sufficiently smooth functions, changing the order of partial derivatives does not change the outcome. For example, $f_{xxy} = f_{xyx} = f_{yxx}$.

Exercise 1.12. Let $f(x, y, z) = e^{2x}\sin(xy^3)$.

(i) Calculate $\frac{\partial f}{\partial x}$, $\frac{\partial f}{\partial y}$ and $\frac{\partial f}{\partial z}$.

(ii) Verify that $\frac{\partial^2 f}{\partial x \partial y} = \frac{\partial^2 f}{\partial y \partial x}$.

(iii) Find $\nabla f|_{(0,2,0)}$.

Exercise 1.13. Show that

$$\nabla(fg) = (\nabla f)g + f(\nabla g).$$

◆

1.3.2 Taylor series

An important mathematical tool to make approximations and check limiting cases is the Taylor series approximations.

Definition 1.3.3: Taylor and Maclaurin series

A *sufficiently smooth*[1] function $f(x)$ can be approximated in terms of a **Taylor expansion** around a point $x = x_0$:

$$f(x) = f(x_0) + f'(x_0)(x - x_0) + \frac{f''(x_0)}{2}(x - x_0)^2 \tag{1.25}$$

$$+\frac{f'''(x_0)}{3!}(x - x_0)^3 + \cdots \tag{1.26}$$

$$= f(x_0) + \sum_{k=1}^{\infty}\frac{1}{k!}\frac{\mathrm{d}^k f}{\mathrm{d}x^k}\bigg|_{x=x_0}(x - x_0)^k, \tag{1.27}$$

where the derivatives are all to be evaluated at the point $x = x_0$ as indicated. For the important special case $x_0 = 0$, this expansion is called the **Maclaurin series**,

$$f(x) = f(0) + f'(0)(x) + \frac{f''(0)}{2}x^2 + \frac{f'''(0)}{3!}x^3 + \cdots \tag{1.28}$$

$$= f(0) + \sum_{k=1}^{\infty}\frac{1}{k!}\frac{\mathrm{d}^k f}{\mathrm{d}x^k}\bigg|_{x=0} x^k. \tag{1.29}$$

Even if the function is equal to the sum of the infinite number of polynomials for any value of x, the practical use of these expansions mainly lies in approximating the functions with low-order polynomials in the vicinity of the point of expansion. It is

[1] A function may be considered to be *smooth* if it is differentiable everywhere, or up to some desired order over some domain, and hence it is there continuous.

easy to see that, unless higher-order derivatives are very large, the presence of the factorial at the denominator and of higher powers of $(x - x_0)$, supposed to be a small number, are such that the contribution of the higher-order terms becomes smaller and smaller.

Remark 1.11: Factorials!
Recall that a **factorial** $n!$ is defined for a positive integer n as

$$n! \equiv n \cdot (n-1) \cdot (n-2) \cdot \ldots \cdot 2 \cdot 1.$$

For example,

$$3! = 3 \cdot 2 \cdot 1 = 6. \tag{1.30}$$

By definition, $0! \equiv 1$.

Remark 1.12. The function $f(x)$ must be differentiable at the point $x = x_0$ of the expansion. For example, $f(x) = \sqrt{x}$ has no Maclaurin series at $x = 0$. It can still be Taylor-expanded for any other $x > 0$.

Example 1.11. Maclaurin series for e^x
Set $f(x) = e^x$, we have

$$f^{(n)}(x) = e^x \quad \text{for } n = 0, 1, 2, \ldots$$

and thus

$$f^{(n)}(0) = 1 \quad \text{for } n = 0, 1, 2, \ldots$$

Therefore, the Maclaurin series for e^x is

$$\sum_{n=0}^{\infty} \frac{f^{(n)}(0)}{n!} x^n = \sum_{n=0}^{\infty} \frac{x^n}{n!}.$$

Hence, it can be shown that, for all x,

$$e^x = \sum_{n=o}^{\infty} \frac{x^n}{n!}.$$

Example 1.12. Maclaurin series for $(1 + x)^\alpha$
Let $f(x) = (1 + x)^\alpha$. Then,

$$\begin{aligned} f'(x) &= \alpha(1 + x)^{\alpha-1} \\ f''(x) &= \alpha(\alpha - 1)(1 + x)^{\alpha-2}, \ \ldots \\ f^{(n)}(x) &= \alpha(\alpha - 1)\ldots(\alpha - n + 1)(1 + x)^{\alpha-n}. \end{aligned}$$

Therefore, $f(0) = 1$ and, for $n > 0$, we have

$$f^{(n)}(0) = \alpha(\alpha - 1)\ldots(\alpha - n + 1).$$

So the Maclaurin series is

$$\sum_{n=0}^{\infty} \frac{f^{(n)}(0)}{n!} x^n = 1 + \sum_{n=1}^{\infty} \frac{\alpha(\alpha - 1)\ldots(\alpha - n + 1)}{n!} x^n.$$

It can be shown that for all x such that $-1 < x < 1$, we have

$$(1 + x)^\alpha = \sum_{r=0}^{\infty} \binom{\alpha}{r} x^r, \quad \binom{\alpha}{0} = 1 \quad \text{and} \quad \binom{\alpha}{r} = \frac{\alpha(\alpha - 1)\ldots(\alpha - r + 1)}{r!}, \quad r > 0.$$

Example 1.13. $E = mc^2$

In Einstein's theory of Special Relativity, the energy E of a particle is given by the famous formula

$$E = mc^2,$$

where c is the speed of light and m is the relativistic mass of the particle with respect to some observer, which, in turn, is given by

$$m = \frac{m_0}{\sqrt{1 - \frac{v^2}{c^2}}},$$

where m_0 is the rest mass, v is the speed of the particle measured by the observer.

Let us now consider the case in which $v/c \ll 1 \iff v \ll c$, i.e. the non-relativistic limit. Thus, one has

$$E = mc^2 = \frac{m_0 c^2}{\sqrt{1 - \frac{v^2}{c^2}}}.$$

Since we are considering the case $v/c \ll 1$, we want to put the above in a convenient form to Taylor expand. One may thus write

$$E = m_0 c^2 \left(1 - \frac{v^2}{c^2}\right)^{-1/2},$$

which is indeed in a similar form to the previous example, but now our x is $-(v/c)^2$.

Therefore,

$$E = m_0 c^2 \left(1 - \frac{v^2}{c^2}\right)^{-1/2} = m_0 c^2 \left[1 - \frac{1}{2}\left(-\frac{v^2}{c^2}\right) + \cdots\right],$$

where we are ignoring terms of higher order. Eventually, one obtains

$$E \approx m_0c^2 + \frac{1}{2}m_0v^2.$$

After going through the following chapters of this book, a careful reader may realise that the equation we obtained simply consists of the particle's rest energy plus its *classical* kinetic energy ($\frac{1}{2}m_0v^2$).

Exercise 1.14. Verify the following Maclaurin series:

$$\sin x = \sum_{k=0}^{\infty} \frac{(-1)^k}{(2k+1)!}x^{2k+1}; \qquad \cos x = \sum_{k=0}^{\infty} \frac{(-1)^k}{(2k)!}x^{2k}.$$

Exercise 1.15. Derive the Maclaurin series for $\ln(1 + x)$. Hence or otherwise find the Maclaurin series up to the x^4 term for the following:

(i) $\cos(2x) \cdot \ln(1 + x)$.
(ii) $\ln(\cos x)$.

Exercise 1.16. The low velocity limit of the relativistic Doppler shift
The Doppler shift (or Doppler effect) is the change in frequency of a wave in relation to an observer who is moving relative to the wave source. A common example of Doppler shift is the change of pitch heard when an ambulance approaches and recedes from an observer. The Doppler effect for sound is given by the formula:

$$f_{\text{observed}} = f_{\text{source}}\left(\frac{1}{1 \pm \frac{v_{\text{source}}}{v_{\text{wave}}}}\right) \qquad (*)$$

where the plus sign is taken for waves travelling away from the observer. For light and electromagnetic waves, the Doppler shift is given by

$$f_{\text{observed}} = f_{\text{source}}\sqrt{\frac{1 + \frac{v}{c}}{1 - \frac{v}{c}}},$$

where v is the relative velocity of source and observer and v is considered positive when the source is approaching.

Show that for low speeds, i.e. $v \ll c$, the first two terms of the expansion of the relativistic Doppler shift give a good approximation of the Doppler shift, i.e. equation (*) is restored.

◆

1.3.3 Integration

Integration can be seen as the reverse of differentiation, and is connected to finding the area under the graph of the function $f(t)$.

Definition 1.3.4: Definite integral

The **definite integral** from a to b is the area of the shape defined by the curve (between $t = a$ and $t = b$), two vertical lines at $t = a$ and $t = b$, and the horizontal axis. We can thus write

$$\int_a^b f(t)\,\mathrm{d}t = \lim_{\delta t \to 0} \sum_i f(t_i)\delta t \tag{1.31}$$

where the function f is evaluated at a set of equally spaced points t_i separated by intervals δt.

Definition 1.3.5: Indefinite integral

The indefinite integral, or primitive, of a function is defined by

$$F(t) = \int_a^t f(u)\,\mathrm{d}u \tag{1.32}$$

where a is an arbitrary value and u is a *dummy variable*. The indefinite integral is a function of t, the upper limit of the integration.

Remark 1.13: What is a 'dummy variable'?

It is a variable that appears in a calculation only as a placeholder and which disappears completely in the final result. Thus, it doesn't really matter what it is named. It needs, however, to be different from the extreme of the integration. So if we are integrating between time 0 and t, the dummy variable cannot be t itself, even if we are integrating over time. In these cases $\tilde{t}$, or t' is often used, to indicate that it is indeed an integration over time, but that this variable is still different from the integration boundary.

Theorem 1.3.1. Fundamental theorem of calculus

The fundamental theorem of calculus links differentiation and integration of functions, and can be stated as follows:

$$\frac{\mathrm{d}F(t)}{\mathrm{d}t} \equiv \frac{\mathrm{d}}{\mathrm{d}t}\left[\int_a^t f(u)\,\mathrm{d}u\right] = f(t). \tag{1.33}$$

So, the derivative of an indefinite integral is the function itself. The indefinite integral is the inverse operation of the derivative. Equivalently, we can also write

$$\int f(t)\,\mathrm{d}t = F(t) + c \tag{1.34}$$

where c is a constant of integration (subject to initial conditions). The second part of the theorem states that

$$\int_a^b f(t)\,\mathrm{d}t = F(b) - F(a) \tag{1.35}$$

so the definite integral of a function between a and b is the difference between the value of the primitive of the function in point b and that in point a. So, the value of the area under a function can be calculated from its primitive, that is the inverse operation to the derivative.

Example 1.14. Since velocity is the time derivative of the position, the distance travelled in one dimension between times t_1 and t_2 by a particle whose velocity is $v(t)$ is given by its integral

$$s = \int_{t_1}^{t_2} v(t)\,\mathrm{d}t.$$

Example 1.15. Consider a particle of mass m moving along the positive x-axis as in the figure below.

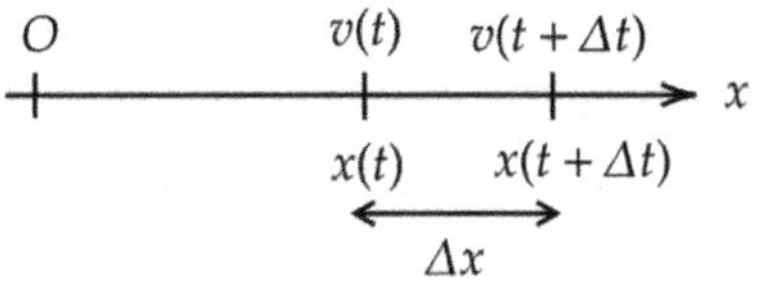

The velocity is positive for a motion in the direction of increasing x, and negative for the direction of decreasing x.

In the infinitesimal time interval $\mathrm{d}t$, the distance travelled by the particle is $\mathrm{d}x = v\mathrm{d}t$. In the finite time interval between t_1, when the position of the particle is x_1, and time t_2, when the position is x_2, the distance travelled is

$$s = x_2 - x_1 = \int_{t_1}^{t_2} v\,\mathrm{d}t.$$

This is represented by the shaded area in the figure below. We would need to know how v varies with t in order to calculate s.

Similarly, for acceleration a, the change in velocity in time interval between time t_1 and t_2 is

$$v_2 - v_1 = \int_{t_1}^{t_2} a\,\mathrm{d}t,$$

and is represented by the shaded area in the second diagram below. As before, we need to know how a varies as a function of time in order to calculate v.

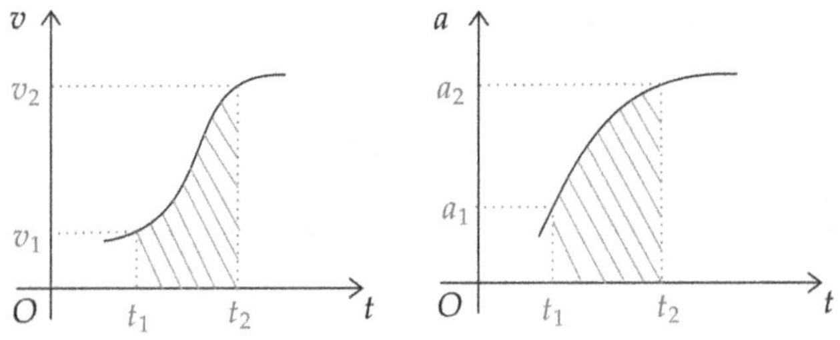

◆

1.3.4 Integration by parts

The method of integration by parts allows us to compute the indefinite integral of a product of two functions, and it is the integration analogue to the product rule for differentiation.

Definition 1.3.6: Integration by parts
In shorthand notation,

$$\int uv' \, \mathrm{d}x = uv - \int u'v \, \mathrm{d}x.$$

For a definite integral, the integration by parts relation reads:

$$\int_a^b u(x)v'(x) \, \mathrm{d}x = [u(x)v(x)]|_a^b - \int_a^b u'(x)v(x) \, \mathrm{d}x.$$

Given a product of two functions, we assign the first term as u and the second as v', where the former should be easy to differentiate and the latter easy to integrate. For the method to be successful, the remaining integral of $u'v$ should be easier to determine.

Example 1.16.

$$I = \int xe^x \, \mathrm{d}x.$$

We identify u and v as follows:

$$u = x \iff u' = 1,$$
$$v = e^x \iff v' = e^x.$$

Thus,

$$I = xe^x - \int 1 \cdot e^x \, \mathrm{d}x = (x - 1)e^x + c.$$

Example 1.17.

$$I = \int \ln x \, \mathrm{d}x.$$

We apply the same trick and insert a factor of 1. We then identify u and v as follows:

$$u = \ln x \iff u' = \frac{1}{x},$$
$$v = x \iff v' = 1.$$

Thus,

$$I = x \ln x - \int \frac{x}{x}\,\mathrm{d}x = x \ln x - \int \mathrm{d}x = x(\ln x - 1) + c.$$

Exercise 1.17.

(i) Find $\int x \sin(x)\,\mathrm{d}x$.

(ii) Find $\int \sin(x)\sin[\cos(x)]\,\mathrm{d}x$.

(iii) Find $\int x^2 e^{3x}\,\mathrm{d}x$.

Exercise 1.18. For all $x > 0$, define

$$\Gamma(x) = \int_0^\infty t^{x-1} e^{-t}\,\mathrm{d}t.$$

Warning: *Do not attempt to evaluate the integral, it cannot be expressed simply in terms of elementary functions!*

1. Show that for all $x > 0$,

$$\Gamma(x + 1) = x\Gamma(x).$$

2. Hence, show that for all positive integers n,

$$\Gamma(n) = (n - 1)!$$

◆

1.3.5 Integration with vectors

The fundamental theorem of calculus also works for vectors, since both integration and derivation are linear operations, and they act independently on the three axes of the coordinate system. For example, if the velocity vector is the rate of change of the position vector with time (a scalar),

$$\boldsymbol{v} = \frac{\mathrm{d}\boldsymbol{r}}{\mathrm{d}t}. \tag{1.36}$$

Then the change in position from an initial time t_1 to a final time t_2 is the integral of the velocity vector with respect to time between t_1 and t_2:

$$\boldsymbol{r}(t_2) - \boldsymbol{r}(t_1) = \lim_{\delta t \to 0} \sum_i \boldsymbol{v}(t_i)\delta t = \int_{t_1}^{t_2} \boldsymbol{v}(t)\,\mathrm{d}t. \tag{1.37}$$

We will also sometimes need a different kind of integral involving vectors (for instance, when we look at the work done by a force): this involves integrating *along a path*, instead of with respect to a scalar variable like time in the previous example. The integration along a path involves taking at each point the dot product of the vector representing the function to be integrated (say, $\boldsymbol{v}$) with the small displacement vector $\delta\boldsymbol{r}$ corresponding to a motion along the path. The result is written as

$$\int_{\text{path}} \boldsymbol{v} \cdot \mathrm{d}\boldsymbol{r} = \lim_{\delta r \to 0} \sum_{\text{path}} \boldsymbol{v} \cdot \delta\boldsymbol{r}. \tag{1.38}$$

The two most common ways to perform path integrals are choosing as a path either the straight line between the two extremes of integration, or the sum of three paths, each of them parallel to one of the axes of the coordinate system. IMPORTANT: the result of a path integral is in general dependent on the integration path chosen.

Example 1.18. Consider the integral of the vector field

$$\boldsymbol{F} = x^2\boldsymbol{i} + xy\boldsymbol{j} + y^2\boldsymbol{k}$$

along a straight-line path from $(x, y, z) = (0, 0, 0)$ to $(x, y, z) = (1, 1, 1)$.

The integral can be expanded as

$$\int_{\text{path}} \boldsymbol{F} \cdot \mathrm{d}\boldsymbol{r} = \int_{\text{path}} [x^2\mathrm{d}x + xy\mathrm{d}y + y^2\mathrm{d}z]. \tag{1.39}$$

Now, we need to put the information about the path. Everywhere along this particular path the values of x, y and z are equal, as are the changes δx, δy and δz, when moving along any path segment. We can choose any of the variables to work with. Let's choose x, for the sake of the argument. So we will replace both y and z with x, and δy and δz with δx. With this substitution we can write the integral as

$$\int_{\text{path}} [x^2\mathrm{d}x + xy\mathrm{d}y + y^2\mathrm{d}z] = 3\int_{x=0}^{1} x^2\,\mathrm{d}x = 1. \tag{1.40}$$

(We would have obtained the same result by working in terms of any of the other variables.) Now let's integrate over the sum of three paths, each of them parallel to one axis. This has the advantage that while moving parallel to an axis, the increments with respect to the other one are zero, e.g. while integrating over $\mathrm{d}x$, the components in the directions $\boldsymbol{j}$ and $\boldsymbol{k}$ do not contribute. For the above example, to go from $(0, 0, 0)$ to $(1, 1, 1)$ we first go to $(1, 0, 0)$ along the x-axis, then to $(1, 1, 0)$ moving parallel to the y axis (at constant $(x = 1, z = 0)$), then parallel to the z-axis (at constant $(x = 1, y = 1)$, then sum all the three contributions.

$$\int_{(0,0,0)}^{(1,0,0)} x^2\mathrm{d}x = \left.\frac{x^3}{3}\right|_0^1 = \frac{1}{3} \tag{1.41}$$

$$\int_{(1,0,0)}^{(1,1,0)} xy\mathrm{d}y = \left.\frac{x^2}{2}\right|_0^1 = \frac{1}{2} \tag{1.42}$$

$$\int_{(1,\,1,\,0)}^{(1,\,1,\,1)} y^2 \mathrm{d}z = z\Big|_0^1 = 1. \tag{1.43}$$

And the sum is $\frac{11}{6}$, different from the integral computed over the straight line.

◆

1.4 Differential equations

Differential equations are different from the ordinary equations we are used to. In a differential equation a relation is expressed between a function and its derivatives, and the unknown is not the value of a variable, but a function itself. Differential equations are widely used in physics, and they are a very powerful tool, especially since several physical laws connect the force acting on a body (and the resulting acceleration) to its position or velocity.

In this section, we will only consider linear differential equations applicable to the mechanics that will be dealt with in this book (mostly for oscillatory motion). We have the following two types of equations:

1.4.1 Separable first order ordinary differential equations

Definition 1.4.1: Separable first order ODEs

They are ordinary differential equations of the form:

$$\frac{\mathrm{d}x(t)}{\mathrm{d}t} = -bx(t) \tag{1.44}$$

where b is a constant.

In order to solve this kind of equation, we can separate the two sides, such that on the left we have the x-dependent variables and on the right only t. To be noticed that this operation involves multiplying by an infinitesimal differential quantity $\mathrm{d}t$, so it is not strictly speaking mathematically correct; however, it can be demonstrated that a mathematically accurate approach leads to the same result.

$$\frac{\mathrm{d}x(t)}{x(t)} = -b\ \mathrm{d}t.$$

Integrating,

$$\int \frac{\mathrm{d}x(t)}{x(t)} = \int -b\ \mathrm{d}t$$

such that

$$\ln x = -bt + c.$$

Finally, taking exponentials on both sides:

$$x(t) = e^{-bt}e^{c} = Ae^{-bt}$$

where e^c has been incorporated into the constant A, which can be determined through boundary conditions, for instance the position or velocity of the system at a given time.

1.4.2 First order ordinary differential equation. Integrating factor

The above case shows a homogeneous equation, namely one where all terms contain the function $x(t)$ or one of its derivatives. An equation in the more general form is in general not separable:

$$\frac{\mathrm{d}x}{\mathrm{d}t} + p(t)x = f(t) \tag{1.45}$$

for some functions p and f.

We look to find a function $I(t)$, called **integrating factor**, such that multiplying equation (1.45) by $I(t)$, the left side becomes an exact derivative. So, let's multiply equation (1.45) by a function $I(t)$ and then decide what the right choice for $I(t)$ would be:

$$I(t)x'(t) + p(t)I(t)x(t) = f(t)I(t).$$

The left side of the equation looks like the product rule expansion, so:

$$\frac{\mathrm{d}}{\mathrm{d}t}(I(t)y(t)) = I(t)x'(t) + I'(t)x(t).$$

From the equation above, we can see that we must pick $I(t)$ to be any non-zero solution of:

$$\frac{\mathrm{d}I(t)}{\mathrm{d}t} = p(t)I(t)$$

which has the general solution

$$I(x) = \exp\left(\int p(t)\,\mathrm{d}t\right). \tag{1.46}$$

Remark 1.14: Practical use
For practical purposes we often set the constant of integration to be zero.

1.4.3 Second order homogeneous linear differential equations

Definition 1.4.2: Second order homogeneous linear differential equation
They have the form:

$$a\frac{\mathrm{d}^2x}{\mathrm{d}t^2} + b\frac{\mathrm{d}x}{\mathrm{d}t} + cx = 0.$$

It is linear as the coefficients a, b and c are constant, and homogeneous as once all terms related to the function x are placed on the left-hand side, the right-hand side is 0.

We consider an ansatz[2]: $x(t) = e^{rt}$, where $r \in \mathbb{C}$. Substituting into the differential equation, we obtain:

$$e^{rt}(ar^2 + br + c) = 0. \tag{1.47}$$

For the left-hand side to be zero, considering that the exponential is always positive, the algebraic equation inside the parenthesis (called characteristic equation) has to be zero. There are two quadratic solutions: r_1 and r_2, such that:

$$r_{1,2} = \frac{-b \pm \sqrt{\Delta}}{2a}, \quad \Delta = b^2 - 4ac$$

where Δ is the discriminant.

Remark 1.15: Characteristic equation
In equation (1.47) we call the equation in parenthesis the **characteristic equation**, and it allows one to solve any linear differential equation translating it into the equivalent algebraic equation where derivatives are replaced by corresponding powers of a real variable. The above method shows how to obtain it, however, in practice, the exercise can be started from the characteristic equation and then following the procedure described below.

Thus, the general solution will be given by the linear combination of the exponential of each of the two roots:

[2] **Ansatz:** educated guess that is verified later by result.

$$x(t) = Ae^{r_1 t} + Be^{r_2 t}$$

where A and B are constants to be determined through boundary conditions.

We encounter three different cases depending on the nature of the constants:

- **Both roots are real and different.** $\Delta > 0$. We keep the general solution:

$$x(t) = Ae^{r_1 t} + Be^{r_2 t}.$$

- **Double real root.** $\Delta = 0$. So we have $r_1 = r_2 = r$ and the general solution becomes:

$$x(t) = (A + B)e^{rt}.$$

- **Both roots are imaginary and different.** $\Delta < 0$. We can easily see that the roots are the complex conjugate of each other: $r_1 = a + ib$ and $r_2 = a - ib$. So, the two solutions can also be written as:

$$x_1(t) = e^{rt} = e^{(a+ib)t} = e^{at}[\cos(bt) + i\sin(bt)],$$
$$x_2(t) = e^{at}[\cos(bt) - i\sin(bt)]$$

where we have used Euler's formula $e^{ibt} = \cos(bt) + i\sin(bt)$.

Then, since physical functions must be real numbers, we can take real combinations of these two solutions:

$$X_1 = \frac{1}{2}(x_1 + x_2) = e^{at}\cos(bt),$$
$$X_2 = \frac{1}{2i}(x_1 - x_2) = e^{at}\sin(bt).$$

So that we can write the general solution as:

$$x(t) = AX_1 + BX_2 = e^{at}[A\cos(bt) + B\sin(bt)]$$

with A and B being real numbers.

1.4.4 Second order inhomogeneous linear differential equations

It has the form:

$$a\frac{d^2x}{dt^2} + b\frac{dx}{dt} + cx = f(t).$$

Its solution involves several steps:

- First, we need to find the **general solution of the homogeneous equation**, $x_h(t)$, as detailed above.
- Then, we seek a general solution in the form of an ansatz consisting of the solution of the homogeneous equation and the **particular solution** of the inhomogeneous equation, $x_p(t)$.

$$x(t) = x_h(t) + x_p(t).$$

In general, we determine the particular solution by guessing. However, there are a few standard results that can be applied:

$f(t)$	Particular solution $y_p(t)$
ce^{at}	Ae^{at}
$c_1 \sin(at) + c_2 \cos(at)$	$A \sin(at) + B \cos(at)$
Polynomial of degree n	Polynomial of degree n
$c_1 e^{at} \sin(bt) + c_2 e^{at} \cos(bt)$	$Ae^{at} \sin(bt) + Be^{at} \cos(bt)$

Note: even if $f(t)$ is only a function of *sine, cosine* should be included in the choice of a particular solution. This is also the case for polynomial functions, if there is a polynomial of degree n: $x^n + x^{n-1} + \cdots + c$, all degrees should be included in your choice.

- Finally, the constants can be determined by substituting the ansatz into the differential equation and through boundary conditions.

Exercise 1.19. Find the general solution of the following equations.

(i) $\frac{dx}{dt} = t^2(x^2 + 1)$.

(ii) $\frac{dx}{dt} = t^2(x + 1)$.

Exercise 1.20. Solve the following equations with boundary conditions.

(i)

$$\frac{d^2x}{dt^2} + 2\frac{dx}{dt} + 2x = \sin^2 t$$

with boundary conditions: $x(0) = 2$ and $\frac{dx}{dt} = -1$

(ii)

$$\frac{d^2x}{dt} - 2\frac{dx}{dt} + x = \cosh t \cosh 2t$$

with boundary conditions: $x(0) = 2$ and $\frac{dx}{dt}(0) = -1$

(iii)

$$\frac{d^2x}{dt^2} - x = 7t^2 + 3t$$

with boundary conditions: $x(0) = 2$ and $\frac{dx}{dt}(0) = -1$.

Exercise 1.21.

(i) Look for a solution of the equation

$$\frac{d^2x}{dt^2} - 2\alpha\frac{dx}{dt} + \alpha^2 x = 0 \qquad (*)$$

of the form $x(t) = e^{\lambda t}$ and show that the characteristic equation has the repeated root $\lambda = \alpha$. Hence $x(t) = e^{\alpha t}$ is a solution of (*).

(ii) Verify (by substitution) that $x(t) = te^{\alpha t}$ is a solution of (*) with the initial conditions $x(0) = 0$ and $x'(0) = 1$.

(iii) Next, we will consider an initial value problem *close* to the one above.

(a) Find the solution $X(t)$ of the initial value problem

$$\frac{d^2X}{dt^2} - (2\alpha + \varepsilon)\frac{dX}{dt} + \alpha(\alpha + \varepsilon)X = 0, \quad X(0) = 0, \quad X'(0) = 1,$$

where $\varepsilon \neq 0$.

(b) Show that

$$\lim_{\varepsilon \to 0} X(t) = te^{\alpha t}.$$

[*Hint:* recall that for small z, $e^z \approx 1 + z$.]

Chapter 2

Newton's laws

2.0 Introduction

In this chapter, we will introduce Newton's laws of motion. The discovery of the *laws of dynamics*, or *laws of motion*, was a turning point in the history of physics, and, loosely speaking, in the history of science. Objects in motion were a mystery, but Newton uncovered the underlying laws that govern them.

Galileo first realised that an object would continue to move with a constant speed in a straight line unless disturbed. Galileo's intuition was named *principle of inertia*.

What about objects changing speed due to *something* affecting them? The answer can be found in Newton's laws of motion. Newton would refer to them as *axioms* on which he would build what we nowadays call *Newtonian or classical mechanics.*

We will thus explore the concept of force and motion under a constant force. Motions under gravity and projectiles will be analysed with and without air resistance.

Towards the end of the chapter, we will look at momentum and its conservation for isolated systems.

2.1 Newton's laws of motion

In 1687 Sir Isaac Newton formulated his three laws in his *Principia Mathematica*. They are the basis of classical mechanics and, indeed, axioms of our study.

> **Law 2.1.1: Newton's first law: inertia**
> A body remains in a state of rest ($v = 0$) or uniform motion ($v =$ const.), i.e. not accelerating, unless acted on by an external force.

doi:10.1088/978-0-7503-2690-2ch2

In formulae,

$$\frac{\mathrm{d}\boldsymbol{v}}{\mathrm{d}t} = 0, \text{ i.e. } \boldsymbol{v} = \text{constant} \quad \text{if} \quad \boldsymbol{F}_{\text{ext}} = 0.$$

Law 2.1.2: Newton's second law: motion

The time rate of change of momentum of a body equals the force acting on the body, i.e.

$$\boldsymbol{F} = \frac{\mathrm{d}\boldsymbol{p}}{\mathrm{d}t},$$

where $\boldsymbol{p} = m\boldsymbol{v}$ is the (*linear*) momentum of a body of mass m moving with velocity $\boldsymbol{v}$. If m is constant, then

$$\boldsymbol{F} = \frac{\mathrm{d}\boldsymbol{p}}{\mathrm{d}t} = \frac{\mathrm{d}(m\boldsymbol{v})}{\mathrm{d}t} = m\frac{\mathrm{d}\boldsymbol{v}}{\mathrm{d}t} = m\boldsymbol{a}$$

with acceleration $\boldsymbol{a} = \frac{\mathrm{d}\boldsymbol{v}}{\mathrm{d}t}$.

Focus on the nature of the second law: it simultaneously defines a quantity (force) and says something about it (that it causes a change in momentum).

Remark 2.1. Solving differential equations with Newton's second law

Newton's second law is a differential equation, since it connects acceleration (i.e. the second derivative of a position) to a quantity (i.e. the force) that in general depends on the position itself. More details on this can be found in appendix **A**.

Law 2.1.3: Newton's third law: action and reaction

To every force (action) there is an equal but opposite reaction.

In other words, if a body A exerts a force on body B of $\boldsymbol{F}$, then force of B on A is $-\boldsymbol{F}$.

These laws are only valid in an **inertial** (i.e. non-accelerating) **frame of reference**.

Definition 2.1.1: Inertial frame
A frame of reference in which Newton's first law holds true is said to be an **inertial frame**.

Although this definition may sound tautological, the actual physical significance is represented by the very existence of inertial frames, i.e. this special collection of frames of reference where a body is not moving if no force is applied to it, and accelerates according to the second law.

The most common *practical* reference frame is the surface of the Earth. Even though this is strictly speaking not correct since the planet rotates around its axis, we consider it sufficiently close to an inertial frame for the sake of observing most phenomena occurring on the human scale where the effect of the Earth's rotation can be neglected.

Remark 2.2. The first law is a special case of the second
When $\boldsymbol{F} = 0$, then the acceleration is

$$\boldsymbol{a} = \frac{\mathrm{d}\boldsymbol{v}}{\mathrm{d}t} = 0.$$

Thus, velocity is constant.
In order to convince yourself, you may recall that zero derivative implies constant function. This is the converse of the derivative of a constant. Hence, any function f is a constant *if and only if*, for all x, $f'(x) = 0$.

Remark 2.3. Derivation of the third law from the second
We can also derive the third law from the second law as follows.
Apply a force $\boldsymbol{F}$ to body 1, which is in contact with a second body 2. Body 1 pushes on body 2 with force $\boldsymbol{F}_2$ and body 2 pushes on body 1 with force $\boldsymbol{F}_1$ as shown below.

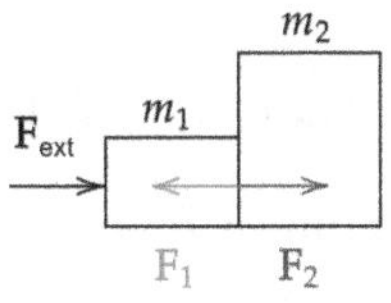

Applying Newton's second law, for the combined system,

$$\boldsymbol{F} = (m_1 + m_2)\boldsymbol{a}$$

If instead we look at body 1 only, the equation becomes

$$\boldsymbol{F} + \boldsymbol{F}_1 = m_1\boldsymbol{a}$$

and if we look at body 2

$$F_2 = m_2\boldsymbol{a}$$

Adding the last two equations, and considering the result of the first:

$$\boldsymbol{F} + \boldsymbol{F}_1 + \boldsymbol{F}_2 = (m_1 + m_2)\boldsymbol{a} = \boldsymbol{F}$$
$$\boldsymbol{F}_1 + \boldsymbol{F}_2 = 0$$

and, hence,

$$\boldsymbol{F}_1 = -\boldsymbol{F}_2$$

Clearly, the forces $\boldsymbol{F}_1$ and $\boldsymbol{F}_2$ between the two subsets of the system are equal in magnitude but opposite in direction.

Example 2.1. Three-dimensional motion

A particle of mass m moves such that its position vector as a function of time is given by

$$\boldsymbol{r}(t) = 2\cos(t)\hat{\boldsymbol{i}} + 2\sin(t)\hat{\boldsymbol{j}} + t^2\hat{\boldsymbol{k}}$$

Calculate the force $\boldsymbol{F}$ which must be acting on this particle to cause this motion. Furthermore, determine the direction of the force on the particle at time $t = 0$.

Solution. We previously found that the acceleration is

$$\boldsymbol{a}(t) = \ddot{\boldsymbol{r}}(t) = -2\cos(t)\hat{\boldsymbol{i}} - 2\sin(t)\hat{\boldsymbol{j}} + 2\hat{\boldsymbol{k}}.$$

Then,

$$\boldsymbol{F} = m\boldsymbol{a} = -2m\cos(t)\hat{\boldsymbol{i}} - 2m\sin(t)\hat{\boldsymbol{j}} + 2m\hat{\boldsymbol{k}}$$

by Newton's second law.

At $t = 0$, we have

$$\boldsymbol{a}(t = 0) = -2\hat{\boldsymbol{i}} + 2\hat{\boldsymbol{k}}$$

Therefore, at $t = 0$,

$$\boldsymbol{F} = m\boldsymbol{a} = m(-2\hat{\boldsymbol{i}} + 2\hat{\boldsymbol{k}})$$

The unit vector defining the direction of $\boldsymbol{F}$ is

$$\frac{\boldsymbol{F}}{|\boldsymbol{F}|} = \frac{m(-2\hat{\boldsymbol{i}} + 2\hat{\boldsymbol{k}})}{\sqrt{m^2(2^2 + 2^2)}} = -\frac{1}{2}\hat{\boldsymbol{i}} + \frac{1}{2}\hat{\boldsymbol{k}}$$

◆

> **Remark 2.4. TOP TIP**
> Remember when you are asked for direction of a force, you are supposed to give the unit vector(s).

Exercise 2.1.

(i) State Newton's laws.
(ii) In what type of frame is Newton's second law valid?
(iii) Why is the first law a special case of the second?
(iv) Derive Newton's third law from Newton's second law.

Exercise 2.2. Three-dimensional motion

Determine the unit vector defining the direction of the force on a particle of mass $m = 1$ kg at time $t = 0$, for each of the following position vectors $\boldsymbol{r}$:

(i) $\boldsymbol{r}(t) = t^2\hat{\boldsymbol{i}} + 2t\hat{\boldsymbol{j}} + 3t^2\hat{\boldsymbol{k}}$.
(ii) $\boldsymbol{r}(t) = \cos(t)\hat{\boldsymbol{i}} + \sin(t\sqrt{3})\hat{\boldsymbol{j}} + te^{10t}\hat{\boldsymbol{k}}$.
(iii) $\boldsymbol{r}(t) = \cos(2t)\hat{\boldsymbol{i}} + 2t^2\hat{\boldsymbol{j}} + e^{t\sqrt{2}}\hat{\boldsymbol{k}}$.
(iv) $\boldsymbol{r}(t) = \sin(2t)\hat{\boldsymbol{i}} + \cos(t\sqrt{3})\hat{\boldsymbol{j}} + te^{2t}\hat{\boldsymbol{k}}$.

◆

2.2 The concept of force

2.2.1 The vector nature of force

Everyone has an intuitive understanding of the concept of force from everyday experience. Typically, we identify a force with a push or a pull. A force is a vector with a magnitude and direction and uses the parallelogram law for addition. In many cases, the point of application of a force is also relevant, since it matters where the force acts. In that case, if a reference frame is present, in addition to the vector representing the force, another vector must be given, connecting the origin to the point of application of the force with respect to the origin.

If a number of forces $\boldsymbol{F}_1, \boldsymbol{F}_2, \ldots, \boldsymbol{F}_n$ act on a point particle, the total or *resultant*, also referred to as *net*, force is just the vector sum of these forces:

$$\boldsymbol{F}_{\text{resultant}} = \boldsymbol{F}_1 + \cdots + \boldsymbol{F}_n = \sum_{j=1}^{n} \boldsymbol{F}_j$$

A single force that has the same magnitude and direction as the calculated net force would then have the same effect as all the individual forces. This fact, called the **principle of superposition** for forces, is connected to the linearity of derivations and integrations, since the acceleration resulting from a sum of forces is the same as the sum of the accelerations due to each of the forces.

2.2.2 Types of forces

Forces can be divided into two types: **contact forces** and **field forces** (a.k.a. *non-contact forces*).

> Definition 2.2.1: Contact force
> A **contact force** is any force which involves physical contact between two objects (or bodies).

Examples of contact forces include:

Tension
Tension is the force that a rope, wire or a stick etc, exert to resist another, external force that is pulling on them. Tension balances the force of the pull until it cannot bear it anymore and the object exerting it breaks. For example, an object tied to a rope hanging from a ceiling. The object will experience a downward force due to gravity which will be balanced by the tension in the rope. If the object increases its weight, so will the tension of the rope increase, and the body will still not move, until the rope, or the ceiling, or the object breaks. Microscopically, the rope is made of molecules that have an electrostatic attraction between them, and pulling them apart results in this attraction getting stronger, compensating the external force.

Normal force
When an object is pushed against a flat surface (for instance, a person standing on a floor under the influence of gravity), it will stay at its position. The push (in this case, of gravity) is balanced by a **normal force**. This normal force is taken to be perpendicular to the surface the object is resting on. Microscopically, the normal force comes from the atomic or molecular lattice of the surface; the external push will produce small deformations to this structure, and the lattice will tend to resist them. So it is also electromagnetic in nature.

Friction
When two objects (or an object and a surface) are in contact, there will be a resistance to them sliding with respect to each other. This resistance depends on the fact that the contact surface between the two bodies is not perfectly flat, and it acts parallel to the surface of the object. The force is applied by the surface the object is resting on, and will be proportional to the normal force between the two surfaces, multiplied by a coefficient μ.

In the approximation we use in this book, friction is:

- not dependent on the contact area;
- not dependent on the relative speed between the surfaces (but different between static and kinetic);
- proportional to the normal reaction force between the surfaces;

Friction is therefore written as:

$$F_f = \mu|N|, \tag{2.1}$$

where μ is the coefficient of friction between the two surfaces, depending on how rough the contact surface is. Indeed, also friction has an electromagnetic nature, since it depends on the repulsion of the external electrons of the two non-flat surfaces when two objects are forced to slide.

Static friction. When an object is at rest on a rough surface the friction between them is labelled as *static* friction. Static friction is written as $F_f \leqslant \mu_s|N|$. This means when a force is applied to the object at rest, the static friction balances this force until the magnitude of the force exceeds $\mu_s|N|$. At this point the object starts sliding and kinetic friction takes over.

Kinetic friction. When the object is in motion on a surface, the friction between them will usually be smaller than the static one, and is called *kinetic* friction. Unlike static friction, kinetic friction is independent of the force pulling the object, and its coefficient is μ_K.

Remark 2.5. When examined in the atomic level, all contact forces are caused by electric forces. The contact surface of both bodies is not perfectly smooth, and friction comes from the electrostatic repulsion between the external layers of electrons of the two surfaces that cannot smoothly slide with respect to each other due to their irregularities.

Definition 2.2.2: Field forces

Field forces, also known as **non-contact** forces, do not involve physical contact, but rather act on objects through open space. A vector field exists around the origin of the force and the direction and magnitude of the force depends on the position of the object in the vector field.

Examples of field forces include the *fundamental forces of nature*:

1. **gravitational forces** between objects,
2. **electromagnetic forces** between electric charges,
3. **strong forces** that keep protons and neutrons together inside nuclei,
4. **weak forces** that arise in certain radioactive decay processes.

Remark 2.6. In classical physics, we are concerned only with gravitational and electromagnetic forces.

2.2.3 Analysis of a physical system using Newton's second law

- In many problems, all you are given is a physical situation (for example, a block resting on a plane, strings connecting masses, etc), and it is up to you to find all the forces acting on all the objects, using $\boldsymbol{F} = m\boldsymbol{a}$. The forces generally point in various directions, so one has to isolate the objects and draw all the forces acting on each of them. Since forces are vectors, it is usually convenient to separate the various components, projecting into a convenient reference frame.
- In other problems, you are given the force explicitly as a function of time, position, or velocity, and the task immediately becomes the mathematical one of solving the $F = ma = m\ddot{x}$ equation.

Statics
If the acceleration of an object modelled as a particle is zero, the object is in equilibrium, and the net force on the object is zero:

$$\sum \boldsymbol{F} = 0 \tag{2.2}$$

If the body is at rest, it stays at rest; if it is moving, it continues to move at constant velocity. The two cases are not so different, because a body moving with constant speed will be at rest in its own reference frame. In such cases, any forces on the body balance one another, and both the forces and the body are said to be in equilibrium. Commonly, the forces are also said to cancel one another, but the term 'cancel' does not mean that they cease to exist.

Dynamics
If an object experiences an acceleration, its motion can be analysed with the particle under a net force model. The appropriate equation for this model is Newton's second law,

$$\sum \boldsymbol{F} = m\boldsymbol{a} \tag{2.3}$$

The net force on an object is the vector sum of all forces acting on the object. (We sometimes refer to the net force as the total force or the resultant force.) In solving a problem using Newton's second law, one must determine all components of the force vector.

The vector expression for Newton's second law is equivalent to three component equations:

$$\sum F_x = ma_x; \quad \sum F_y = ma_y; \quad \sum F_z = ma_z. \tag{2.4}$$

To solve problems with Newton's second law, we often draw a free-body diagram in which the only body shown is the one for which we are summing forces. Since the point-like approximation is made, the forces applied to the body are acting on the same point. Each force is drawn as a vector arrow with its tail anchored on the body. A coordinate system is usually superimposed, oriented so as to simplify the solution, and the acceleration of the body is sometimes shown with a vector arrow (labelled as an acceleration). This whole procedure is designed to focus our attention on the body of interest.

Definition 2.2.3: Free–body diagram
The term *free-body* diagram is used to denote a diagram with all the forces drawn on a given object.

After drawing such a diagram for each object in the set-up, it could be either convenient to write down all the $\boldsymbol{F} = m\boldsymbol{a}$ equations they imply (and solve them as a system of equations), or sum up the forces and get the resultant.

Example 2.2. Consider a chandelier suspended from the ceiling by a chain of negligible mass, as depicted in the figure below. The system can be considered in two dimensions, on the x–y plane, where x is the horizontal and y the vertical direction.

The force diagram shows the two forces acting on the chandelier: the downward pointing gravitational force $m\boldsymbol{g}$ and the upward pointing force $\boldsymbol{T}$ (tension) exerted by the chain.

There are no forces in the $\hat{\imath}$-direction, $\sum F_x = 0$, so there are no further contributions to the system.

From the condition $\sum F_y = 0$, we have

$$\sum F_y = T - mg = 0 \Longrightarrow T = mg.$$

So the tension of the chain perfectly balances the gravitational force of the chandelier, and if the chain does not break (or the ceiling collapse!), it will be able to sustain any weight.

Example 2.3. Equation of motion for a block on a rough horizontal surface
Consider the motion of a block of mass m moving with an initial positive horizontal velocity on a rough horizontal surface.

The external forces acting on it are the gravitational force $\boldsymbol{mg}$, the normal force $\boldsymbol{N}$, and the force of kinetic friction $\boldsymbol{F}_k$.

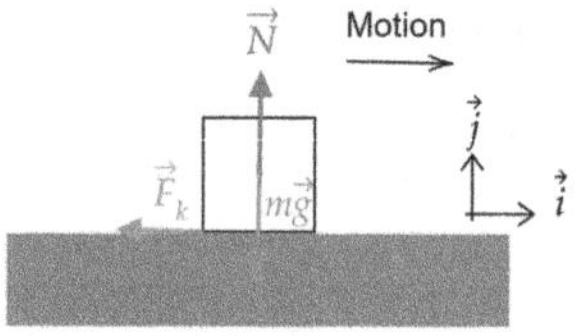

The particle will feel the friction force along the $\hat{\boldsymbol{i}}$-direction:

$$\sum F_x = -F_k = ma$$

While on the $\hat{\boldsymbol{j}}$-direction gravity and the normal reaction will compensate:

$$\sum F_y = N - mg = 0 \Longrightarrow N = mg$$

Substituting $N = mg$ and $F_k = \mu_k N$ into the first equation, we get:

$$-\mu_k N = -\mu_k mg = ma_x$$
$$\Longrightarrow a_x = -\mu_k g.$$

Example 2.4. Roller-coaster physics
A roller-coaster car of mass m is on a part of its track inclined at an angle α that can be approximated as an inclined plane.

(i) If the track is icy, we may assume it to be frictionless. Find the acceleration of the car. What happens if $\alpha = 90°$?

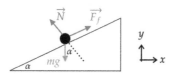

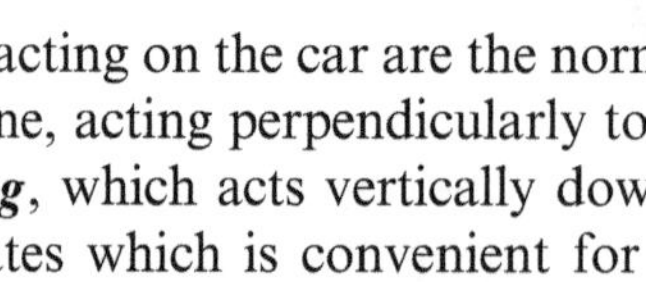

From everyday experience, we know that since the car is on an icy incline, it will accelerate down the incline. (The same thing happens to a car down a hill with its brakes not set!)

The only forces acting on the car are the normal reaction force N exerted by the inclined plane, acting perpendicularly to the plane, and the gravitational force $\boldsymbol{F} = \boldsymbol{mg}$, which acts vertically downwards. We make a clever choice of coordinates which is convenient for our set-up: the coordinate axes with x along the incline and y perpendicular to it as in the diagram.

We find the components of the gravitational force to be one of magnitude $mg \sin \alpha$ along the positive x-axis and one of magnitude $mg \cos \alpha$ along the negative y-axis. We can run a very quick check and we would find out that the magnitude of the gravitational force is indeed $|mg|$ since $(\cos^2\alpha + \sin^2\alpha) = 1$.

Therefore, we have the following system of equations

$$\sum F_x = mg\sin\theta = ma_x$$
$$\sum F_y = N - mg\cos\theta = 0.$$

Solving the equation for $_{ax}$ yields

$$a_x = g\sin\theta.$$

We also note that the acceleration component $_{ax}$ is independent of the mass of the roller-coaster car! It depends only on the angle of inclination and on g. Clearly, for $\alpha = 90°$, the inclined plane becomes vertical, and the car is an object in free fall and $a_x = g$. We find $a_x = g$ rather than $a_x = -g$ because we have chosen positive x to be downward.

(ii) In summertime, ice melts and therefore friction is restored. Find the friction coefficient μ_s needed to prevent the car from moving down the incline.

By definition of friction force, $F_f = \mu|N|$. In this case, $|N| = mg\cos\alpha$. Thus, the equation for the forces acting along x-axis is

$$\sum F_x = mg\sin\alpha - \mu mg\cos\alpha$$

In equilibrium, $\sum F_x$ must be zero. Therefore, combining leads to

$$mg\sin\alpha - \mu mg\cos\alpha = 0 \Longrightarrow \mu = \tan\alpha.$$

Therefore, if $\mu_s \geqslant \tan\alpha$, the car doesn't move.

(iii) Assume that the body starts moving and the friction coefficient is reduced to μ_k, find the acceleration of the mass along its direction of motion.

We are considering the case in which $\mu_k < \tan\alpha$, and the car moves. We now recover an equation for motion along our x-axis which is

$$\sum F_x = mg\sin\alpha - \mu mg\cos\alpha = ma_x$$

Thus,

$$a_x = g(\sin\alpha - \mu_k\cos\alpha)$$

is the acceleration of the mass along its direction of motion.

Example 2.5. Atwood machine

Two blocks, having masses m_1 and m_2 (with $m_1 > m_2$), are connected by a light, inextensible string of length L which hangs over a light, frictionless pulley. The two blocks hang vertically under writing. This set-up is often referred to as *Atwood machine*.

(i) Find the acceleration of the blocks indicating the forces acting on each block and writing down the equation of motion for each. Remembering the constraint that the length L of the string should remain constant, combine your two equations into one and solve it to find the acceleration of each block and the tension in the string.

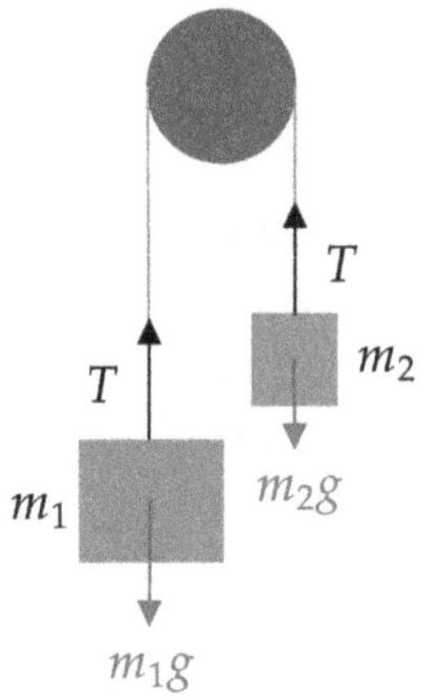

Let the tension in the string be denoted by T. The string is *inextensible*, hence its length is constant. Thus, the accelerations of the two blocks are equal in magnitude but opposite in direction. Let the acceleration of the first block be denoted by a, which is taken to be positive downwards. From the diagram, one may easily notice that each block is acted upon by two forces: the upward pointing tension of the string T, and the downward pointing gravitational force. Then, for the first block

$$m_1g - T = m_1a.$$

Whilst, for the second block,

$$m_2g - T = -m_2a.$$

Subtracting the two equations yields

$$(m_1 - m_2)g = (m_1 + m_2)a,$$

thus

$$a = \frac{m_1 - m_2}{m_1 + m_2}g.$$

In order to check the physical validity of the formula we have just derived, one may consider some limiting cases.

- One may notice that if we have equal masses, i.e. $m_1 = m_2 \equiv m$, then the system would be in equilibrium and would not (and should not) accelerate. Indeed, we would have that

  $$a = \frac{m - m}{m + m}g = 0 \Longrightarrow a = 0,$$

  as required. ✓

- One may also go on and further investigate what happens if we have, say, $m_1 \gg m_2$. We would therefore be able to neglect the effect of m_2, and the larger mass would indeed be falling freely under gravity. From the formula we derived, we indeed get

$$a = \frac{m_1 - m_2}{m_1 + m_2} g \xrightarrow{m_1 \gg m_2} \frac{m_1}{m_1} g = g \Longrightarrow a = g.$$

Remember that we considered a to be positive downwards and the same would apply for g, consistently with the physical world. ✓

Substituting in the initial equation for the first block (or the second) gives

$$T = m_1 g - m_1 a = m_1 g\left(1 - \frac{m_1 - m_2}{m_1 + m_2}\right) = 2\frac{m_1 m_2}{m_1 + m_2} g.$$

(ii) If the blocks are realised from rest when m_2 is a distance ℓ below the pulley, how long is it before the pulley jams?

When released from rest, mass 2 accelerates upwards at

$$a = \frac{m_1 - m_2}{m_1 + m_2} g.$$

It reaches an upward displacement ℓ (and hence jams the pulley) when

$$\frac{1}{2} a t^2 = \ell \Longrightarrow t = \sqrt{\frac{2\ell}{a}} = \sqrt{\frac{2\ell(m_1 + m_2)}{(m_1 - m_2) g}}.$$

Exercise 2.3. A mass of 1 kg is on an inclined plane with an angle α with respect to the horizontal. The plane has a static friction coefficient of $\mu = 0.2$.

(i) Calculate the components of the gravitational force which are parallel and perpendicular to the plane.

(ii) Calculate the minimal value of α for which the parallel component of the gravitational force is stronger than friction, and the body starts moving.

(iii) Assuming that once the body is in motion the friction coefficient reduces to 0.1, calculate the acceleration of the mass along its direction of motion.

Exercise 2.4. A mass m is at rest on a frictionless plane, inclined with an angle α with respect to the horizontal direction. At time $t = 0$, the mass starts sliding under the influence of gravity.

(i) Calculate the forces acting on the mass, the velocity and the distance travelled (computed with respect to a coordinate parallel to the plane) as a function of time t in the direction parallel to the plane, assuming the plane to be frictionless.

(ii) Consider now two frictions: (1) the friction of the plane, acting with friction coefficient $\mu < \tan\alpha$; (2) the air friction acting on the mass, such that resistance can be described by a force $F = -\beta v$, where v is the velocity of the mass. Calculate the velocity as a function of time in the presence of component (1) alone, and of the sum of components (1) and (2).

Exercise 2.5. A point-like body with mass m is placed on a plane, inclined with an angle α to the horizontal. Between the body and the plane there is a friction coefficient $\mu < \tan\alpha$. The body is left free to slide on the plane under the influence of the Earth's gravitational field, starting at a distance d from the bottom of the plane (measured along the plane direction, so the initial height will be $d \sin\alpha$).

(i) Sketch the system and the forces acting on it.

(ii) Show that the time taken for the body to reach the bottom of the plane is

$$t = \sqrt{\frac{2d}{g(\sin\alpha - \mu\cos\alpha)}}.$$

Exercise 2.6. A body with mass m_1 is laying on an inclined plane, that has an angle α with respect to the horizontal (see figure).

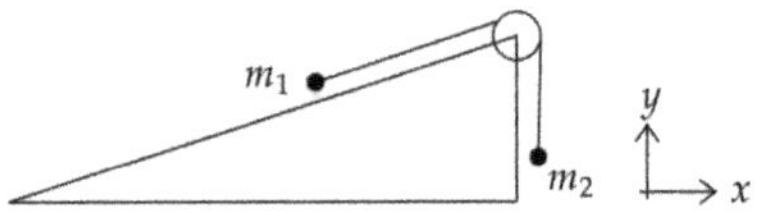

Assume initially no friction between the body and the plane. The body is attached to a massless rope, connected to a massless and frictionless pulley, placed at the top of the plane. On the other side of the pulley, the rope is attached to another body of mass m_2, hanging vertically below it. Both masses are subject to the Earth's gravitational field, and the plane is initially not moving.

(i) Give the expressions for the forces acting on both masses, as well as the tension of the rope and the mass ratio needed to have the system in equilibrium.

(ii) Consider now the case where there is a friction coefficient μ between mass m_1 and the inclined plane. Write the forces acting on the system, clearly separating the case when the rope is moving towards the left and when it is moving towards the right.

(iii) Show that while in the frictionless case the equilibrium between the forces is only possible for a specific value of m_1/m_2, in the presence of friction, due to the opposite sign of the friction force in the two directions, the system is static over a range of mass ratios m_1/m_2, and give the expression for this range.

(iv) Neglect friction again, and assume that the mass ratios are such that the system moves in the direction of mass m_1 (to the left in the figure). Now consider that the inclined plane and the masses feel a constant acceleration in the negative x-direction. Find the acceleration of the plane needed to keep both masses in the same position with respect to the plane (neglect any change in the angle of the rope holding mass m_2). [*Hint: think about the horizontal component of the acceleration of m_1 when the plane is not moving.*]

◆

2.3 Motion under a constant force

2.3.1 General theory

If the force is constant, $\frac{\mathrm{d}\boldsymbol{F}}{\mathrm{d}t} = 0$, if the mass does not change so is the acceleration by Newton's second law:

$$\frac{\mathrm{d}\boldsymbol{F}}{\mathrm{d}t} = m\frac{\mathrm{d}\boldsymbol{a}}{\mathrm{d}t} = 0 \iff \frac{\mathrm{d}\boldsymbol{a}}{\mathrm{d}t} = 0.$$

General theory
We can think of the change in velocity from the area under a graph of constant acceleration:

$$\boldsymbol{v}(t_2) = \boldsymbol{v}(t_1) + \boldsymbol{a}(t_2 - t_1),$$

or by integrating the very simple differential equation

$$\frac{\mathrm{d}\boldsymbol{v}}{\mathrm{d}t} = \boldsymbol{a};$$

$$\int \mathrm{d}\boldsymbol{v} = \int \boldsymbol{a}\mathrm{d}t;$$

$$\boldsymbol{v} = \boldsymbol{a}t + \boldsymbol{C}$$

where $\boldsymbol{C}$ is an arbitrary (vector) constant of integration.

Suppose $\boldsymbol{v} = \boldsymbol{u}$ at $t = 0$ such that $\boldsymbol{u}$ is the initial velocity. This is a **boundary** or **initial condition**, so

$$\boldsymbol{v} = \frac{\mathrm{d}\boldsymbol{r}}{\mathrm{d}t} = \boldsymbol{u} + \boldsymbol{a}t \qquad (2.5)$$

this can be separated and integrated again to give:

$$\int \mathrm{d}\boldsymbol{r} = \int (\boldsymbol{u} + \boldsymbol{a}t)\mathrm{d}t \qquad (2.6)$$

$$\boldsymbol{r} = \boldsymbol{u}t + \frac{1}{2}\boldsymbol{a}t^2 + \boldsymbol{C}_2 \qquad (2.7)$$

where $\boldsymbol{C}_2$ is another vector constant of integration, that can be determined by another boundary (initial) condition. It is easy to see that C_2 corresponds to the value of position $\boldsymbol{r}_0$ at $t = 0$, so the equation can be written as:

$$\boldsymbol{r} = \boldsymbol{r}_0 + \boldsymbol{u}t + \frac{1}{2}\boldsymbol{a}t^2$$

and the displacement from the origin up to a time t is

$$\Delta \boldsymbol{r} = \boldsymbol{r} - \boldsymbol{r}_0 = \boldsymbol{u}t + \frac{1}{2}\boldsymbol{a}t^2 = \frac{1}{2}(\boldsymbol{u} + \boldsymbol{v})t. \tag{2.8}$$

Now notice that $\boldsymbol{a} = \frac{\Delta \boldsymbol{v}}{\Delta t} = \frac{(\boldsymbol{v} - \boldsymbol{u})}{t - 0}$, therefore if we take the dot product between $\boldsymbol{a}$ and $\Delta \boldsymbol{r}$ we get:

$$\begin{aligned}
\boldsymbol{a} \cdot \Delta \boldsymbol{r} &= \boldsymbol{a} \cdot \left(\boldsymbol{u}t + \frac{1}{2}\boldsymbol{a}t^2\right) \\
&= \boldsymbol{u} \cdot (\boldsymbol{v} - \boldsymbol{u}) + \frac{1}{2}(\boldsymbol{v} - \boldsymbol{u})^2 \\
&= \boldsymbol{u} \cdot \boldsymbol{v} - u^2 + \frac{1}{2}v^2 - \boldsymbol{v} \cdot \boldsymbol{u} + \frac{1}{2}u^2 \\
&= \frac{1}{2}(v^2 - u^2), \\
\Rightarrow \quad v^2 &= u^2 + 2\boldsymbol{a} \cdot \Delta \boldsymbol{r}.
\end{aligned}$$

Now we can collect together the equations of motion for linear motion under a constant acceleration (or force)

$$\boldsymbol{v} = \boldsymbol{u} + \boldsymbol{a}t, \tag{2.9}$$

$$\Delta \boldsymbol{r} = \boldsymbol{u}t + \frac{1}{2}\boldsymbol{a}t^2, \tag{2.10}$$

$$v^2 = u^2 + 2\boldsymbol{a} \cdot \Delta \boldsymbol{r}, \tag{2.11}$$

$$\Delta \boldsymbol{r} = \frac{1}{2}(\boldsymbol{u} + \boldsymbol{v})t. \tag{2.12}$$

2.3.2 One dimension

In the special case of one-dimensional motion (i.e. where the velocity, acceleration and displacement are all co-linear) we define s as the distance travelled and then have

$$v = u + at \tag{2.13}$$

$$s = ut + \frac{1}{2}at^2 \tag{2.14}$$

$$v^2 = u^2 + 2as \tag{2.15}$$

$$s = \frac{1}{2}(u + v)t. \tag{2.16}$$

These results should be familiar from school. But beware, it is only applicable to the simple case of constant acceleration, and a typical mistake is using it in cases when it is not appropriate.

Example 2.6. The Department of Transport recommended safe stopping distances (i.e. the total distance travelled from when the driver first sees a reason to stop, to when the vehicle stops) for cars as a function of speed contain two components: a thinking distance s_{think} (i.e. the distance travelled between the moment when you first see a reason to stop, to the moment when you use the brake) and a braking distance s_{brake} (i.e. the distance travelled from the time the brake is applied until the vehicle stops). Let the initial speed of the car be u.

The distance covered while thinking (since during this time we can assume the car's speed to be constant) is

$$s_{\text{think}} = ut_{\text{think}}.$$

The distance covered while braking (with constant deceleration) can be found from $v^2 = u^2 + 2as$; putting $v = 0$ gives

$$s_{\text{brake}} = \frac{u^2}{2(-a)}.$$

(Note that the acceleration will be negative since the car is braking.)

Hence, the total distance is

$$s = ut_{\text{think}} + \frac{u^2}{2(-a)}.$$

The published distances fit this curve with $t_{\text{think}} = 0.68$ s and $a = -13$ m s^{-2}.

Example 2.7. A non-constant force is acting on a body with mass m, at rest, at the origin of a reference frame, according to the time dependence

$$F = \frac{m}{(t + c)^2}$$

where c is an arbitrary constant. Notice that c has to be positive and non-zero otherwise the acceleration at time $-c$ would be infinite. Calculate velocity and position as a function of time.

Solution. Since the force is not constant, also the acceleration is time-dependent, and the simple quadratic equation used above cannot be applied.

$$a = \frac{1}{(t + c)^2}.$$

Then, for velocity

$$v = \int_0^t \frac{d\tilde{t}}{(\tilde{t}+c)^2} = \left.\frac{-1}{\tilde{t}+c}\right|_0^t = \frac{1}{c} - \frac{1}{t + c}.$$

So the velocity starts at zero, but will tend to a value of $1/c$ as time goes on (which is expected since acceleration tends to zero, then velocity becomes a constant).

$$s = \frac{\tilde{t}}{c} - \ln(\tilde{t}+c)\Big|_0^t = \frac{t}{c} + \ln\left(\frac{c}{t+c}\right).$$

The position has a linear component with the terminal velocity, minus another that accounts for the fact that velocity is actually always smaller than the asymptotic value. However, the linear infinity of the first term is stronger than the logarithmic one of the second term.

Exercise 2.7.

1. Given the acceleration vector $\boldsymbol{a}(t)$ of an object at time t, state how you would determine
   (i) the velocity vector $\boldsymbol{v}(t)$,
   (ii) and the displacement vector $\Delta\boldsymbol{r}(t)$ of the object from its starting point.
2. An object has initial velocity $\boldsymbol{u}$ at a time $t = 0$ and has a constant acceleration vector $\boldsymbol{a}$.
   (i) Derive an expression for its velocity $\boldsymbol{v}$ at a later time t.
   (ii) Derive also an expression for its displacement $\Delta\boldsymbol{r}$ at time t, and show that

$$\Delta\boldsymbol{r} = \frac{1}{2}(\boldsymbol{u} + \boldsymbol{v})t.$$

3. An object of mass m is acted on by a constant force $\boldsymbol{F}$. At time $t = 0$ its velocity is $\boldsymbol{u}$ and its position vector is $\boldsymbol{r}_0$. Give expressions for its velocity $\boldsymbol{v}$ and position vector $\boldsymbol{r}$ at a subsequent time t.
4. Show that

$$v^2 = u^2 + \frac{2}{m}\boldsymbol{F} \cdot (\boldsymbol{r} - \boldsymbol{r}_0)$$

5. Show that the position at time t is

$$\boldsymbol{r} = \boldsymbol{r}_0 + \boldsymbol{u}t + \frac{\boldsymbol{F}}{2m}t^2.$$

◆

2.3.3 Free fall under gravity—one dimension

We have seen the general theory of motion under a constant force. We shall now consider the case in which the 'constant force' is gravity.

> **Definition 2.3.1: Free fall**
> We generally refer to any motion of a body which is being acted upon **only by gravity** as *free fall.*

Remark 2.7. Free fall
It might be helpful to note that, when using the term *free fall*, one may not necessarily refer to an object actually falling down, but also to an object being thrown upwards under the action of gravity only.

Let us consider a one-dimensional rectilinear motion happening along, without loss of generality, the z-axis. We consider quantities to be positive if the motion is pointing upwards. Hence, the vector equation

$$\boldsymbol{F} = ma\hat{\boldsymbol{k}} = m\ddot{z}\hat{\boldsymbol{k}}$$

can be replaced by the scalar equation

$$F = ma = m\ddot{z},$$

where a is the acceleration of the body, z its position and F the one-dimensional force by which the body is acted upon. They are all measured in the positive z-direction.

We shall first look at a very simple yet instructive problem, namely the **vertical motion** of a particle under **gravity**, neglecting air resistance. A tennis player, whilst serving, throws a ball of mass m vertically upwards with speed u. Neglecting the ball's spin, and air resistance, we seek the maximum height achieved by the ball (figure 2.1).

As before, our equation of motion is

$$F = m\ddot{z}. \tag{2.17}$$

The only force acting upon the ball is gravity, which points vertically downwards, and we are neglecting air resistance. Therefore one can merrily write

$$F = -mg. \tag{2.18}$$

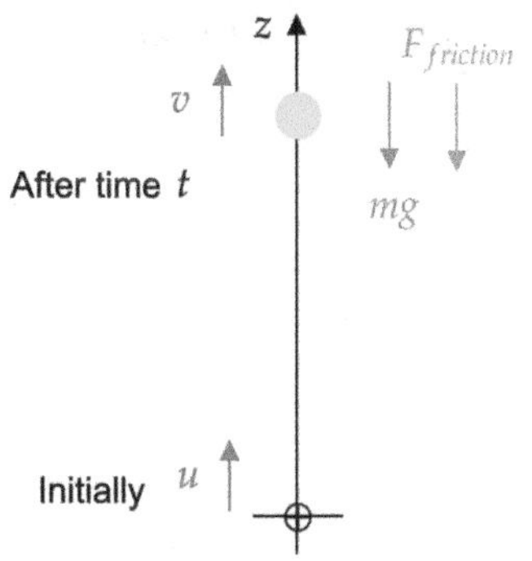

Figure 2.1. The ball is initially at the origin and is thrown vertically upwards in a straight line (axis Oz) with speed u. It is acted upon by the gravity force mg and possibly an air resistance (or drag) force D. At time t the ball has upward velocity v.

Equating equation (2.17) and equation (2.18) one has the equation of motion

$$m\ddot{z} = -mg,$$

which can be identically expressed as

$$m\frac{d^2z}{dt^2} = -mg.$$

Letting $v = \dot{z} = \frac{dz}{dt}$, one can be express the equation of motion as

$$m\frac{dv}{dt} = -mg$$
$$\Longrightarrow \frac{dv}{dt} = -g.$$

Multiplying both sides by dt and integrating gives

$$\int dv = -\int g dt$$
$$v = -gt + c_1$$

where c_1 is a constant of integration. However, by employing the fact that the ball was thrown at some initial velocity $v = u$ for $t = 0$,[1] one may get rid of the integration constant and obtain $c_1 = u$. Hence, the velocity v at time t is given by

$$v = u - gt.$$

In order to obtain the upward displacement z at time t, one can express v as

$$v = \frac{dz}{dt} = u - gt \Longrightarrow \frac{dz}{dz} = u - gt.$$

Again, multiplying both sides by dt and integrating gives

$$\int dz = \int (u - gt)dt$$
$$\Longrightarrow z = ut - \frac{1}{2}gt^2 + c_2$$

where c_2 is a second integration constant. Since we consider the ball to be thrown from the origin of our reference frame, or rather z-axis in this case, we have the initial condition $z = 0$ when $t = 0$, which yields $c_2 = 0$. Hence, the position of the body at time t is given by

$$z = ut - \frac{1}{2}gt^2.$$

[1] This is also defined to be an *initial condition*.

Remark 2.8. Parabolic motion?
Recall that a parabola has an equation of the form:

$$y = f(x) = ax^2 + bx + c,$$

with a, b, c constants. Clearly x and y are variables. In our case, we have $x \mapsto t$ and $y \mapsto z$, further we notice that $c = 0$, $b = u$ and $a = -\frac{1}{2}g$. One may also recall that when $a < 0$, as it is in our case, the parabola opens downwards.

One may thus note that the *shape* of the path is a parabola. Therefore the maximum height is at its vertex. To find such a point, one may find it helpful to recall that the maximum height z_{max} is hence achieved when $\frac{\mathrm{d}z}{\mathrm{d}t} = 0$, i.e. when $v = 0$. For v to be zero, one should have

$$v = u - gt = 0 \Longrightarrow u = gt \Longrightarrow t = \frac{u}{g}.$$

Thus z_{max} is achieved at time $t = u/g$ and is given by

$$z_{\text{max}} = u\left(\frac{u}{g}\right) - \frac{1}{2}g\left(\frac{u}{g}\right)^2 = \frac{u^2}{2g}.$$

After throwing the ball, however, the tennis player realises he did not like his toss, therefore decides not to hit the ball, but rather to catch it, back at its initial position. We now seek how much time the ball takes to return to its initial position. One may note that the ball thus returns to the origin O of our coordinate system, which is simply the x-axis in this instance, when $x = 0$, i.e. when

$$\begin{aligned} x = ut - \frac{1}{2}gt^2 = 0 \\ \Longrightarrow t\left(u - \frac{1}{2}gt\right) = 0 \\ \Longrightarrow t = 0 \quad \text{or} \quad t = \frac{2u}{g}. \end{aligned}$$

Clearly, at $t = 0$ the ball was indeed at its initial position, where it returns after a time $2u/g$.

2.3.4 Particle motion with air friction

We shall now consider a similar problem where we neglect the gravitational force, and our ball is now moving in a fluid (for instance, air). It will feel a resistive friction force that slows down its motion. In the case of laminar flux (i.e. without the formation of vortexes), this force will be proportional and opposite to the particle's velocity:

$$F_{\text{friction}} = -\beta v \tag{2.19}$$

where the coefficient β is the product of the body's area transverse to the motion and a shape-dependent coefficient. Our ball is moving in air with an initial speed u, with no additional external forces acting on it (for instance, moving horizontally on a frictionless plane). Under the only influence of air friction, Newton's second law will be

$$m\frac{dv}{dt} = -\beta v.$$

This first order differential equation can be solved by separation of variables once again, i.e. multiplying both sides by $\frac{dt}{mv}$ and integrating,

$$\int \frac{dv}{v} = -\frac{\beta}{m}dt$$
$$\Longrightarrow \ln(v) = -\frac{\beta}{m}t + \widetilde{C_1}$$
$$\Longrightarrow v = \exp\left(-\frac{\beta}{m}t + \widetilde{C_1}\right) = \exp\left(-\frac{\beta}{m}t\right)\exp(\widetilde{C_1}).$$

One may now set $\exp(\widetilde{C_1}) = C_1$ which yields

$$v = C \exp\left(-\frac{\beta}{m}t\right)$$

where parameter C_1 has to be determined by the initial condition, namely speed at time zero, $v(t = 0) = A = u$. Therefore,

$$v = u \exp\left(-\frac{\beta}{m}t\right).$$

The position as a function of time can be found by integrating the velocity function:

$$x = \int u \exp\left(-\frac{\beta t}{m}\right) dt = -\frac{um}{\beta} \exp\left(-\frac{\beta}{m}t\right) + C_2.$$

One may set as initial condition that at $t = 0$, $x(t = 0) = 0$, so $C_2 = \frac{um}{\beta}$, and the solution is

$$x = \frac{um}{\beta}\left[1 - \exp\left(-\frac{\beta}{m}t\right)\right].$$

Let us now consider the more complete case when the body feels the Earth's gravitational field along with air resistance. Let us consider a body which is left falling at time $t = 0$ from position $x(t = 0) = 0$ and with initial speed $v(t = 0) = 0$. Assuming the positive direction of the x-axis to be pointing downwards, the relation between acceleration and force becomes:

$$m\frac{\mathrm{d}v}{\mathrm{d}t} = mg - \beta v = -(\beta v - mg).$$

If we divide both sides by $m(v - mg/\beta)$ and multiply by $\mathrm{d}t$, we obtain:

$$\frac{\mathrm{d}v}{v - \frac{mg}{\beta}} = -\frac{\beta}{m}\mathrm{d}t. \tag{2.20}$$

Integrating

$$\int \frac{\mathrm{d}v}{v - \frac{mg}{\beta}} = -\int \frac{\beta}{m}\mathrm{d}t$$
$$\Longrightarrow \ln\left(v - \frac{mg}{\beta}\right) = -\frac{\beta}{m}t + \tilde{C}$$
$$\Longrightarrow v - \frac{mg}{\beta} = \exp\left(-\frac{\beta}{m}t + \tilde{C}\right)$$
$$\Longrightarrow v = \frac{mg}{\beta} + \exp\left(-\frac{\beta}{m}t\right)\exp(\tilde{C}).$$

Letting $\exp(\tilde{C}) = C$ and imposing the initial condition that the speed $v = 0$ at $t = 0$, one obtains

$$C = -\frac{mg}{\beta}.$$

Therefore,

$$v = \frac{mg}{\beta}\left[1 - \exp\left(-\frac{\beta}{m}t\right)\right].$$

Let us consider what happens for large values of t, i.e. $t \gg 1$. Because of the minus sign, the exponential term is going to tend to zero and the final velocity will attain a constant value

$$v_f = v(t \gg 1) = \frac{mg}{\beta}.$$

It might be helpful to notice that this is the speed at which the air friction perfectly equates the gravitational pull.

Exercise 2.8. An object with mass m falls through the air towards the Earth.

(i) Assuming that the only forces acting on the object are gravity and air resistance $\boldsymbol{F}_f = -\beta \boldsymbol{v}$, draw a diagram (hence, an appropriate reference frame) and write Newton's second law for the system.

(ii) Now, by writing the acceleration as $a = dv/dt$, verify that the solution for the above equation is of the form

$$v = \frac{mg}{\beta} - \frac{A}{\beta}e^{-\beta t/m} \quad (*)$$

(iii) Assuming the body starts from rest, find coefficient A in equation (*) for $v(0) = 0$.

Exercise 2.9. Two identical bodies with mass m are set free-falling at time $t = 0$, position $z = 0$ and initial speed $v_0 = 0$ in the Earth's gravitational field, with acceleration $-g$ along the z-axis. The first is in vacuum, the second is in air and feels a resistance opposite in direction to its speed: $\boldsymbol{F} = -\beta \boldsymbol{v}$.

(i) Give the expression for the terminal velocity v_t of the body in air at which the air resistance exactly compensates the gravitational force. Show that the time t_t that it takes for the first body (falling in vacuum) to reach the same velocity v_t is $t_t = m/\beta$.

(ii) Write down the forces acting on the body in air at a generic time t, and from that the differential equation connecting velocity and its first derivative (the acceleration). Solve this equation to find the expression of the velocity as a function of time.

(iii) Show that the velocity that the body in air reaches at the time t_t (when the body in vacuum has velocity v_t) is equal to $v_t(1 - 1/e)$.

Exercise 2.10. The baking cup in air

If a light body falls through air, it encounters some degree of air resistance such that its velocity increases and asymptotically approaches a limit value v_L.

The graph below represents the mathematical model according to which the friction force due to air resistance increases when the velocity of the body increases.

Assume that a baking cup of mass 0.7 g and diameter 10 cm falls through air. Its velocity as a function of time varies as in the graph.

Notice from the graph that the velocity of the baking cup, which is initially at rest, increases and asymptotically approaches the limit value $v_L = 1.4\ \mathrm{m\ s^{-1}}$.

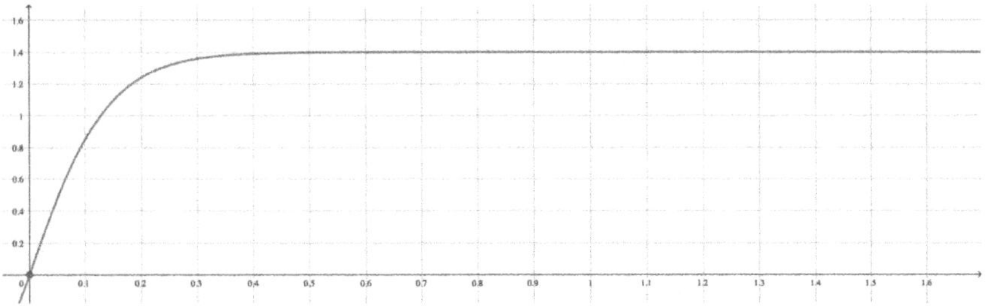

This graph describes a function of the form

$$v(t) = \alpha \cdot \frac{e^{\beta t} - 1}{e^{\beta t} + 1}$$

where α and β are positive real values; t is time in seconds and v is velocity in metres per second.

(i) Describe the variation of the forces acting on the baking cup during the fall and explain why the function $v(t)$ must satisfy the following conditions:
- the value of the function $v(t)$ must asymptotically approach the limit value $v_L = 1.4 \text{ m s}^{-1}$;
- the acceleration $a = dv/dt$ of the baking cup at time $t = 0$ coincides with the acceleration of gravity $g = 9.8 \text{ m s}^{-2}$.

Determine the values of parameters α and β, along with their respective units of measurement, such that the function $v(t)$ satisfies the above mentioned conditions.

Find the coordinates of some points of the function $v(t)$ for $t \geqslant 0$ and verify that they coincide with those of the function in the graph above. Particularly, compare them for $t = 0.1$ s; $t = 0.2$ s; $t = 0.3$ s; $t = 0.4$ s.

(ii) The mathematical model used is based on the assumption that the intensity of the friction force acting on the falling baking cup increases with velocity v according to the relation $F_f = kv^2$, where k is the friction coefficient.

From the graph of $v(t)$ one deduces that, for $t \geqslant 0.5$ s, the velocity of the baking cup can be considered to be constant.

By applying the principle of inertia (Newton's first law) to this stage of the fall and using again the fact that acceleration at time $t = 0$ must equal g, express α and β as a function of the mass m of the baking cup, friction coefficient k and acceleration of gravity g.

(iii) From point (ii) one deduces that the velocity depends on both time t and friction coefficient k. Precisely, given

$$z = \sqrt{k} \quad \text{and} \quad b = 2t\sqrt{\frac{g}{m}}$$

the function of the fall velocity can be written in the form:

$$v = \frac{\sqrt{mg}}{z} \cdot \frac{e^{bz} - 1}{e^{bz} + 1}.$$

Determine how the function v is modified if the parameter k (and, consequently, z) decreases until it vanishes. Interpret the physical meaning of the result obtained.

(iv) Given the following function

$$F(t) = A \ln\left(\frac{e^{14t} + 1}{2}\right) + Bt$$

where A and B are real values, verify that it can be a primitive (anti-derivative) of the function

$$v(t) = 1.4 \cdot \frac{e^{\beta t} - 1}{e^{\beta t} + 1}.$$

Determine the units of measurement of the factor 1.4 in the expression for $v(t)$ and the value of coefficients A, B and β.

Calculate the average function value for the function $v(t)$ in the time interval (expressed in seconds) $0 \leqslant t \leqslant 1$. Give a physical meaning to the value obtained. [*Hint: You may recall that the average value of a function* $f(x)$ *over the interval* $a \leqslant x \leqslant b$ *is given by:* $f_{\text{avg}} = \frac{1}{b-a}\int_a^b f(x)\,\mathrm{d}x$.]

◆

2.4 Projectiles

In 1604–8 Galileo identified the trajectory of a projectile to be parabolic. However, it is remarkable that he was able to achieve a mathematical description of ballistic motion without the mathematical knowledge we possess today (figure 2.2).

> **Definition 2.4.1: Projectile**
> A point-like object which is acted upon only by uniform gravity, and possibly air resistance, is called a **projectile**.

Projectile motion is very common. As a child, the reader might have been pervaded by a sense of fascination when, in ball games, one would throw the ball upwards or forwards, at different angles, and see where it would end up. A ball may indeed be described as a projectile. Clearly, the word *projectile* itself suggests its wider applications in artillery. However, be aware that rocket propulsion in missiles does not allow them to be considered projectiles.

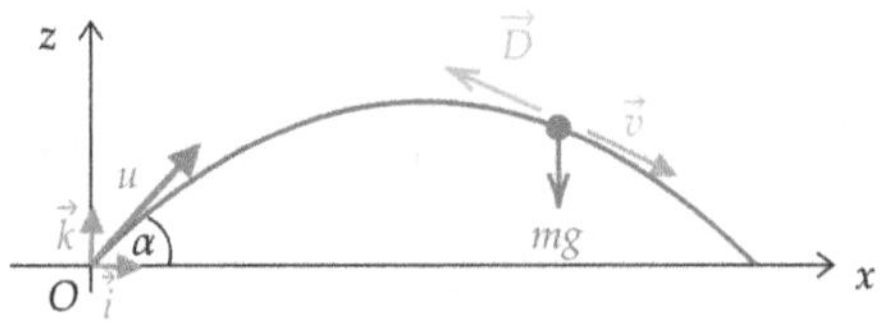

Figure 2.2. Ballistic trajectory.

Throughout our treatment, we will assume that air may (or may not) exert a drag force opposing the velocity of the projectile. Clearly, projectile motion will occur in a vertical plane containing the initial position of the projectile and parallel to its initial velocity.

2.4.1 Projectiles without air resistance

Motion

We will now consider the problem which is equivalent to free fall in gravity, but in three dimensions. Consider a body projected with an initial speed $\boldsymbol{u}$ which makes an angle α with the horizontal (sometimes called *the angle of elevation*).

Suppose that the velocity lies in the (x, z)-plane. In the absence of the drag force (neglecting air resistance),

$$\boldsymbol{u} = u(\cos\alpha\hat{\boldsymbol{i}} + \sin\alpha\hat{\boldsymbol{j}}).$$

The vector equation of motion is therefore

$$m\frac{\mathrm{d}\boldsymbol{v}}{\mathrm{d}t} = m\left(\frac{\mathrm{d}v_x}{\mathrm{d}t}\hat{\boldsymbol{i}} + \frac{\mathrm{d}v_z}{\mathrm{d}t}\right) = -mg\hat{\boldsymbol{k}}, \tag{2.21}$$

whence

$$\frac{\mathrm{d}\boldsymbol{v}}{\mathrm{d}t} = \boldsymbol{a} = -g\hat{\boldsymbol{k}}.$$

Immediately, one can see that

$$\boldsymbol{v} = \boldsymbol{u} + \boldsymbol{a}t = u(\cos\alpha\hat{\boldsymbol{i}} + \sin\alpha\hat{\boldsymbol{j}}) - gt\hat{\boldsymbol{k}} = u\cos\alpha\hat{\boldsymbol{i}} + (u\sin\alpha - gt)\hat{\boldsymbol{k}}.$$

One may now notice that v_x, i.e. the $\hat{\boldsymbol{i}}$-component of $\boldsymbol{v}$, is constant, equal to its initial value

$$v_x = u\cos\alpha,$$

whilst

$$v_y = u\sin\alpha - gt.$$

The position of the particle at time t can now be obtained by integrating the expressions for v_x and v_z, and applying the initial condition of the object being projected from the origin, i.e. $x = 0$ and $z = 0$ when $t = 0$. This gives

$$x = ut\cos\alpha \tag{2.22}$$

$$z = ut\sin\alpha - \frac{1}{2}gt^2 \tag{2.23}$$

the solution for the trajectory of the particle.

An alternative derivation considers the displacement vector equation

$$\begin{aligned}\boldsymbol{r} - \boldsymbol{r}_0 &= \boldsymbol{u}t + \frac{1}{2}\boldsymbol{a}t^2 \\ &= \boldsymbol{u}t - \frac{1}{2}gt^2\hat{\boldsymbol{k}} \\ &= ut\cos\alpha\hat{\boldsymbol{i}} + \left(ut\sin\alpha - \frac{1}{2}gt^2\right)\hat{\boldsymbol{k}},\end{aligned}$$

whence one can read off

$$\begin{aligned}x(t) &= ut\cos\alpha, \\ z(t) &= ut\sin\alpha - \frac{1}{2}gt^2.\end{aligned}$$

Finally, one has

$$|\boldsymbol{v}|^2 = v^2 = v_x^2 + v_y^2 = u^2 + 2\boldsymbol{a}\cdot\Delta\boldsymbol{r} = u_x^2 + u_z^2 - 2gz.$$

Since $u_x^2 = v_x^2$, this implies that

$$v_z^2 = u^2\sin^2\alpha - 2gz.$$

Shape of the path
Let us begin by considering

$$x = ut\cos\alpha \Longrightarrow t = \frac{x}{u\cos\alpha}. \tag{2.24}$$

To find the path taken by the particle we can substitute from equation (2.24) into equation (2.23). This gives a time independent function for z as a function of x:

$$\begin{aligned}z &= u\left(\frac{x}{u\cos\alpha}\right) - \frac{1}{2}g\left(\frac{x}{u\cos\alpha}\right)^2 \\ &= x\tan\alpha - \frac{g}{2u^2\cos^2\alpha}x^2.\end{aligned}$$

Once again, this is indeed the equation of a vertical **parabola** which opens downwards.

The range
The range R on the horizontal plane, or level ground, is found by setting $z = 0$, that is

$$\begin{aligned}z &= x\tan\alpha - \frac{g}{2u^2\cos^2\alpha}x^2 = 0 \\ &\Longrightarrow x\left(\tan\alpha - \frac{gx}{2u^2\cos^2\alpha}\right) = 0.\end{aligned}$$

This equation has two roots, or solutions, i.e. $x = 0$, and the other

$$x = R = \frac{2u^2}{g} \sin\alpha \cos\alpha$$

Recalling that $\sin(2\alpha) = 2 \sin\alpha \cos\alpha$, it can be re-written as

$$R = \frac{u^2}{g} \sin(2\alpha).$$

Time of flight
The time of flight T is the time taken to reach $z = 0$ and is found from equation (2.23)

$$z = ut \sin\alpha - \frac{1}{2} g t^2 = 0 \tag{2.25}$$

whence one has either $t = 0$, which is the initial position, or, for the whole range,

$$T = \frac{2u \sin\alpha}{g}.$$

Since the shape of the trajectory is parabolic, the time of flight T is twice the time $t_{\max}$ taken to reach the maximum height.

Maximum range
R is maximum when

$$\sin(2\alpha) = 1, \quad \text{i.e.} \quad \alpha = \frac{\pi}{4} \tag{2.26}$$

in which case

$$R_{\max} = \frac{u^2}{g}. \tag{2.27}$$

Alternatively, we could have calculated the maximum point of the R function:

$$\frac{\mathrm{d}R}{\mathrm{d}\alpha} = 0$$

$$2\frac{u^2 \cos(2\alpha)}{g} = 0$$

$$\cos(2\alpha) = 0$$

$$\Longrightarrow \alpha = \frac{\pi}{4}.$$

> **Remark 2.9. Projectile motion with air resistance**
> In exercise **2.19**, you will be asked to derive the position and velocity equations for a projectile considering the drag force exerted by air, i.e. air resistance. It is strongly recommended to attempt it as it is indeed highly instructive.

Exercise 2.11. A projectile is launched from ground level at an angle α to the horizontal with initial velocity $\boldsymbol{u} = u(\cos\alpha\hat{\boldsymbol{i}} + \sin\alpha\hat{\boldsymbol{k}})$.

(i) Neglecting air resistance, show that its horizontal displacement and height z at any instant are related by

$$z = x\tan\alpha - \frac{g}{2u^2}x^2\sec^2\alpha.$$

(ii) Find the maximum height attained by the projectile and the horizontal distance travelled before it falls back to height $z = 0$.

(iii) For a given speed u, what launch angle α should be chosen to maximise the horizontal range?

(iv) A human cannonball is a circus act in which a performer is launched from a specially designed compressed-air 'cannon' and lands on a safety net some distance away. The world record distance for a human cannonball flight is approximately 60 m, measured from the mouth of the 'cannon' to the net. Assuming that the safety net was at the same height as the cannon mouth and that the optimum launch angle was chosen for the record attempt, estimate the launch speed of the performer. (You may again neglect the effects of air resistance.)

(v) The length of the cannon barrel in the record attempt was 8 m. Assuming that the performer was accelerated uniformly from rest along the length of the barrel, find the acceleration that would have been necessary to attain the required speed. Hence, find also the magnitude and direction of the apparent g-force (i.e. the apparent weight per unit mass) experienced by the performer during the acceleration.

Exercise 2.12. A child at a fair throws a coconut at a plate on a stall; hitting the plate wins a prize. The child throws the coconut from a starting height h above the ground, with initial speed u and at an angle of elevation α to the horizontal.

(i) Neglecting air resistance, find the height above the ground and the horizontal distance moved by the coconut as a function of time t.

(ii) Hence show also that the vertical height z above the ground is related to the horizontal displacement x by the quadratic equation

$$\frac{gx^2}{2u^2}\tan^2\alpha - x\tan\alpha + z - h + \frac{gx^2}{2u^2} = 0.$$

(iii) If $u = 4$ m s^{-1} and $h = 1$ m, while the plate is a distance $d = 2$ m away and at a height of 0.3 m above the ground, find two possible values of the angle α which would enable the child to hit the plate and win the prize.

Exercise 2.13. A clown stands on a stage directly in front of a person who throws an egg directly at him. Sadly the egg is launched with the wrong combination of initial speed and launch angle to hit his jacket and it goes over the clown's head and the egg lands behind him on the stage.

The egg is launched with an initial speed, u (in m s^{-1}), from a horizontal distance, d (in m), from the stage and a vertical height, h (in m), below the stage at an angle of α to the horizontal.

(i) Neglecting air resistance, the internal dynamics of the egg and the egg-spin, show that the maximum height the egg attains (relative to the launch point), $y_{\max}$, is:

$$y_{\max} = \frac{u^2 \sin^2 \alpha}{2g},$$

where g is the acceleration due to gravity.

(ii) Hence show that, x, the distance (in m) from the edge of stage where the egg lands is given by:

$$x = \frac{u \cos \alpha}{g}\left(u \sin \alpha + \sqrt{u^2 \sin^2 \alpha - 2gh}\right) - d.$$

Exercise 2.14. In an airdrop of supplies, a crate of mass $m = 100$ kg is released at time $t = 0$ from an aircraft moving at a horizontal velocity $\boldsymbol{u} = 100\hat{\boldsymbol{i}}$ m s^{-1} at a height $h = 200$ m above the ground. If air resistance can be neglected, find

(i) the time taken to hit the ground,

(ii) the crate's velocity vector at the moment of impact, and

(iii) the horizontal distance it has travelled from the point of release when the impact occurs. [*Work in a coordinate system where the z-axis points vertically upwards. You may assume that the ground is horizontal.*]

(iv) In the actual drop, air resistance does not remain negligible because a parachute is used. Assume the effect of the parachute is to introduce an extra velocity-dependent force

$$\boldsymbol{F}_{\text{para}} = -\lambda \boldsymbol{v},$$

where $\lambda = 500$ kg s^{-1}. Write down the vector differential equation for the crate's velocity $\boldsymbol{v}$ representing Newton's second law of motion.

(v) Verify that its solution is of the form

$$\boldsymbol{v} = \boldsymbol{A}e^{-\lambda t/m} - \frac{mg}{\lambda}\hat{\boldsymbol{k}}$$

where $\boldsymbol{A}$ is a constant vector and $\hat{\boldsymbol{k}}$ is the unit vector in the z-direction.

(vi) Find an expression for $\boldsymbol{A}$ if the velocity at time $t = 0$ is $\boldsymbol{u}$.

(vii) Describe the evolution of the velocity with time after the parachute has opened and show that during a very long drop the crate's velocity will tend to a constant (the *terminal velocity*).

(viii) Find the magnitude and direction of the terminal velocity in the present case.

Exercise 2.15. A rescue aeroplane flying in a straight line at constant height H and speed U needs to drop supplies to isolated explorers at ground level. Given that the aeroplane eventually flies directly over the explorers, what distance before the explorers does it need to release the supply package if it is to hit them? (Neglect air resistance.)

Exercise 2.16. Safety zone

An anti-aircraft gun is placed at the origin and it fires shells at a fixed speed u, but at any angle α. An air-plane wishes to fly over the gun. Show that the shape of the area that it must avoid is a parabola, called **parabola of safety** (since points *above* the parabola are *safe* from shells).

Exercise 2.17. Squash after lecture

A physics professor enjoys playing squash after his lectures. Squash is a game where you have to hit the ball in such a way that the opponent is not able to play a valid return.

At some point during the match, he hits the ball at a height h above the ground with an angle of 45° with respect to the horizontal. The ball leaves the point located at a distance d from the wall with initial velocity u.

Neglecting air resistance, collisions, spinning and rolling, find how far from the wall the ball hits the ground again.

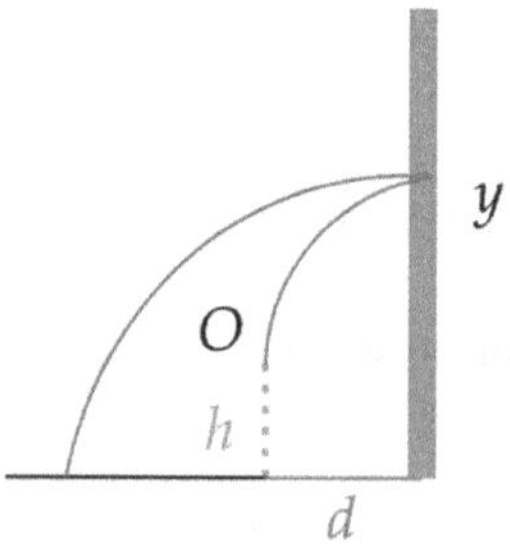

Exercise 2.18. A heavy particle is fired with speed U from the base of an inclined plane. The plane is inclined at an angle α to the horizontal. Find the distance up the plane that the particle lands as a function of the angle θ between the inclined surface of the plane and the initial direction of the particle. (Neglect air resistance.)

Exercise 2.19. Projectile motion with air resistance
Assuming a projectile is shot with an initial velocity v under gravity and a linear resistance force

$$F_D = -\beta v,$$

where, for algebraic convenience, β can be expressed as $\beta = m\kappa$. Assuming the initial speed of the particle u makes an angle α with the horizontal, as in figure 2.2. Derive the equations of the velocity components of the projectile as well as the position components at time t. One may assume that the initial position is the origin, i.e. $x = 0$ and $z = 0$ at $t = 0$. Is the shape of the trajectory still a parabola?

◆

2.5 Momentum and impulse

Definition 2.5.1: Impulse
Consider a force $\boldsymbol{F}$ which acts on a body of mass m during a time interval from t_1 to t_2. The impulse (a vector quantity) of $\boldsymbol{F}$ during this time interval is defined as

$$\boldsymbol{I} = \int_{t_1}^{t_2} \boldsymbol{F}\, \mathrm{d}t.$$

But from Newton's second law, the force at each instant is the rate of change of momentum:

$$\boldsymbol{F} = \frac{\mathrm{d}\boldsymbol{p}}{\mathrm{d}t} \tag{2.28}$$

so

$$\boldsymbol{I} = \int_{t_1}^{t_2} \frac{\mathrm{d}\boldsymbol{p}}{\mathrm{d}t}\, \mathrm{d}t = \int_{p_1}^{p_2} \mathrm{d}\boldsymbol{p} = \boldsymbol{p}_2 - \boldsymbol{p}_1 \tag{2.29}$$

where $\boldsymbol{p}_1 = m\boldsymbol{v}_1$, the momentum of particle at time t_1, and, similarly, $\boldsymbol{p}_2 = m\boldsymbol{v}_2$ at time t_2. Hence, Newton's second law can also be stated as: impulse = change in momentum,

$$\boldsymbol{I} = \Delta\boldsymbol{p} = \boldsymbol{p}_2 - \boldsymbol{p}_1. \tag{2.30}$$

If force, $\boldsymbol{F}$, is constant (in direction and magnitude) throughout the time interval $\Delta t = (t_2 - t_1)$, then

$$\boldsymbol{I} = \boldsymbol{F}(t_2 - t_1) = \boldsymbol{F}\Delta t. \tag{2.31}$$

An important special case is when the force, $\boldsymbol{F}$, is very large and rapidly varying, and Δt is very small, for instance when a ball bounces on a wall. $\boldsymbol{I} = \Delta\boldsymbol{p}$ is finite, and is usually easier to calculate from the momentum difference rather than trying to integrate the force.

Note as $\boldsymbol{I} = \int_{t_1}^{t_2} \boldsymbol{F}\, \mathrm{d}t$, then

$$\boldsymbol{F} = \frac{\mathrm{d}\boldsymbol{p}}{\mathrm{d}t} = \frac{\mathrm{d}\boldsymbol{I}}{\mathrm{d}t}. \tag{2.32}$$

Example 2.8. Particle of mass m bouncing off a wall elastically (i.e. with no loss of kinetic energy (K.E.)).

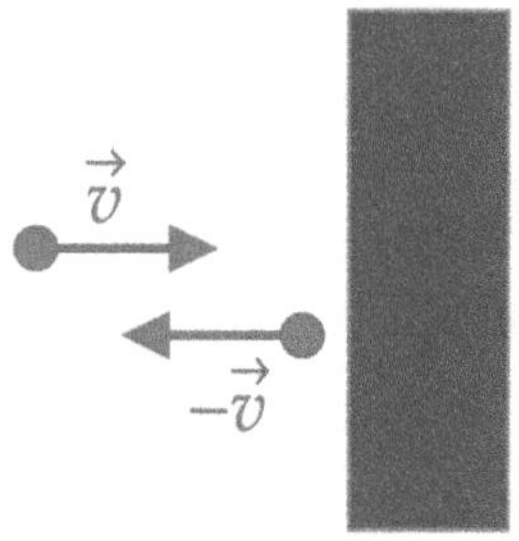

Change in momentum of particle is

$$\Delta p = (-mv) - mv = -2mv$$

which is the impulse on the particle.

Therefore, by Newton's third law, the impulse on the wall is $+2mv$.

◆

2.6 Conservation of momentum for isolated systems

From Newton's second law for a single body,

$$\boldsymbol{F} = \frac{\mathrm{d}\boldsymbol{p}}{\mathrm{d}t}, \quad \text{if } \boldsymbol{F} = 0 \Longrightarrow \boldsymbol{p} \text{ is constant.} \tag{2.33}$$

The same applies to a collection of arbitrarily many bodies: let body i have mass m_i. Let it experience an external force $\boldsymbol{F}_{\mathrm{ext},i}$, and several internal forces $\boldsymbol{F}_{i,j}$ acting from each particle i to to each other particle j. Then, Newton's second law gives

$$\frac{\mathrm{d}\boldsymbol{p}_i}{\mathrm{d}t} = \boldsymbol{F}_{\mathrm{ext},i} + \sum_j \boldsymbol{F}_{i,j}. \tag{2.34}$$

But from Newton's third law,

$$\boldsymbol{F}_{j,i} = -\boldsymbol{F}_{i,j} \tag{2.35}$$

hence the sum of the forces over all values of i and j cancels out, i.e. $\sum_{i,j} \boldsymbol{F}_{i,j} = 0$, therefore adding the equations together,

$$\sum_i \frac{\mathrm{d}\boldsymbol{p}_i}{\mathrm{d}t} = \sum_i \boldsymbol{F}_{\mathrm{ext},i} + \sum_{i,j} \boldsymbol{F}_{i,j} = \sum_i \boldsymbol{F}_{\mathrm{ext},i} \tag{2.36}$$

or

$$\frac{\mathrm{d}\boldsymbol{P}}{\mathrm{d}t} = \boldsymbol{F}_{\mathrm{ext,tot}} \tag{2.37}$$

where

$$\boldsymbol{P} = \sum_i \boldsymbol{p}_i; \qquad \boldsymbol{F}_{\mathrm{ext,tot}} = \sum_i \boldsymbol{F}_{\mathrm{ext},i}. \tag{2.38}$$

The rate of change of the total momentum is equal to the total external force.

> Law 2.6.1: Conservation of momentum
>
> If there is no external force (i.e. if only internal forces act), then
>
> $$\frac{\mathrm{d}\boldsymbol{P}}{\mathrm{d}t} = 0 \tag{2.39}$$
>
> and the total momentum $\boldsymbol{P}$ is conserved.

In terms of impulses, the change ΔP in the total momentum from time t_1 to t_2 is equal to the total impulse $\boldsymbol{I}_{\mathrm{ext}}$ from the external force:

$$\Delta \boldsymbol{P} = \boldsymbol{I}_{\mathrm{ext}} \quad \text{where} \quad \boldsymbol{I}_{\mathrm{ext}} = \int_{t_1}^{t_2} F_{\mathrm{ext},i}\, \mathrm{d}t. \tag{2.40}$$

Example 2.9. Consider a body of mass M moving with velocity V and not subject to any forces. Its momentum remains constant, i.e. $\boldsymbol{p} = M\boldsymbol{V} = \text{constant}$. The body then explodes into several pieces with masses m_i, $i = 1, 2, 3,\ldots$ travelling with velocities v_i, $i = 1, 2, 3,\ldots$. Then.

$$\boldsymbol{p}_1 + \boldsymbol{p}_2 + \boldsymbol{p}_3 + \cdots = m_1 v_1 + m_2 v_2 + m_3 v_3 + \cdots$$

For simplicity, let us consider the case where the explosion only produces three pieces, even if the argument can easily be extended to any number, and gives an impulse $\boldsymbol{I}_i$ to each of them.

Because of Newton's third law (action and reaction are equal and opposite). If impulse on particle 1 due to particle 2 is $\boldsymbol{I}_{12}$ and that due to particle 2 on particle 1 is $\boldsymbol{I}_{21}$, then

$$\boldsymbol{I}_{12} = -\boldsymbol{I}_{21}.$$

For particle 1 the total impulse on it due to the other particles is

$$\boldsymbol{I}_1 = \boldsymbol{I}_{12} + \boldsymbol{I}_{13}$$

and for the whole system we have

$$\begin{aligned} \boldsymbol{I}_1 + \boldsymbol{I}_2 + \boldsymbol{I}_3 &= (\boldsymbol{I}_{12} + \boldsymbol{I}_{13}) + (\boldsymbol{I}_{21} + \boldsymbol{I}_{23}) + (\boldsymbol{I}_{31} + \boldsymbol{I}_{32}) \\ &= (\boldsymbol{I}_{12} + \boldsymbol{I}_{21}) + (\boldsymbol{I}_{23} + \boldsymbol{I}_{21}) + (\boldsymbol{I}_{31} + \boldsymbol{I}_{32}) \\ &= 0 + 0 + 0 \end{aligned}$$

Thus total impulse is zero and so the total momentum is the **same before and after** the explosion,

$$\begin{aligned} \boldsymbol{p} &= \boldsymbol{p}_1 + \boldsymbol{p}_2 + \boldsymbol{p}_3 + \cdots \\ M\boldsymbol{V} &= m_1\boldsymbol{v}_1 + m_2\boldsymbol{v}_2 + m_3\boldsymbol{v}_3 + \cdots \end{aligned}$$

that is, **total momentum is conserved**.

In this case (non-relativistic) mass is conserved,

$$M = m_1 + m_2 + m_3 + \cdots.$$

On the other hand **kinetic energy is not conserved,**

$$\frac{1}{2}MV^2 \neq \frac{1}{2}m_1v_1^2 + \frac{1}{2}m_2v_2^2 + \frac{1}{2}m_3v_3^2 + \cdots$$

but the extra kinetic energy comes from chemical potential energy or energy stored in the springs.

Example 2.10. Two particles, one initially at rest, collide and stick together (this is a fully inelastic collision, where the maximal mechanical energy is converted into heat). The initial momentum is

$$\boldsymbol{p}_i = m_1\boldsymbol{v}_1 + 0.$$

Final momentum is

$$\boldsymbol{p}_f = M\boldsymbol{V} = (m_1 + m_2)\boldsymbol{V}.$$

Since total momentum is conserved, $\boldsymbol{p}_i = \boldsymbol{p}_f$, so $m_1\boldsymbol{v}_1 = (m_1 + m_2)\boldsymbol{V}$ and

$$\boldsymbol{V} = \frac{m_1}{m_1 + m_2}\boldsymbol{v}_1$$

and the combined particle moves in the same direction as the incoming one, but at a reduced speed.

The initial kinetic energy is

$$K_i = \frac{1}{2}m_1v_1^2 + 0.$$

The final kinetic energy

$$K_f = \frac{1}{2}(m_1 + m_2)V^2 = \frac{1}{2}(m_1 + m_2)\left(\frac{m_1}{m_1 + m_2}v_1\right)^2 = \frac{1}{2}\frac{m_1^2}{(m_1 + m_2)}v_1^2 = \frac{m_1}{m_1 + m_2}K_i.$$

Thus there is a **loss** of kinetic energy in a 'sticking' collision. The energy is converted into heat and/or sound.

Exercise 2.20.

(i) Write the expression for the impulse, $\boldsymbol{I}$, received by an object when acted on by a force, $\boldsymbol{F}$.

(ii) Using Newton's second law, find an expression relating $\boldsymbol{I}$ to the momentum of the object.

(iii) Express Newton's third law for two particles in terms of the impulses acting on the particles, and show, in the absence of an external force, that the combined momentum of the two particles is constant.

(iv) Define the momentum of a particle of mass m.

(v) Show, using Newton's laws, that the total momentum of a system of particles is conserved in the absence of external forces.

(vi) Starting from Newton's laws, derive the law of conservation of total momentum $\boldsymbol{P}$ of a system of N interacting particles and state clearly what condition has to be satisfied in order for it to apply.

Exercise 2.21. A particle of mass $m = 2$ kg starts from rest at time $t = 0$. It then experiences, for 2 s, a time-dependent force given by

$$\boldsymbol{F}(t) = [t(2 - t)\hat{\boldsymbol{i}} + 4\hat{\boldsymbol{j}}]\,\text{N},$$

where t is the time measured in seconds. After time $t = 2$ the force is zero.

(i) Calculate the total impulse imparted to the body by the force. Hence, find the body's velocity at time $t = 2$.

(ii) Show that after two seconds the object has moved approximately 4.06 m from its original position. In what direction has it moved?

(iii) Describe the particle's motion after the force has ceased to act. Hence, find its displacement vector after a total time $t = 5$ s has elapsed (i.e. 3 s after the force ceased to act).

Exercise 2.22. A ball with mass m falls under gravity, starting at time $t = 0$ from height h and initial horizontal velocity v_0. At the point of impact with the ground, there is a movable vertical wall with negligible mass, attached to a horizontal spring with elastic constant k.

(i) Calculate the horizontal position x_s of the ball when it hits the ground (and the wall).

(ii) Immediately after the ball bounces upward, it will hit the wall. Its vertical upward movement will not be perturbed by the presence of the wall, assumed frictionless. However, horizontally, it will feel the elastic force slowing down the horizontal component of the movement. Write down the forces acting on the ball before and after hitting the wall.

(iii) Further, indicate the forces acting on the system in that moment, in the case the wall has air friction, producing a resistance $F = -\beta v$.

(iv) In the frictionless case, determine the condition for which the horizontal and vertical velocities reach zero exactly at the same time (the first slowed down by the elastic force, the second by gravity).

Exercise 2.23. An end-of-examinations party has gotten out of hand, and a student A of mass m is running towards the edge of a fenced swimming pool at a horizontal velocity of $v_0\hat{\boldsymbol{i}}$, intending to jump in an attempt to clear the fence.

(i) Assuming that the impulse $I_z\hat{\boldsymbol{k}}$ that A receives from the ground on jumping is purely vertical, show that the velocity vector of the student at time t after take-off is

$$\boldsymbol{v}(t) = v_0\hat{\boldsymbol{i}} + \left(\frac{I_z}{m} - gt\right)\hat{\boldsymbol{k}}.$$

(ii) Find also an expression for the student's displacement vector $\boldsymbol{r}$ from the launch point at time t. [*You may neglect the effects of air resistance.*]

(iii) If the fence has a height h, find the minimum vertical impulse $I_{\min}$ that A will need to achieve in the jump to raise the centre of mass by an amount h in order to just clear the fence.

(iv) If the vertical impulse is only just sufficient to clear the fence, find also the horizontal distance d from the fence from which A will have to jump in order to clear it.

(v) If $h = 1$ m, $m = 60$ kg and $v_0 = 5$ m s^{-1}, evaluate your expressions for $I_{\min}$ and d.

Chapter 3

Kinematic relations

3.0 Introduction: what is energy?

The concept of energy is something with which we are very familiar from everyday life.

Example 3.1. Student: I don't have enough energy to study classical mechanics, ugh ···

◆

What the student is referring to in this example is the chemical energy that our body gains by breaking apart sugar molecules in our cells. Or maybe even something more intangible, rather like 'goodwill'. However, there are many other types of energy and in this chapter we will focus on kinetic and potential energy. Roughly speaking, kinetic energy is linked to the movement of bodies, while mechanical potential energy is linked to their position in the presence of a conservative field (we will see what this is later on), so the ability of producing a movement.

We will thus begin by introducing the concept of work and see how it is related to energy. Secondly, we will define potential energy, first considering the 1D case and then extending our analysis to 3D. Thirdly, we will speak about the conservation of mechanical energy. Finally, we will extend the work–energy theorem to systems of many particles.

3.1 Work and energy

Definition 3.1.1: Work

Work is the application of a force to a body through a displacement, and it leads to energy transfer to that object.

The infinitesimal work done by a force $\boldsymbol{F}$ through an infinitesimal displacement $d\boldsymbol{r}$ is

doi:10.1088/978-0-7503-2690-2ch3

$$dW = \boldsymbol{F} \cdot d\boldsymbol{r}. \tag{3.1}$$

Notice that work is a scalar, while both force and displacement are vectors. To obtain the total work done going from point r_A to point r_B we need to integrate over all the infinitesimal displacements between the two extremes of the motion (figure 3.1):

$$\begin{aligned} W &= \int_{r_A}^{r_B} \boldsymbol{F} \cdot d\boldsymbol{r} = \int_{r_A}^{r_B} F \cos\alpha \, dr \\ &= \int_{r_A}^{r_B} [F_x \, dx + F_y \, dy + F_z \, dz] \end{aligned} \tag{3.2}$$

where α is the angle between the directions of the force and of the displacement. The S.I. unit of work (and energy) is the joule (J) and 1 J = 1 N m.

Remark 3.1. There are a few things to notice here:

(i) The scalar product between force and displacement converts vector quantities into a scalar, i.e. a number.

(ii) Only the component of the force in the direction of the displacement matters; this is given by $F \cos\alpha$.

(iii) Remember that the scalar product $\boldsymbol{F} \cdot d\boldsymbol{r}$ can also be expressed as $F_x \, dx + F_y \, dy + F_z \, dz$.

(iv) If the force is constant (in both magnitude and direction), work is given by

$$W = \boldsymbol{F} \cdot (\boldsymbol{r}_B - \boldsymbol{r}_A).$$

If you have any doubt on any of these remarks, go back to chapter 1 on the mathematical preliminaries of this course. In particular, line integrals are considered in section 1.3.5.

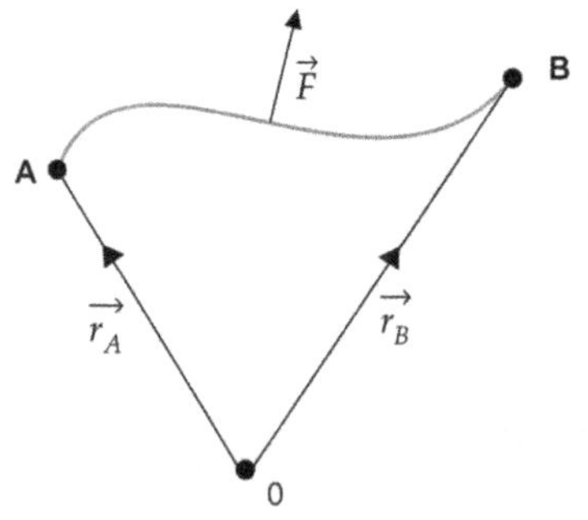

Figure 3.1. Work done from point r_A to point r_B.

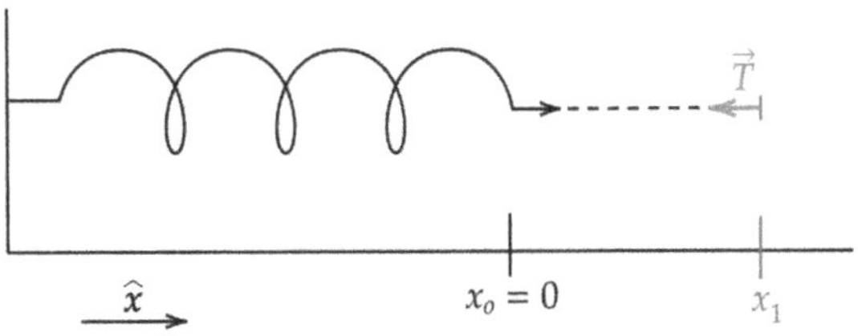

Figure 3.2. A spring is stretched through a distance x in the positive x direction.

Example 3.2. Work done on a spring

Consider an unstretched spring as shown in figure 3.2. We define an x-axis pointing to the right and we set $x = 0$ at the position at which the non-fixed end of the spring is found at equilibrium, i.e. where the spring is if unstretched and not compressed.

We know from experience that, if we stretch or compress a spring, a restoring force will act to oppose the movement and take the spring back to its equilibrium position (as long as the extension or compression is not too big).

Let's now try to see what is the work done when we stretch a spring from its equilibrium position $x_0 = 0$ to a position $x_1 = x$ in the positive x direction. Clearly, the distance between the two points is x.

Hooke's law tells us that the restoring force of the spring is proportional to the distance through which the spring is stretched (or compressed), i.e. $\boldsymbol{F}_{\text{spring}} = \boldsymbol{T} = -k\boldsymbol{x} = -kx\hat{\boldsymbol{x}}$, where k is the spring constant and the negative sign comes from the fact that the force opposes the motion. Notice that, if we were compressing the spring, then the force would be acting in the positive x direction, always trying to restore the equilibrium position.

In this case, with the spring stretched to the right, we have that $F_{\text{spring}} = -kx$ and thus to stretch the spring we have to apply a force $F_{\text{ext}} = +kx$. What is the work *done by the external force*?

$$W = \int_{x_0}^{x_1} F_{\text{ext}}\, \mathrm{d}x = \int_0^x kx'\, \mathrm{d}x' = \frac{1}{2}kx^2,$$

where x' is a dummy variable—see chapter 1, remark 1.13.

Note that the work *done by the spring* is equal but opposite to that done by the external force. Since the work done by the spring is negative, we say that *the spring has work done on it.*

Example 3.3. Work done by gravity

Consider a particle of mass m falling due to gravity in the negative z direction from z_1 to z_2 through a distance h, such that $\Delta z = z_2 - z_1 = -h$ (figure 3.3).

The force applied to the particle is thus $F_g = -mg$ in the z direction. The work *done on the particle* is

$$W = \int_{z_1}^{z_2} F_g\, \mathrm{d}z = -\int_{z_1}^{z_2} mg\, \mathrm{d}z = -mg(z_2 - z_1) = -mg(-h) = mgh.$$

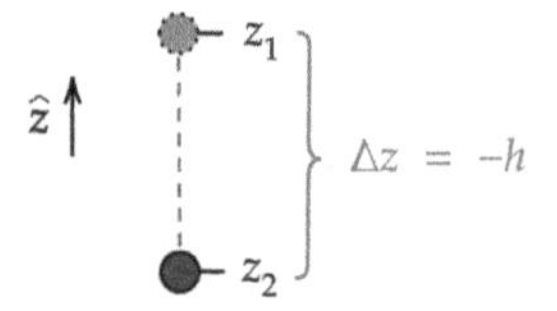

Figure 3.3. A particle falling due to gravity in the negative z direction from z_1 to z_2 through a distance h.

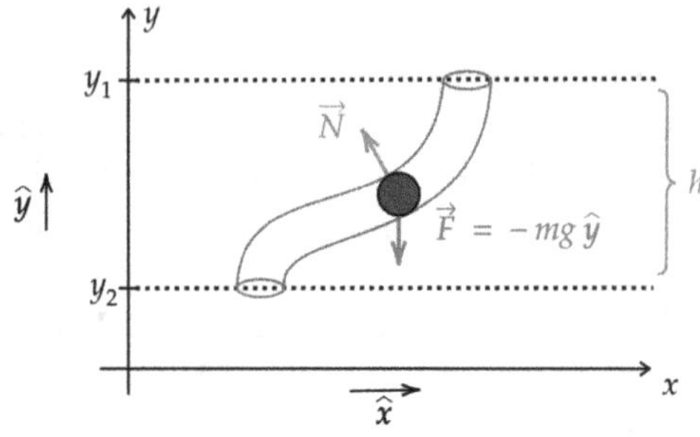

Figure 3.4. A particle inside a smooth tube falling due to gravity.

Example 3.4. Work done by gravity on a particle in a smooth tube

Consider a setup as shown in figure 3.4. A particle of mass m is falling due to gravity inside a *smooth* tube. The tube is in the x–y plane and goes from height y_1 to y_2 through a distance h, such that $\Delta y = y_2 - y_1 = -h$.

The forces acting on the particle are just the gravitational force and the normal force of the tube (at every point perpendicular to it). The gravitational force is given by $\boldsymbol{F}_g = -mg\hat{\boldsymbol{y}}$ and the resulting force acting on the particle is given by $\boldsymbol{F}_R = \boldsymbol{N} + \boldsymbol{F}_g$. Therefore, the infinitesimal work done on the particle when it moves through a distance $\mathrm{d}\boldsymbol{r}$ in the tube is

$$\mathrm{d}W = (\boldsymbol{N} + \boldsymbol{F}_g) \cdot \mathrm{d}\boldsymbol{r} = \boldsymbol{N} \cdot \mathrm{d}\boldsymbol{r} + \boldsymbol{F}_g \cdot \mathrm{d}\boldsymbol{r}.$$

However, as already mentioned, $\boldsymbol{N}$ is perpendicular to the tube, and thus to the direction of motion $\mathrm{d}r$, at every point. Therefore, $\boldsymbol{N} \cdot \mathrm{d}\boldsymbol{r} = 0$ at every point of the tube and we are left with $\mathrm{d}W = \boldsymbol{F}_g \cdot \mathrm{d}\boldsymbol{r}$.

Since the motion is in the x–y plane, $\mathrm{d}\boldsymbol{r}$ can be written as $\mathrm{d}\boldsymbol{r} = \mathrm{d}x\hat{\boldsymbol{x}} + \mathrm{d}y\hat{\boldsymbol{y}}$. Then it follows that

$$\mathrm{d}W = \boldsymbol{F}_g \cdot \mathrm{d}\boldsymbol{r} = -mg\hat{\boldsymbol{y}} \cdot (\mathrm{d}x\hat{\boldsymbol{x}} + \mathrm{d}y\hat{\boldsymbol{y}}) = -mg\mathrm{d}y,$$

since $\hat{\boldsymbol{y}} \cdot \hat{\boldsymbol{x}} = 0$ and $\hat{\boldsymbol{y}} \cdot \hat{\boldsymbol{y}} = 1$.

Notice that this setup has thus been simplified to the same situation as the previous example, even though this one seemed more complicated. Let's show that the work done on the particle is the same in both cases:

$$W = \int_{y_1}^{y_2} -mg\,\mathrm{d}y = -mg(y_2 - y_1) = -mg(-h) = mgh.$$

This example shows that purely constraining forces (like normal forces, tensions etc) do no work.

Notice that the fact that the tube is here set to be smooth is a crucial piece of information, because, if it weren't, there would be friction forces acting on the particle, and work would be lost to win this friction.

◆

3.2 Relationship between work and kinetic energy

By manipulating equation (3.2), we can obtain an important relation between work and kinetic energy.

Since $\boldsymbol{F} = m\boldsymbol{a} = m\frac{\mathrm{d}\boldsymbol{v}}{\mathrm{d}t}$, we have

$$W = \int_{r_A}^{r_B} \boldsymbol{F} \cdot \mathrm{d}\boldsymbol{r} = m \int_{r_A}^{r_B} \boldsymbol{a} \cdot \mathrm{d}\boldsymbol{r} = m \int_{r_A}^{r_B} \frac{\mathrm{d}\boldsymbol{v}}{\mathrm{d}t} \cdot \mathrm{d}\boldsymbol{r}.$$

Rearranging the previous equation and remembering that velocity is just the derivative of position with respect to time, i.e. $\boldsymbol{v} = \frac{\mathrm{d}\boldsymbol{r}}{\mathrm{d}t}$, we obtain

$$W = m \int_{r_A}^{r_B} \mathrm{d}\boldsymbol{v} \cdot \frac{\mathrm{d}\boldsymbol{r}}{\mathrm{d}t} = m \int_{v_A}^{v_B} \boldsymbol{v} \cdot \mathrm{d}\boldsymbol{v}. \tag{3.3}$$

By applying the definition of scalar product, we observe:

$$\boldsymbol{v}\mathrm{d}\boldsymbol{v} = v_x \mathrm{d}v_x + v_y \mathrm{d}v_y + v_z \mathrm{d}v_z = \frac{1}{2}\mathrm{d}\left(v_x^2 + v_y^2 + v_z^2\right) = \frac{1}{2}\mathrm{d}(v^2). \tag{3.4}$$

If you are confused, try working through the passages backwards. For instance, $d(v_x^2) = 2v_x d(v_x)$, so we have that $v_x d(v_x) = \frac{1}{2}d(v_x^2)$. Note that we cannot integrate until we have transformed the scalar product $\boldsymbol{v} \cdot \mathrm{d}\boldsymbol{v}$ into the scalar $\frac{1}{2}\mathrm{d}(v^2)$.

Therefore, by combining equations (3.3) and (3.4) we obtain:

$$W = \frac{1}{2} m \int_{v_A}^{v_B} \mathrm{d}(v^2) = \frac{1}{2} m \left(v_B^2 - v_A^2\right) = \frac{1}{2} m\, v_B^2 - \frac{1}{2} m\, v_A^2.$$

Definition 3.2.1: Kinetic energy
We define the kinetic energy K as

$$K = \frac{1}{2} m\boldsymbol{v} \cdot \boldsymbol{v} = \frac{1}{2} m v^2. \tag{3.5}$$

Note that energy is a scalar, as expected.

By defining kinetic energy in such a way, we obtain an important result.

Theorem 3.2.1: Work–energy theorem
The work done by an external force on a particle through a path that goes from point A to point B equals the change in kinetic energy of the particle between points A and B. That is,

$$W = \Delta K. \tag{3.6}$$

This theorem allows us to define energy as the capacity of a body to perform work.

3.3 Power

Definition 3.3.1: Instantaneous power
Instantaneous power P is the rate of change of work per unit time:

$$P = \frac{\mathrm{d}W}{\mathrm{d}t}. \tag{3.7}$$

The S.I. unit is the joule/second (J s^{-1}) or watt (W).

Remark 3.2. Other units for power
There exist other units of power, although obsolete or not officially accepted anymore. For instance, car manufacturers use *horsepower* as a unit for the power of their engines. One metric horsepower (hp) is defined as the power to raise a mass of 75 kg against the Earth's gravitational field over a distance of 1 m in 1 s. This is equivalent to 735.499 W. We could as well use *duckpower* as a unit. Since a duck can lift 250 g at 1 m s^{-1}, one horsepower is equal to 300 duckpower[1]. To match the power of a Ferrari SF90's engine (990 hp) we would need 297 000 ducks.

Since $\mathrm{d}W = \boldsymbol{F} \cdot \mathrm{d}\boldsymbol{r}$,

$$P = \frac{\mathrm{d}W}{\mathrm{d}t} = \boldsymbol{F} \cdot \frac{\mathrm{d}\boldsymbol{r}}{\mathrm{d}t} = \boldsymbol{F} \cdot \boldsymbol{v}.$$

[1] Reddit, *Converting horsepower to duckpower* https://www.reddit.com/r/theydidthemath/comments/4oyzjm/converting_horsepower_to_duckpower/.

From this last equation, and using the usual $\boldsymbol{F} = m\boldsymbol{a}$, we have

$$P = \boldsymbol{F} \cdot \boldsymbol{v} = m\frac{\mathrm{d}\boldsymbol{v}}{\mathrm{d}t} \cdot \boldsymbol{v} = m\frac{\boldsymbol{v} \cdot \mathrm{d}\boldsymbol{v}}{\mathrm{d}t}.$$

But we have already seen in equation (3.4) that $\boldsymbol{v} \cdot \mathrm{d}\boldsymbol{v} = \frac{1}{2}\mathrm{d}(v^2)$. Therefore,

$$P = \frac{1}{2}m\frac{\mathrm{d}(v^2)}{\mathrm{d}t} = \frac{\mathrm{d}\left(\frac{1}{2}mv^2\right)}{\mathrm{d}t} = \frac{\mathrm{d}K}{\mathrm{d}t}.$$

Note that constants can be taken in or out of differentials without any problems.

Let's summarize these results by giving an alternative definition for power:

Definition 3.3.2: Instantaneous power

Instantaneous power P is the rate of change of kinetic energy per unit time, i.e.

$$P = \boldsymbol{F} \cdot \boldsymbol{v} = \frac{\mathrm{d}K}{\mathrm{d}t}. \tag{3.8}$$

For now, we have been discussing instantaneous power. What about average power?

Definition 3.3.3: Average power

The average power $\bar{P}$ over a time interval Δt is given by

$$\bar{P} = \frac{W}{\Delta t}$$

where W is the work done in the time interval.

Exercise 3.1. State the infinitesimal amount of work done by a force $\boldsymbol{F}$ when its application point moves a distance $\mathrm{d}\boldsymbol{r}$. Moreover, define the instantaneous power P and prove that when the application point of the force is moving at a velocity $\boldsymbol{v}$ the power can be written as $P = \boldsymbol{F} \cdot \boldsymbol{v}$.

Determine the power required to pull a body of mass m at a constant sped v_0 up an inclined smooth plane with an angle α. Why is it not necessary to specify the direction of the force applied in this case?

Exercise 3.2. Give the equation for the kinetic energy of a particle moving at velocity v, with mass m.

Show, using Newton's laws of motion, that the rate of change of the kinetic energy is equal to the power developed by the force acting on the particle.

◆

3.4 Potential energy and conservative forces

To introduce potential energy, let's start by looking at the simplest possible case, i.e. a one-dimensional space.

3.4.1 Potential energy in 1D

Consider a one-dimensional force acting on a particle in the positive x direction. We have that $\boldsymbol{F} = F\,\hat{\boldsymbol{x}}$ and $\mathrm{d}\boldsymbol{x} = \mathrm{d}x\,\hat{\boldsymbol{x}}$. As we have seen, the work done by the force on the particle in going from point x_1 to point x_2 is

$$W = \int_{x_1}^{x_2} F\,\mathrm{d}x.$$

> Remark 3.3. Note that since we are in a one-dimensional problem, the force and the displacement must either be parallel or anti-parallel. As in this situation the force is acting in the positive x direction, they are parallel and the scalar product $\boldsymbol{F}\cdot\mathrm{d}\boldsymbol{x}$ is simply equal to $F\,\mathrm{d}x$.

Now suppose $F \equiv F(x)$, i.e. the force depends only on the position of the particle (and not, for instance, on its velocity). Then, we can define the potential energy V in the following way.

> **Definition 3.4.1: Potential energy (1D)**
> In 1D, given a force $F \equiv F(x)$, which depends only on the position of the particle on which it is acting, there exists a quantity V—the potential energy—such that its change from point x_1 to point x_2 corresponds to the work that would need to be done by an external force F_{ext}, which opposes F, to move the particle between those two points. In formulae,
>
> $$V(x_2) - V(x_1) = W_{\mathrm{ext}} = \int_{x_1}^{x_2} F_{\mathrm{ext}}\,\mathrm{d}x = -\int_{x_1}^{x_2} F\,\mathrm{d}x = -W. \tag{3.9}$$
>
> It follows that
>
> $$V(x) = -\int F\,\mathrm{d}x. \tag{3.10}$$

> Remark 3.4. Note that *only differences in potential energy matter*. In fact, we can express the potential energy in terms of an indefinite integral, but it will imply a constant of integration to which we need to assign a value. To do so, we need to set an (arbitrary)

reference potential energy with respect to which the potential energy $V(x)$ of a general point x is calculated. In cases in which the potential energy is inversely proportional to distance, it is common to set the potential energy to zero at infinity, i.e. $V(x) = 0$ as $x \to \infty$ in 1D, because in this way the constant of integration becomes zero. We give an example of this situation later, in the case of the gravitational potential energy between two point masses. Another similar case is the electric potential energy between two stationary, electrically charged particles. When on the other hand the potential energy is proportional to the height of an object in gravitational field assumed to be locally constant (see following example), it makes sense to choose the ground as the point of zero potential. For the potential energy due to an elastic force, the equilibrium position (for instance, when a spring is at rest) is usually chosen.

Example 3.5. Constant gravitational force

We'll give a very simple example to show what we mean when we say that only differences in potential energy matter.

Consider a particle of mass m falling due to Earth's gravity in the negative z direction (figure 3.5). Since the displacement involved is much smaller than the distance with respect to the centre of the Earth, the gravitational force acting on the particle is considered to be constant: $\boldsymbol{F}_g = -mg\hat{\boldsymbol{z}}$. What's the particle's potential energy? Well, we have just seen that $V(z) = -\int F \, \mathrm{d}z$, so

$$V(z) = -\int (-mg) \, \mathrm{d}z = mgz + c, \tag{3.11}$$

where c is a constant of integration. The presence of this constant is crucial, because it could take any value and therefore we cannot uniquely define the particle's potential energy.

However, say we want to find the particle's potential energy at position z_1 with respect to position z_0 (with $z_1 > z_0$). Then,

$$\begin{aligned}\Delta V = V(z_1) - V(z_0) &= mgz_1 + c - (mgz_0 + c)\\ &= mgz_1 - mgz_0 + \cancel{c} - \cancel{c} = mg(z_1 - z_0).\end{aligned}$$

Now, we can assign a specific value to the potential energy of the first position. We say that the particle at position z_1 has potential energy ΔV with respect to position z_0.

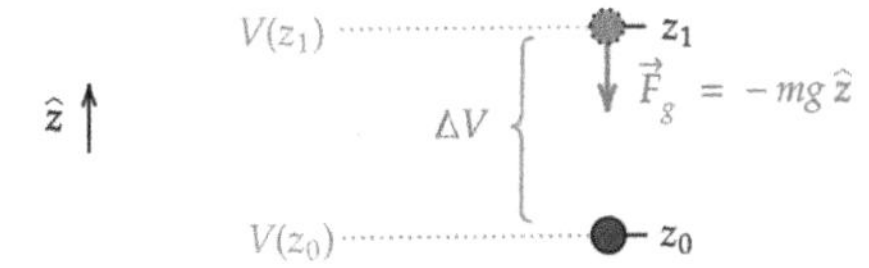

Figure 3.5. A particle of mass m falling due to gravity in the negative z direction.

Let's try to see this in another way. We assign an arbitrary reference value to the potential at the second position, for instance $V(z_0) = 0$. Then, equation (3.11) must be equal to zero at position z_0, i.e.

$$V(z_0) = mgz_0 + c = 0.$$

It follows that $c = -mgz_0$. Therefore, equation (3.11) becomes

$$V(z) = mgz - mgz_0 = mg(z - z_0).$$

Therefore, we have again that the potential of the first position z_1 is $V(z_1) = mg(z_1 - z_0)$ with respect to position z_0. Fixing an arbitrary value of the potential at an arbitrary point has allowed us to calculate the potential energy of a particle at any point.

Example 3.6. Gravitational potential energy between two point masses

In the general case, when the displacement of a particle is not much smaller than the distance from the source of the gravitational force, gravity cannot be approximated as a constant anymore, but we need to use its full radial dependence, given by

$$F = \frac{GMm}{r^2}, \tag{3.12}$$

where G is the universal gravitational constant, M and m are the values of the masses of the two particles, and r is the distance between them. The force acts along the line connecting the two masses.

It follows that the gravitational potential is (figure 3.6):

$$V(r) = \int F \, \mathrm{d}r = GMm \int \frac{\mathrm{d}r}{r^2} = -\frac{GMm}{r} + c,$$

where c is a constant of integration.

Since the potential decreases as the distance between the masses increases, it is convenient to set the potential to zero when the distance is infinitely large, i.e. $\lim_{r\to\infty} V(r) = 0$. Therefore, we have

$$\lim_{r\to\infty} V(r) = \lim_{r\to\infty}\left(-\frac{GMm}{r} + c\right) = 0 + c = 0.$$

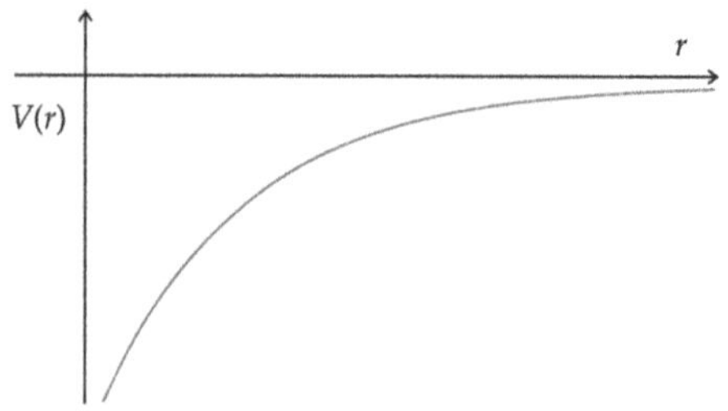

Figure 3.6. The dependence of gravitational potential energy on radial distance.

It follows that $c = 0$ and so the potential energy is simply

$$V(r) = -\frac{GMm}{r}. \tag{3.13}$$

Example 3.7. Gravitational potential energy near the Earth

In the previous examples, we have used two different definitions of the gravitational potential energy, namely equations (3.11) and (3.13). There is no contradiction: the first is an approximation that can be used for bodies near the Earth, while the second is more general and can also be used far away from the Earth's surface. Let's see how to obtain the gravitational potential energy near the Earth using equation (3.13).

Suppose there is a particle of mass m at a height h above the surface of the Earth, such that $h \ll R_E$, where R_E is the radius of the Earth. Let's take as a reference the potential energy felt by the particle on the surface of the Earth. Then, we want to find the difference in potential energy between points $z_2 = R_E + h$ and $z_1 = R_E$. This is given by

$$V(R_E + h) - V(R_E) = -\frac{GM_E\, m}{(R_E + h)} + \frac{GM_E\, m}{R_E} = -\frac{GM_E m}{R_E}\left(\frac{1}{1 + h/R_E} - 1\right), \tag{3.14}$$

where M_E is the mass of the Earth.

Consider the function $f(x) = \frac{1}{1+x}$. For $x \ll 1$, we can calculate its Maclaurin expansion (Taylor series around $x = 0$):

$$\frac{1}{1+x} = \sum_{n}^{\infty}(-1)^n x^n = 1 - x + x^2 - x^3 + \cdots$$

If you are not sure of how this expansion is obtained, see chapter 1, section 1.3.2.

Since $x \ll 1$, we can neglect the higher powers and approximate the function as $\frac{1}{1+x} \approx 1 - x$. Then, for $x = h/R_E$, we have that $\frac{1}{1+h/R_E} \approx 1 - \frac{h}{R_E}$. Substituting into equation (3.14), we obtain

$$V(R_E + h) - V(R_E) = -\frac{GM_E\, m}{R_E}\left(\cancel{1} - \frac{h}{R_E} - \cancel{1}\right) = \frac{GM_E}{R_E^2}mh = mgh,$$

where $g \equiv GM_E/R_E^2$ is the gravitational acceleration on Earth.

◆

From equation (3.9) it follows immediately that

$$W = -[V(x_2) - V(x_1)] = V(x_1) - V(x_2) \tag{3.15}$$

Remark 3.5. Signs
From equation (3.15), we see that the work done by the internal force is positive if it decreases the potential energy, i.e. $V(x_2) < V(x_1)$. Vice versa, if the potential energy increases, it means the internal force is doing negative work, i.e. the external force is doing work on the system.

From equation (3.10), using the fundamental theorem of calculus, we have that the force in a one-dimensional system can be expressed as minus the derivative of the potential with respect to position.

$$F = -\frac{\mathrm{d}V}{\mathrm{d}x}. \tag{3.16}$$

We will see in the next section how this can be generalised to three dimensions.

Exercise 3.3. In one dimension, give the expression for the potential energy corresponding to a position-dependent force $F(x)$. Show how the work done by this force when moving a particle from $x = x_1$ to $x = x_2$ can be related to the particle's potential energy in these points.

Exercise 3.4. A mass m moving in one dimension has a potential energy $V(x) = \frac{1}{2}kx^2$, where k is a constant and x the position of the particle.
Find the force F corresponding to the potential V.

Exercise 3.5. The *jet d'eau* (water spout) in Lake Geneva is a gigantic fountain where water is pushed vertically upwards at a speed of 60 m s^{-1}.

(i) Neglecting air resistance, calculate the maximum height reached by a droplet of water in the fountain, and the total time it takes after launch to fall back into the lake.

(ii) Horizontal wind produces a constant force of 10^{-3} N on a water droplet with a mass of 1 g. Calculate the horizontal distance between the bottom of the jet, where the water is ejected, and the position at which it falls back into the lake.

(iii) Is the shape described by the water in the fountain a parabola? Justify your answer.

◆

3.4.2 Potential energy in 3D

Having studied the one-dimensional problem, we now need to extend the concept of potential energy to three dimensions.

By analogy, we write

Definition 3.4.2: Potential energy (3D)
The change in potential energy from point $\boldsymbol{r}_1$ to point $\boldsymbol{r}_2$ is defined as

$$W = \int_{\boldsymbol{r}_1}^{\boldsymbol{r}_2} \boldsymbol{F}(\boldsymbol{r}) \cdot \mathrm{d}\boldsymbol{r} = V(\boldsymbol{r}_1) - V(\boldsymbol{r}_2) = -\Delta V \tag{3.17}$$

and

$$V(\boldsymbol{r}) = -\int_{\boldsymbol{r}_0}^{\boldsymbol{r}} \boldsymbol{F}(\boldsymbol{r}') \cdot \mathrm{d}\boldsymbol{r}', \tag{3.18}$$

where the integral is defined over a path from $\boldsymbol{r}_0$ to $\boldsymbol{r}$, and we define $V(\boldsymbol{r}_0) = 0$.

There is still an important issue that needs to be addressed: namely, whether the potential energy even exists. In 1D we have seen that the potential energy $V(x)$ can be defined if and only if $F \equiv F(x)$, i.e. the force only depends on the position on the particle on which it is acting. However, in three dimensions this condition is not sufficient anymore, although necessary. Now, we need to define conservative forces.

Definition 3.4.3: Conservative force
A force is called **conservative** if the work done by it on a system is independent of the path taken, and only depends on the initial and final positions of the system.
If a force is conservative, then we can associate it with a potential energy function $V(x, y, z)$.

Remark 3.6. *Potential energy can only be defined for a conservative force.* For instance, we can talk about electric potential energy, because the electrostatic force is conservative. However, we cannot talk about a potential energy function associated with friction, because friction forces are not conservative.

We can give a more rigorous mathematical definition of a conservative force:

Definition 3.4.4: Conservative force
A force field $\boldsymbol{F} \equiv \boldsymbol{F}(\boldsymbol{r})$ defined over all space[2] is conservative, and its associated potential energy is V, if and only if it satisfies any of these equivalent conditions:

[2] Technically, it can also be defined within a simply-connected volume of space.

(i)

$$\oint_{\text{path}} \boldsymbol{F} \cdot \mathrm{d}\boldsymbol{r} = 0 \tag{3.19}$$

(ii)

$$\boldsymbol{F} = -\nabla V \tag{3.20}$$

(iii)

$$\nabla \times \boldsymbol{F} = 0, \tag{3.21}$$

where the Nabla operator is defined as

$$\nabla \equiv \left(\frac{\partial}{\partial x}, \frac{\partial}{\partial y}, \frac{\partial}{\partial z}\right) = \frac{\partial}{\partial x}\hat{\boldsymbol{i}} + \frac{\partial}{\partial y}\hat{\boldsymbol{j}} + \frac{\partial}{\partial z}\hat{\boldsymbol{k}}.$$

When the ∇ operator is applied to a scalar field to produce a vector (like in example (ii), the operation is called *gradient*. When the scalar product between this operator and a vector field is taken, like in the expression $\nabla \cdot \boldsymbol{E} = 0$, it is called *divergence*. The cross-product of this operator with a vector field, like in example (iii) is called *curl*.

Remark 3.7. Note that condition (iii) can be obtained through the following reasoning. If a potential of a force $V(x, y, z)$ exists and is a differentiable function, the order of the second derivatives does not matter, so differentiating it with respect to, e.g. x and y will give the same result as differentiating with respect to y first and then with respect to x:

$$\frac{\partial^2 V(x, y)}{\partial x \partial y} = \frac{\partial^2 V(x, y)}{\partial y \partial x}. \tag{3.22}$$

But $\partial V/\partial x = F_x$, and $\partial V/\partial y = F_y$, so another way to express the condition that a force is conservative is

$$\frac{\partial F_x}{\partial y} = \frac{\partial F_y}{\partial x}. \tag{3.23}$$

Equation (3.23) is simply the z-component of equation (3.21). The other components can be obtained analogously.

Let's now give some examples of conservative and non-conservative forces:

Conservative forces	Non-conservative forces
Electrostatics	Friction
Gravitation	Air resistance
Elastic	Viscous forces

Example 3.8. Proving that a force is conservative

Let's show that the force $\boldsymbol{F} = (-2xy, -x^2, +4)$ is conservative, using equation (3.21).

The vector product $\nabla \times \boldsymbol{F}$, called the *curl* of the force, is thus given by (using the shorthand notation ∂_x to represent $\frac{\partial}{\partial x}$ etc):

$$\nabla \times \boldsymbol{F} = \begin{pmatrix} \partial_x \\ \partial_y \\ \partial_z \end{pmatrix} \times \begin{pmatrix} F_x \\ F_y \\ F_z \end{pmatrix} = \begin{pmatrix} \partial_y F_z - \partial_z F_y \\ \partial_z F_x - \partial_x F_z \\ \partial_x F_y - \partial_y F_x \end{pmatrix}. \tag{3.24}$$

Thus, saying that the curl of the force is zero is equivalent to the following system of equations (which can also be derived, as shown above, from the consideration that second derivatives do not depend on the order of differentiation):

$$\begin{cases} \partial_y F_z = \partial_z F_y \\ \partial_z F_x = \partial_x F_z \\ \partial_x F_y = \partial_y F_x \end{cases} \tag{3.25}$$

All these equations must be true for the force to be conservative.

For the force given at the beginning of the example, we have

$$\begin{cases} 0 = 0 \\ 0 = 0 \\ -2x = -2x \end{cases}$$

Therefore, the force $\boldsymbol{F} = (-2xy, -x^2, +4)$ is conservative.

Another way of showing that a force is conservative is finding a candidate potential function whose gradient gives the force itself. For instance in this case it can be shown that $\boldsymbol{F} = -\nabla(x^2 y - 4z)$. A candidate potential function can be found by integrating the expression of the force from the origin to a generic point (x, y, z) along a path (usually a straight line, or summing three paths parallel to the coordinate axes). However, it may not be trivial to find a potential this way if the function has a complex analytical form, and this potential may be path-dependent, so it is always necessary to verify that its gradient yields the desired expression for the force. In general, if the value of the potential is not explicitly asked for, and the goal is simply to check if a force is conservative or not, it is more convenient to use equation (3.21).

Exercise 3.6. Conservative forces

(a) Give a definition of conservative force in words.

(b) State the three equivalent mathematical conditions for a force field defined over all space to be conservative.

(c) Verify that the force $\boldsymbol{F} = (x^2, xy, y^2)$ is *not* conservative, using equation (3.21).

Exercise 3.7. Again conservative forces

In exercise 3.6, we have shown that the vector field $\boldsymbol{F} = (x^2, xy, y^2)$ is not conservative. To prove this in alternative way, consider example 1.18 of chapter 1. In that example, the same vector field was used and the work done by it on a particle was calculated along a straight line going from (0,0,0) to (1,1,1). The result obtained was

$$W = \int_{\text{path}} \boldsymbol{F} \cdot \mathrm{d}\boldsymbol{r} = 1.$$

Now, repeat the same calculation but along a different path, i.e. first go from (0,0,0) to (1,0,0); then, from (1,0,0) to (1,1,0); and, finally, from (1,1,0) to (1,1,1).

Since the start and end points are the same in both cases, if the force were conservative, you would obtain the same result. In fact, remember that a conservative force is defined to have path integrals independent of the path taken but only dependent on the end points. However, since we have already shown this vector field to be non-conservative, you will obtain a different value for the work done along the two paths.

Exercise 3.8. Define the potential energy function $V(\boldsymbol{r})$ given by a conservative force $\boldsymbol{F}(\boldsymbol{r})$ acting on a particle such that $V = 0$ at $r = r_0$.

Explain how $\boldsymbol{F}$ can be determined if V is known.

A particle moves in the x–y plane under the action of a conservative force $\boldsymbol{F}(x, y)$, which has a potential function of $V = Kx^3y^2$, where K is a constant. Given this information determine $\boldsymbol{F}(x, y)$ as well as the work done on the particle by this force, when moving it from the origin; $x = 0$, $y = 0$ to the point $x = 2$, $y = 4$.

Exercise 3.9. A particle in three dimensions has a potential energy given by

$$V = z^2 + y^3 + 2x^2y^2.$$

Determine the equation of the force $\boldsymbol{F}$ acting on the particle in a general position (x, y, z).

If the particle moves from the origin (0, 0, 0) to the position (1, 1, 2), what is the change in kinetic energy assuming that only the force determined above is acting on the particle.

If the force acting on a second particle is

$$\boldsymbol{F} = -2xy^3\hat{\boldsymbol{x}} - 3x^2y^2\hat{\boldsymbol{y}},$$

then find the potential energy in a general two-dimensional position (x, y) given that $V = 0$ at the origin.

Exercise 3.10. A particle moves in a straight line in three dimensions, starting from the position vector $\boldsymbol{r}_1 = \hat{x} + \hat{y} + 3\hat{z}$ to a point with position vector $\boldsymbol{r}_2 = 2\hat{x} + 2\hat{y} + 5\hat{z}$.

Find the work done during the displacement from $\boldsymbol{r}_1$ to $\boldsymbol{r}_2$ produced by:

(i) a *constant* force $\boldsymbol{F}_1 = 3\hat{x} + 2\hat{y} + \hat{z}$.

(ii) a *position dependent* force $\boldsymbol{F}_2 = 2xy^3\hat{x} + 3x^2y^2\hat{y}$.

Which of the two forces above is conservative and why?

Exercise 3.11. A force is given by the expression

$$\boldsymbol{F} = 3xy\hat{x} + y^2x\hat{y} + zy\hat{z}.$$

(i) Is this force conservative?

(ii) Calculate the work done by this force from point (0, 0, 0) to point (1, 1, 1), first moving along x to (1, 0, 0), then along y to (1, 1, 0) and then along z to (1, 1, 1).

(iii) Then, calculate the work done by this force following a straight line along the diagonal directly from point (0, 0, 0) to (1, 1, 1).

◆

3.4.3 Conservation of mechanical energy

To introduce the concept of mechanical energy, let's now apply the work–energy theorem discussed earlier in this chapter (theorem 3.2.1). We'll do this in 1D. Since $W = \Delta K$, and using equation (3.15),

$$\begin{aligned}\Delta K &= -\Delta V\\ K(x_2) - K(x_1) &= V(x_1) - V(x_2)\\ K(x_2) + V(x_2) &= K(x_1) + V(x_1).\end{aligned}$$

The equation just obtained shows that in the absence of friction the sum of kinetic and potential energy is conserved, i.e. it remains constant. We call this quantity the mechanical energy. In the presence of friction, mechanical energy is not conserved, because it is transformed into other forms of energy, such as heat.

Definition 3.4.5: Mechanical energy

The mechanical energy E of a system is defined as the sum of the kinetic energy and potential energy of the system, i.e.

$$E = K + V. \tag{3.26}$$

It is linked to the motion and to the position of a system.

Example 3.9. Moving in a gravitational potential
If a particle is subject only to gravity and friction can be neglected, its mechanical energy is conserved. Let's use this information to work out the velocity that a particle needs to have in order to escape from Earth's gravitational attraction and go into space.

The Earth has mass M, while the particle has mass m. Call r the distance from the centre of the Earth (of course, $r > R_E$, where R_E is the Earth's radius).

Since the particle's mechanical energy is conserved, we have that

$$E_i = E_f$$

$$\frac{1}{2}mv_e^2 - \frac{GMm}{r} = 0 - \frac{GMm}{r_{\max}}.$$

This equation means that, when the particle has transformed all of its kinetic energy into potential energy, it will be at a distance $r_{\max}$ from the centre of the Earth. Rearranging the terms, we obtain

$$\frac{v_e^2}{2GM} = \left(\frac{1}{r} - \frac{1}{r_{\max}}\right).$$

For $r_{\max} \to \infty$,

$$\frac{v_e^2}{2GM} = \frac{1}{r}.$$

Thus, the velocity needed by a particle of mass m and at position r from the centre of the Earth to escape the planet's gravitational attraction is given by

$$v_e = \sqrt{\frac{2GM}{r}} \tag{3.27}$$

and is called **escape velocity**.

◆

You might be wondering why conservative forces are named like this. It is because they conserve mechanical energy.

Theorem 3.4.1: Conservation of mechanical energy
In an isolated system only subject to conservative forces, mechanical energy is conserved.

Definition 3.4.6: A system is called **isolated** if it does not exchange energy or matter with the surroundings. The only truly isolated system is the Universe (at least, from what we know).

If non-conservative forces, such as friction, are also present, then mechanical energy does not remain constant anymore. For instance, a ball rolling on a non-ideal surface will eventually stop because friction will convert the kinetic energy into heat. Nevertheless, heat is also a form of energy. What remains conserved in this case is the total energy of the system (provided this is isolated).

3.4.4 Work–energy theorem for systems of many particles

It is important here to briefly discuss how the work–energy theorem [theorem 3.2.1] changes when the system is composed of more parts (not only one particle!).

Theorem 3.4.2: General work–energy theorem

The work done on a system by external forces equals the change in energy of the system. In formulae,

$$W_{\text{external}} = \Delta K + \Delta V + \Delta K_{\text{internal}} \tag{3.28}$$

where the terms on the rhs are

- the overall kinetic energy ΔK;
- the internal potential energy ΔV;
- the internal (e.g. thermal) kinetic energy $\Delta K_{\text{internal}}$.

Since a particle doesn't have an internal structure, we are left with only the first term on the rhs, in agreement with theorem 3.2.1.

Example 3.10. Raising a book

This new formulation of the work–energy (W–E) theorem means that we need to choose a system on which work is done. There might be various choices that we can make. Let's have a look at an example of such a situation.

Consider a person lifting a book with constant speed such that its kinetic energy does not vary.

- **System = (book)**: The rhs of equation (3.28) is zero, because there is no change in the energy of the system itself. Both the person and gravity are external forces acting on the system, therefore we have
$$W_{\text{ext}} = W_{\text{person}} + W_{\text{gravity}} = (mgh) + (-mgh) = 0.$$
- **System = (book + Earth)**: Now the person is the only external force acting on the system. Gravity is still present but, instead of being considered an external force doing work, it is accounted for as a change in the internal potential energy of the system. Therefore,
$$W_{\text{person}} = \Delta V_{\text{Earth-book}} \Longleftrightarrow mgh = mgh.$$
- **System = (book + Earth + person)**: Now there is no external force anymore. The change in internal energy is caused by the increase in the potential energy

of the book and the decrease in the person's internal energy (to lift the book, the person needs to burn some calories). Equation (3.28) now gives

$$0 = \Delta V_{\text{Earth-book}} + \Delta V_{\text{person}} \Longleftrightarrow 0 = (mgh) + (-mgh).$$

It should be noted that, since a human body does not have an efficiency of 100%, some heat will be produced and this will increase the system's internal kinetic energy. However, the person's potential energy will decrease by more than mgh to compensate for the heat released. The sum of these two quantities will give $(-mgh)$. So, if an amount of energy ε is released in the form of heat, the W–E theorem gives us

$$0 = \Delta V_{\text{Earth-book}} + \Delta V_{\text{person}} + \Delta K_{\text{internal}} \Longleftrightarrow 0 = mgh + (-mgh - \varepsilon) + \varepsilon.$$

Exercise 3.12. A particle moves in the x–y plane under the action of a force described by the potential energy function

$$V = 8x^2 + 2y^2 + \alpha y^4,$$

where α is a non-negative constant, x and y are measured in metres and $V(x, y)$ is in joules. What is the work done on the object by the force as it moves from the point (2,0) to the point (0,0)?

Find the vector expression for the force acting on the object in a general position (x, y).

Which quantity will be conserved during the motion and why?

IOP Publishing

Classical Mechanics
A professor-student collaboration
Mario Campanelli

Chapter 4

Oscillatory motion

4.0 Introduction

Harmonic motion is key to many concepts in physics, ranging from the description of sound waves produced by guitars to the behaviour of a quantum harmonic oscillator. It is a very important tool for a physicist to learn and understand, as many physical phenomena can be related or simplified to the basic oscillatory motions which will be covered in this chapter.

We will consider damped as well as driven and damped harmonic motion, and finally discuss coupled oscillators and how to solve these more complex systems.

4.1 Simple harmonic motion

Definition 4.1.1: Simple harmonic motion
Simple harmonic motion is an oscillatory motion characterised by a single constant angular velocity. It can be described mathematically as a superposition of sinusoidal functions with the same frequency.

Simple harmonic motion is the simplest form of propagation of a wave. Discussing this first will give us the key tools to help us tackle more complex forms of propagation, with harder equations of motion to be solved.

Let's consider now the motion of an ideal spring with no friction acting on it. When the spring is displaced from its equilibrium position by a distance x, the spring will exert a **restoring force** which will oppose the displacement. This restoring force is defined by **Hooke's law** and takes the form:

$$F = -kx.$$

Remark 4.1. Note that in all the equations regarding *springs* in this chapter the x coordinate indicates the **displacement from the position of equilibrium** of the spring. Therefore, the motion we will be considering is that of the displacement of the body from the equilibrium position.

Let's try to find the solution for the motion of a spring displaced from equilibrium.

To write the equation of motion, we need to first apply **Newton's second law** which is:

$$F_T = ma = m\ddot{x}$$

where F_T represents the total sum of the forces acting on the body.

The force described by Hooke's law will be the only one acting on the spring, therefore our total equation of motion will be:

$$m\ddot{x} = -kx.$$

The equation above is a second order differential equation, whose solution is treated in full in section 1.4.3. It can be demonstrated that a solution is an exponential function in the following form:

$$x = Ae^{\alpha t} \tag{4.1}$$

Remark 4.2. In the exponential solution above, A is a general constant, t is the time variable, upon which the solution depends. And finally, α indicates a general proportionality constant for t.

To calculate the value of the parameter α, the first step is to find the first and second derivatives of the solution

$$\dot{x} = A\alpha e^{\alpha t}$$
$$\ddot{x} = A\alpha^2 e^{\alpha t}.$$

Substituting the solution into the equation of motion we obtain:

$$mA\alpha^2 e^{\alpha t} = -kAe^{\alpha t}.$$

By rearranging we see

$$\cancel{A} m\alpha^2 \cancel{e^{\alpha t}} + \cancel{A} k \cancel{e^{\alpha t}} = 0.$$

We are then left with

$$\alpha^2 = \frac{-k}{m}$$

$$\alpha = \pm\, i\sqrt{\frac{k}{m}}$$

$$\alpha \equiv \pm\, i\omega$$

where ω is the angular velocity of the motion.

We have obtained two solutions and we know from section 1.4.3 that our final solution will be a **linear combination** of the two. The two solutions will be combined using two *constants* which must be determined from the initial conditions of the problem.

$$x(t) = Ae^{i\omega t} + Be^{-i\omega t} \tag{4.2}$$

> Remark 4.3. In general, the constants A and B are complex numbers, but because in classical mechanics we normally deal with physical problems the initial conditions will be such that the constants turn out to be real.

The most general solution to a simple harmonic oscillator can be written in **four equivalent forms**.

$$x(t) = Ae^{i\omega t} + Be^{-i\omega t} \tag{4.3}$$

$$x(t) = C\cos\omega t + D\sin\omega t \tag{4.4}$$

$$x(t) = F\cos(\omega t + \phi) \tag{4.5}$$

$$x(t) = G\sin(\omega t + \phi) \tag{4.6}$$

where the sine and cosine forms are obtained by substituting de Moivre's equation $e^{i\alpha} = \cos\alpha + i\sin\alpha$ *in the exponential solution.*

> Remark 4.4. The capital letters in the equations above, are all *arbitrary constants* which must be determined from the boundary conditions given by the problem. The Greek letter ϕ which appears in equations (4.5) and (4.6) is a general *phase shift angle*, which must also be determined from initial conditions.

Note that in the case of equations (4.5) and (4.6) the constants in front of the cosine and sine functions represent the amplitude of the oscillation.

Now, let's look at an example on how to find the constants from initial conditions. We will see that imposing real numbers as initial conditions will automatically ensure that the constants for amplitudes and phases are real, so the overall functions will also be real.

Example 4.1. Obtaining the constants
Let's consider the motion of a simple spring which at the time $t = 0$ is held in position x_0 with no initial velocity (figure 4.1).

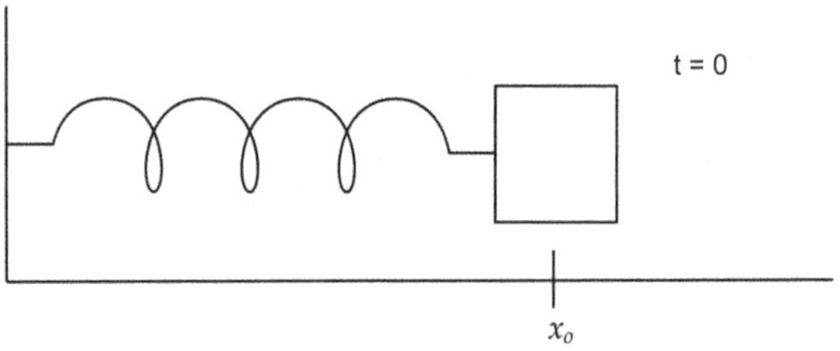

Figure 4.1. An ideal spring with no friction at time $t = 0$.

Such that our boundary conditions become:

$$x(0) = x_0$$
$$\dot{x}(0) = 0.$$

In order to demonstrate the equivalence of the solutions chosen to describe a system, we will be solving to find the constants using both the exponential and sinusoidal solutions to simple harmonic motion. We will start from the **exponential solution**:

1. Substitute $t = 0$ into the general solution to find the conditions which satisfy the first boundary condition
   $x(0) = Ae^0 + Be^0 = x_0.$
2. $x_0 = A + B.$
3. Now take the first derivative of the general solution
   $\dot{x}(0) = Ai\omega e^{i\omega t} - Bi\omega e^{-i\omega t}.$
4. Once again substitute the boundary condition for a time $t = 0$
   $\dot{x}(0) = Ai\omega e^0 - Bi\omega e^0 = 0.$
5. $\omega(A - B) = 0$
   Therefore $A = B.$
6. $A = \dfrac{x_0}{2} = B.$
7. Substituting the values of the constants into the general solution and simplifying we obtain:

$$x(t) = \frac{x_0}{2}e^{i\omega t} + \frac{x_0}{2}e^{-i\omega t}.$$

$$x(t) = \frac{x_0}{2}\cos\omega t + i\frac{x_0}{2}\sin\omega t + \frac{x_0}{2}\cos\omega t - i\frac{x_0}{2}\sin\omega t.$$

8. The final solution is

$$x(t) = x_0 \cos \omega t.$$

Now we can look at the result of applying the boundary conditions to the **sinusoidal solution**.

1. Substitute into the general solution the condition at $t = 0$
   $x(0) = C \cos 0 + D \sin 0.$
2. This gives $C \cos 0 = C = x_0$.
3. We then need to take the first derivative of the general solution and substitute the condition of zero velocity at $t = 0$
   $\dot{x}(0) = -C \sin 0 + D \cos 0.$
4. This gives the condition that $D \cos 0 = D = 0$.
5. Therefore, the final solution is

$$x(t) = x_0 \cos \omega t.$$

We can see how both general forms give the same final solution for motion, given the boundary conditions, as expected.

◆

Here is an exercise for you to have a go at the procedure above:

Exercise 4.1. Let's consider a body attached to an ideal spring with no friction, which at the time $t = 0$ is found in the position x_0 with an initial velocity v_0 (figure 4.2).

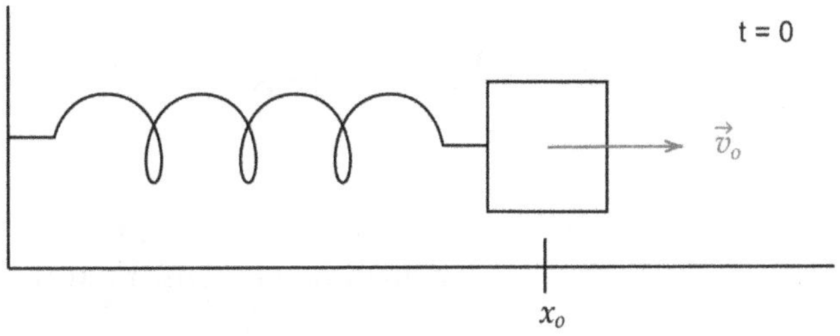

Figure 4.2. An ideal spring with an initial positive velocity at a time $t = 0$.

In this case adopt the general solution in the form

$$x(t) = F \cos (\omega t + \phi).$$

to determine the solution of motion for the body.

Remark 4.5. In the above case there will be a **phase shift** ϕ, this is because when we add a velocity to the body attached to a spring, its maximal displacement will no longer be at $t = 0$. It can be clearly seen in figure 4.3, that if there is an initial velocity, the spring will carry on extending to the maximal amplitude of x_m.

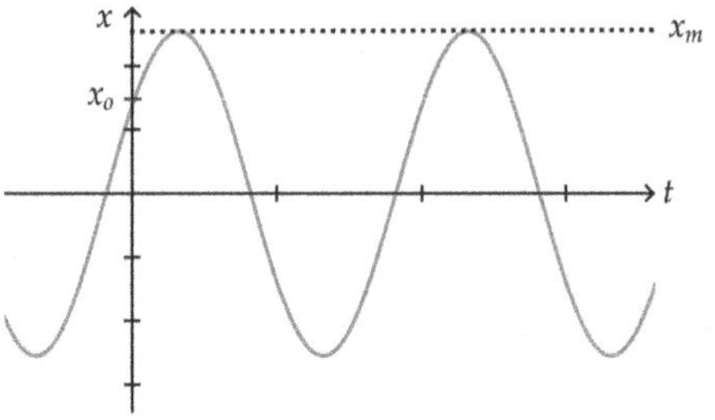

Figure 4.3. The graph of the motion of the ideal spring with an initial velocity.

Exercise 4.2. The simple pendulum
A point mass m hangs from a massless string of length l. The string is displaced through a **small angle** θ from the vertical rest position, as seen in figure 4.4.

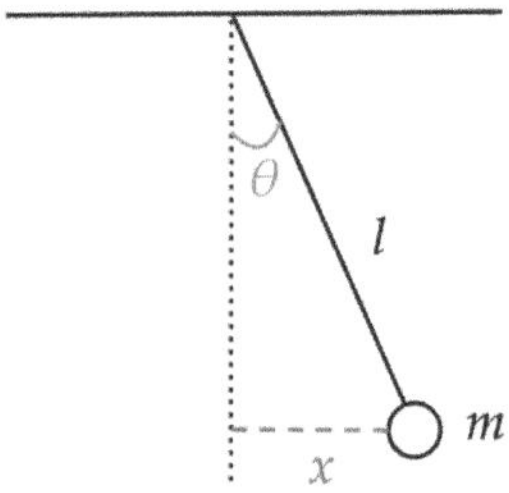

Figure 4.4. The setup of a simple pendulum.

Using the equation of motion for the rotating system, show that the pendulum moves with simple harmonic motion for small oscillations. Find the frequency f of the pendulum.
[*Hint:* You can use the small angle approximation: $\sin \alpha \approx \alpha$.]

Exercise 4.3. A point mass m hangs from a massless string of length l, moving in a horizontal circle as can be seen from figure 4.5 and making an angle θ with the vertical.

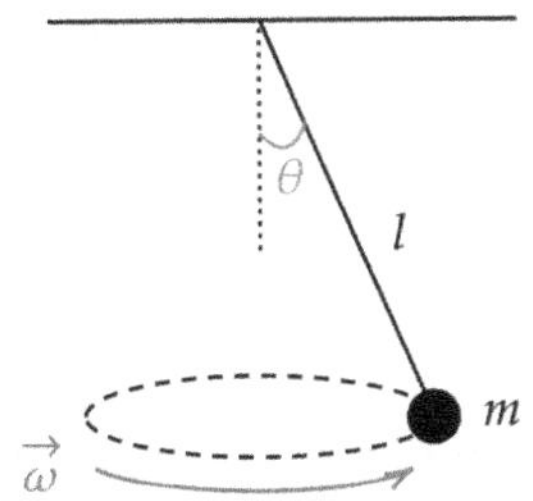

Figure 4.5. A circular pendulum moving under the effect of gravity.

If g is the gravitational acceleration, show that the angular frequency ω of the circular rotation is given by

$$\omega = \sqrt{\frac{g}{l \cos \theta}}.$$

Exercise 4.4. At a time $t = 0$ a ball of mass m is dropped from rest, vertically down from the top of a tunnel dug between the North and South poles. Show that the position r of the ball at a time t satisfies:

$$r(t) = R \cos \omega t.$$

You can neglect air resistance, temperature effects and the mass of the Earth excavated to make the tunnel.
[*Hint:* Write the expression for the fraction of the Earth's mass contained in a sphere of radius $r \leqslant R$.]

◆

4.1.1 Energy of simple harmonic motion

We will apply the **principle of conservation of energy** to the case of a body attached to an ideal spring. In this case we only need to consider two forms of energy; the *potential energy* given by the restoring force, i.e. the spring, and the *kinetic energy* of the body.

Let's look at the kinetic energy first:

$$K = \frac{1}{2} m\dot{x}^2.$$

Taking the first derivative of the harmonic motion solution given by equation (4.5):

$$\dot{x} = -F\omega \sin(\omega t + \phi),$$

$$K = \frac{1}{2} mF^2\omega^2 \sin(\omega t + \phi)^2. \tag{4.7}$$

We have already mentioned that the potential energy is related to the restoring force, and is the elastic potential energy stored in a spring which we know is of the form

$$V = \frac{1}{2}kx^2.$$

By substituting in the equation the position of the spring, given by our general solution we can find the potential energy of a body moving in simple harmonic motion

$$V = \frac{1}{2}kF^2 \cos(\omega t + \phi)^2. \tag{4.8}$$

Therefore, the total energy of simple harmonic motion is given by the sum of the potential and kinetic energies at a given time t:

$$E = \frac{1}{2}kF^2 \cos(\omega t + \phi)^2 + \frac{1}{2}mF^2\omega^2 \sin(\omega t + \phi)^2. \tag{4.9}$$

Since $\omega = \sqrt{\frac{k}{m}}$, and that for any angle α, $\sin^2 \alpha + \cos^2 \alpha = 1$, we see that energy is conserved and takes the form

$$E = \frac{1}{2}kF^2. \tag{4.10}$$

Example 4.2. Finding the maximal amplitude of oscillation using the energy
We can find the maximal amplitude of oscillation very easily by considering the **conservation of the total energy** of the oscillator. At a time $t = 0$ a body of mass m attached to an ideal spring is at the position x_0 with respect to its equilibrium position, and with a velocity v_0.
Let's start by looking at what the initial kinetic and potential energies of the oscillator will be at the time $t = 0$. Using the equations given above and substituting in our initial conditions we get that

$$V_0 = \frac{1}{2}kx_0^2$$
$$K_0 = \frac{1}{2}mv_0^2.$$

Therefore, the total energy will be

$$E_0 = \frac{1}{2}kx_0^2 + \frac{1}{2}mv_0^2.$$

We now need to think about the motion. The oscillator will become stationary when it has reached its maximal amplitude, before reverting direction and returning to equilibrium. Hence when the *total kinetic energy is zero* we know we will have our maximal displacement from the equilibrium position x_m.
Since the total energy is conserved we can say

$$E_0 = K_m + U_m$$

$$\cancel{\tfrac{1}{2}}kx_0^2 + \cancel{\tfrac{1}{2}}mv_0^2 = 0 + \cancel{\tfrac{1}{2}}kx_m^2.$$

Therefore, by rearranging we get

$$x_m = \sqrt{\frac{m}{k}v_0^2 + \frac{\cancel{k}}{\cancel{k}}x_0^2}\,.$$

Substituting in the natural frequency of the oscillator $\omega^2 = k/m$ we get our final solution for the maximal amplitude of the oscillations,

$$x_m = \sqrt{\frac{v_0^2}{\omega^2} + x_0^2}\,.$$

Exercise 4.5. Given that the motion of a friction-less mass attached to an ideal spring is described by the equation

$$x(t) = A\cos\omega t;$$

show that the total mechanical energy of the system, E is

$$E = \frac{1}{2}kA^2.$$

◆

4.2 Damped harmonic motion

> **Definition 4.2.1: Damped harmonic motion**
> Damped harmonic motion is analogous to simple harmonic motion, with the presence of a dragging force, which dissipates the oscillator's energy.
> *Looking at the case considered for the simple harmonic oscillator of the body attached to the end of a spring, the example becomes damped harmonic motion if we introduce the presence of friction between the body and the surrounding air.*

Let's look at how the equation of motion of the simple harmonic oscillator will change with the addition of friction. Here we will be considering **viscous friction** which is described by a drag force *proportional to its velocity* with equation

$$F = -bv = -b\dot{x}, \tag{4.11}$$

where b is the proportionality constant. Let's add this into the full equation of motion:

$$F = -kx - b\dot{x}$$
$$m\ddot{x} = -kx - b\dot{x}$$
$$0 = \ddot{x} + \frac{b}{m}\dot{x} + \frac{k}{m}x.$$

The final differential equation we need to solve is

$$\ddot{x} + 2\gamma\dot{x} + {\omega_0}^2 x = 0, \tag{4.12}$$

where γ is what we will now call the **damping rate** of motion, defined as

$$\gamma = \frac{b}{2m} \tag{4.13}$$

and ω_0 is the **natural angular frequency** of the simple harmonic motion in the absence of damping force, defined in the previous section as

$${\omega_0}^2 = \frac{k}{m}.$$

The solution to the new equation of motion will be in the form

$$x(t) = A\mathrm{e}^{q_1 t} + B\mathrm{e}^{q_2 t}, \tag{4.14}$$

where q_1 and q_2 are the solutions to the *characteristic equation* used to solve such a linear, homogeneous, second order differential equation (for further information see section 1.4.3):

$$q^2 + 2\gamma q + \omega_0^2 = 0,$$

which gives a general solution in the form

$$q_{1,2} = -\gamma \pm \sqrt{\gamma^2 - \omega_0^2}$$
$$\equiv -\gamma \pm \Omega,$$

where Ω is an important parameter whose value can change the form of the damped motion. It is given by:

$$\Omega^2 = \gamma^2 - \omega_0^2. \tag{4.15}$$

We can substitute our findings into our general solution and find the most general motion of the damped harmonic oscillator as

$$x(t) = \mathrm{e}^{-\gamma t}[A\mathrm{e}^{\Omega t} + B\mathrm{e}^{-\Omega t}]. \tag{4.16}$$

The value of Ω will separate our damped harmonic motion into three cases.

Remark 4.6. Remember that γ represents the damping rate of the drag force, whilst ω_0 is the natural frequency which is a property of the oscillator, based on its form and characteristics. Hence the various values that Ω can take depend on the rate at which the system is damped compared to the angular frequency at which it would naturally oscillate.

Underdamping

The first case we will observe occurs when the value of $\Omega^2 < 0$ or, in other words, when the square of the damping rate is smaller than the square of the natural frequency, i.e.

$$\gamma^2 < \omega_0^2. \tag{4.17}$$

When this occurs, Ω will become imaginary, giving rise to an oscillatory solution. We can start by defining a **new angular frequency for damped motion** which is related to Ω.

$$\Omega = i\sqrt{\omega_0^2 - \gamma^2} \equiv i\tilde{\omega} \tag{4.18}$$

$$\tilde{\omega} \equiv \sqrt{\omega_0^2 - \gamma^2}. \tag{4.19}$$

This gives us a solution of motion depending on $\tilde{\omega}$,

$$x(t) = \mathrm{e}^{-\gamma t}[A\mathrm{e}^{i\tilde{\omega}t} + B\mathrm{e}^{-i\tilde{\omega}t}] \tag{4.20}$$

which, in a similar way to the simple harmonic motion, can be written in a more intuitive form using trigonometric functions as seen below:

$$x(t) = F\mathrm{e}^{-\gamma t}[\cos(\tilde{\omega}t + \phi)]. \tag{4.21}$$

Remark 4.7. Note that the angular frequency of damped motion $\tilde{\omega}$, is different from the natural frequency of the simple harmonic oscillator. The new angular frequency depends on the rate of damping whilst the natural frequency ω_0 is an intrinsic property of an oscillator.

The solution is a sinusoidal oscillation, contained inside an exponentially decreasing frame given by the factor of $\mathrm{e}^{-\gamma t}$ in front. The amplitude of the oscillations is therefore modulated, and decreases in time, as one would expect in the presence of friction. This can be observed in figure 4.6.

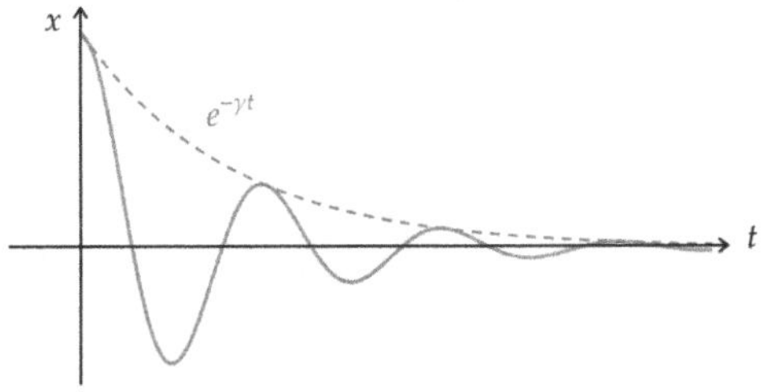

Figure 4.6. The graph showing the magnitude of the oscillations in the case of underdamped motion.

Overdamping

Next we will look at the case where $\Omega^2 > 0$ or in other words the square of the damping rate is larger than the square of the natural frequency, i.e.

$$\gamma^2 > \omega_0^2. \tag{4.22}$$

In this case the exponentials in our general solution will be real;

$$x(t) = Ae^{-(\gamma - i\Omega)t} + Be^{-(\gamma + i\Omega)t}. \tag{4.23}$$

In the above case the real exponentials will just produce damping without oscillations. Here too the factor of $e^{-\gamma t}$ will act as an envelope causing the exponential function inside to decrease, as seen in figure 4.7. This agrees with our expectations, since the damping level will be so high that the system will not have time to complete a full oscillation before all its energy is dissipated.

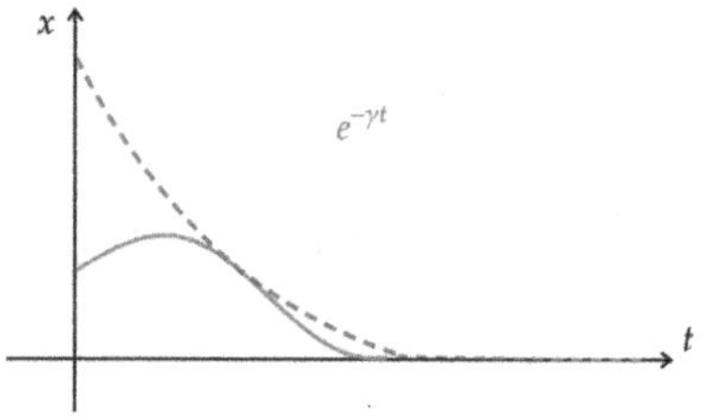

Figure 4.7. The graph showing the magnitude of the oscillations in the case of overdamped motion.

Critical damping

The last case occurs when the value of $\Omega^2 = 0$ or such that the square of the damping rate is identical to the square of the natural frequency.

$$\gamma^2 = \omega_0^2. \tag{4.24}$$

When this occurs our solution will become

$$x(t) = e^{-\gamma t}[Ae^{0t} + Be^{-0t}] = e^{-\gamma t}(A + B).$$

However, this presents a mathematical problem for us. We have just solved a second order differential equation, for which we need to find **two solutions**. The above gives us only one, so a more general form is needed. It is possible to see that inserting an extra term $te^{-\gamma t}$ also returns an acceptable solution;

$$x(t) = e^{-\gamma t}[Ae^{0t} + Bte^{-0t}] = e^{-\gamma t}(A + Bt). \tag{4.25}$$

This particular kind of equation gives us the most rapidly decreasing exponential function of the three cases we have looked at. At **critical damping** the system will return to the equilibrium position with no oscillatory motion whatsoever, as seen in figure 4.8.

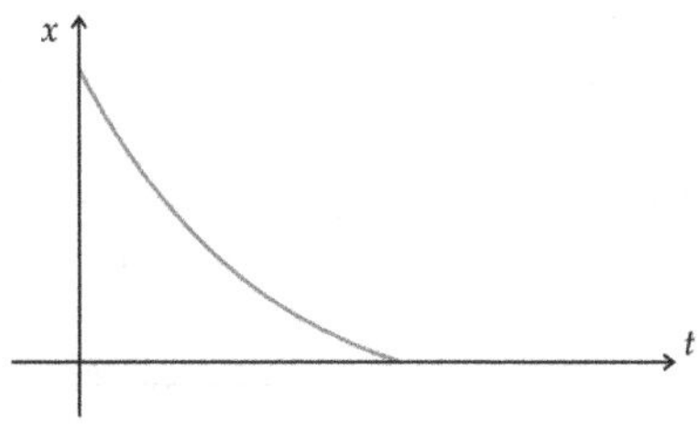

Figure 4.8. The graph showing the amplitude of oscillations in the case of critically damped motion.

Exercise 4.6. When a particle of mass m is displaced from its equilibrium position, it is subject to a force $-kz\hat{k}$. An additional damping force $-2m\alpha\dot{z}\hat{k}$ is applied to the particle. Write down the equation of motion in this case.
Now, consider a solution to the equation of the form $z(t) = Ae^{qt}$ and find q. Show that the maximal displacement from the equilibrium position, will have a time dependent decrease proportional to $e^{-\alpha t}$ provided that $m\alpha^2 < k$

Exercise 4.7. Consider a particle of mass m attached to a spring of constant k, and subject to a damping force proportional to the velocity of the particle, with constant of proportionality λ. In the case of critical damping the general solution to the equation of motion is

$$x = (A + Bt)\exp\left\{\frac{-\lambda t}{2m}\right\}$$

where A and B are arbitrary constants. Suppose the particle is found at rest in a position x_0 from equilibrium at a time $t = 0$. Determine the values of the constants using the boundary condition given and sketch the displacement of x as a function of time.

◆

4.3 Driven and damped harmonic motion

Definition 4.3.1: Driven oscillations
Driven Oscillations, also called **forced oscillations**, occur when a damped oscillator is subject to an external oscillating force.

The new force will drive the oscillator with a constant period described by a **driving angular frequency** ω_d. The driving frequency will remain constant, and is once again distinct from the natural frequency ω_0 of the system. The new equation of motion to be solved, when adding the driving force, is

$$\ddot{x} + 2\gamma\dot{x} + {\omega_0}^2 x = C_0 e^{i\omega_d t}. \tag{4.26}$$

In the equation above, we have used the general mathematical formula where the periodically oscillating driving force is in general a complex number.

> Remark 4.8. By observing our new equation of motion, we can see that it is an inhomogeneous form of the damped harmonic motion, equation (4.12). To solve such a *second order linear inhomogeneous* differential equation, we need the sum of the solution to the homogeneous differential equation and the **particular solution**. See section 1.4.4.

Since we have an exponential on the right-hand side of our equation of motion we will guess that the solution will have a similar form, with the same oscillation frequency. To find the particular solution we will substitute our educated guess and its derivatives into equation (4.26):

$$x(t) = Ae^{i\omega_d t}$$
$$\dot{x}(t) = Ai\omega_d e^{i\omega_d t}$$
$$\ddot{x}(t) = -A\omega_d^2 e^{i\omega_d t}.$$

Since $i^2 = -1$. Substituting into the differential equation we obtain

$$C_0 \cancel{e^{i\omega_d t}} = A\omega_0^2 \cancel{e^{i\omega_d t}} + 2\gamma Ai\omega_d \cancel{e^{i\omega_d t}} - A\omega_d^2 \cancel{e^{i\omega_d t}}$$
$$C_0 = A(\omega_0^2 + 2i\gamma\omega_d - \omega_d^2)$$
$$A = \frac{C_0}{\omega_0^2 + 2i\gamma\omega_d - \omega_d^2}$$

$$x(t) = \frac{C_0}{\omega_0^2 + 2i\gamma\omega_d - \omega_d^2} e^{i\omega_d t}. \tag{4.27}$$

The particular solution to the differential equation found above is also called the **stationary solution**, and has no constants which depend on the initial conditions. The final solution to the driven harmonic oscillator is the sum of the stationary solution and equation (4.16).

$$x(t) = \mathrm{e}^{-\gamma t}[A\mathrm{e}^{\sqrt{\gamma^2-\omega_0^2}\,t} + B\mathrm{e}^{-\sqrt{\gamma^2-\omega_0^2}\,t}] + \frac{C_0}{\omega_0^2 + 2i\gamma\omega_d - \omega_d^2} e^{i\omega_d t}. \tag{4.28}$$

Physical driving forces

In the treatment above we used the most general case of a periodic driving force. However, to extend the treatment to a real physical case, we need the driving motion to be real, this means we cannot describe our driving motion as $C_0 e^{i\omega_d t}$.

Let's look at the example of the block attached to the end of the spring, subject to air friction (figure 4.9). Now, we shall add a driving force to the block, which will have a periodic motion of:

$$\boldsymbol{F_d} = F_d \cos \omega_d t. \tag{4.29}$$

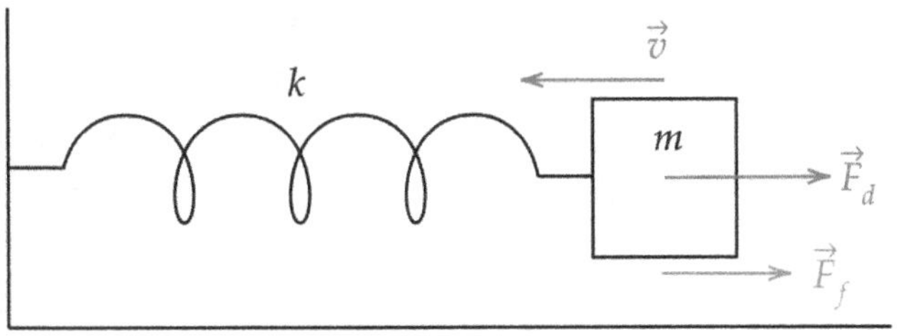

Figure 4.9. An image illustrating a driven and damped oscillator formed by a block of mass m attached to a spring.

Now that we have changed the driving force into a real value, the stationary solution will also change. To make the treatment easier we will first express the cosine in terms of complex exponentials, to help relate the solution to the one found above for the general treatment:

$$F_d \cos \omega_d t = \frac{F_d}{2}(e^{i\omega_d t} + e^{-i\omega_d t}).$$

The equation of motion will become:

$$\ddot{x} + 2\gamma \dot{x} + {\omega_0}^2 x = \frac{F}{2}(e^{i\omega_d t} + e^{-i\omega_d t}) \tag{4.30}$$

where

$$F = \frac{F_d}{m}.$$

> Remark 4.9. Note that in the equation of motion seen above, just as in the previous treatments, ω_d is the driving frequency of the restoring force. One should not mistake this to be the same as the general oscillation frequency ω.

The **Principle of Superposition** allows us to find the solution to the equation of motion of the sinusoidal driving force. By expressing the cosine as the sum of two complex exponentials, we can easily determine what the solution would be with just one exponential, as we found above for equation (4.28). The principle of superposition allows us to then find the full solution by adding the two single exponential ones. Hence our final particular solution will be:

$$x_p(t) = \left(\frac{F/2}{{\omega_0}^2 + 2i\gamma\omega_d - {\omega_d}^2}\right)e^{i\omega_d t} + \left(\frac{F/2}{{\omega_0}^2 - 2i\gamma\omega_d - {\omega_d}^2}\right)e^{-i\omega_d t}. \tag{4.31}$$

Once again the full solution of motion will be the sum of this particular solution and the solution to the homogeneous differential equation (4.16), i.e.

$$x(t) = \left(\frac{F/2}{\omega_0^2 + 2i\gamma\omega_d - \omega_d^2}\right)e^{i\omega_d t} + \left(\frac{F/2}{\omega_0^2 - 2i\gamma\omega_d - \omega_d^2}\right)e^{-i\omega_d t}$$
$$+ \mathrm{e}^{-\gamma t}[A\mathrm{e}^{\sqrt{\gamma^2-\omega_0^2}\,t} + B\mathrm{e}^{-\sqrt{\gamma^2-\omega_0^2}\,t}].$$

This equation can simplified by expressing the complex exponential functions as cosines and sines.

Example 4.3. Transforming between forms of the same solution
Given the form of the particular solution in equation (4.31), find its equivalent expression using sinusoidal functions.
The first step is to rationalize the complex fractions in front of the exponential terms The first fraction becomes

$$\frac{F/2}{(\omega_0^2 - \omega_d^2) + i(2\gamma\omega_d)} \equiv \frac{F/2}{(\omega_0^2 - \omega_d^2) + i(2\gamma\omega_d)} \times \frac{(\omega_0^2 - \omega_d^2) - i(2\gamma\omega_d)}{(\omega_0^2 - \omega_d^2) - i(2\gamma\omega_d)}$$
$$\equiv \frac{F/2(\omega_0^2 - \omega_d^2)}{(\omega_0^2 - \omega_d^2)^2 + 4\gamma^2\omega_d^2} - i\frac{F/2\ 2\gamma\omega_d}{(\omega_0^2 - \omega_d^2)^2 + 4\gamma^2\omega_d^2}.$$

Analogously, the second fraction becomes

$$\frac{F/2}{\omega_0^2 - 2i\gamma\omega_d - \omega_d^2} \equiv \frac{F/2\,(\omega_0^2 - \omega_d^2)}{(\omega_0^2 - \omega_d^2)^2 + 4\gamma^2\omega_d^2} + i\frac{F/2\ 2\gamma\omega_d}{(\omega_0^2 - \omega_d^2)^2 + 4\gamma^2\omega_d^2}.$$

By replacing these into equation (4.31) and collecting the terms with the same fractions in front, we obtain

$$x_p(t) = \left[\frac{F\,(\omega_0^2 - \omega_d^2)}{(\omega_0^2 - \omega_d^2)^2 + 4\gamma^2\omega_d^2}\right]\frac{1}{2}(e^{i\omega_d t} + e^{-i\omega_d t})$$
$$+ \left[\frac{F\ 2\gamma\omega_d}{(\omega_0^2 - \omega_d^2)^2 + 4\gamma^2\omega_d^2}\right]\frac{1}{2i}(e^{i\omega_d t} - e^{-i\omega_d t}).$$

By recalling the complex trigonometric identities

$$\cos\alpha = \frac{1}{2}(e^{i\alpha} + e^{-i\alpha})$$
$$\sin\alpha = \frac{1}{2i}(e^{i\alpha} - e^{-i\alpha}),$$

we can see that the equation above becomes:

$$x_p(t) = \left[\frac{F\,(\omega_0{}^2 - \omega_d{}^2)}{(\omega_0{}^2 - \omega_d{}^2)^2 + 4\gamma^2\omega_d^2}\right]\cos\omega_d t + \left[\frac{F\,2\gamma\omega_d}{(\omega_0{}^2 - \omega_d{}^2)^2 + 4\gamma^2\omega_d^2}\right]\sin\omega_d t. \tag{4.32}$$

◆

We can simplify the particular solution even further than that shown in the example above. To do so, we define the quantity:

$$R \equiv \sqrt{(\omega_0^2 - \omega_d{}^2)^2 + (2\gamma\omega_d)^2}. \tag{4.33}$$

We can then write equation (4.32) as

$$x_p(t) = \frac{F}{R}\left(\frac{\omega_0^2 - \omega_d{}^2}{R}\cos\omega_d t + \frac{2\gamma\omega_d}{R}\sin\omega_d t\right). \tag{4.34}$$

Consider the triangle shown in figure 4.10, we can define an angle ϕ whose cosine and sine can be expressed as seen in equations (4.36) and (4.35).

$$\cos\phi = \frac{\omega_0^2 - \omega_d{}^2}{R} \tag{4.35}$$

$$\sin\phi = \frac{2\gamma\omega_d}{R}. \tag{4.36}$$

We can substitute these identities into our particular solution and obtain

$$x_p(t) = \frac{F}{R}(\cos\phi\cos\omega_d t + \sin\phi\sin\omega_d t).$$

In the equation above, we can clearly recognise the expanded formula for a cosine of a sum of two angles. The equation above can therefore be rewritten as

$$x_p(t) = \frac{F}{R}\cos(\omega_d t + \phi). \tag{4.37}$$

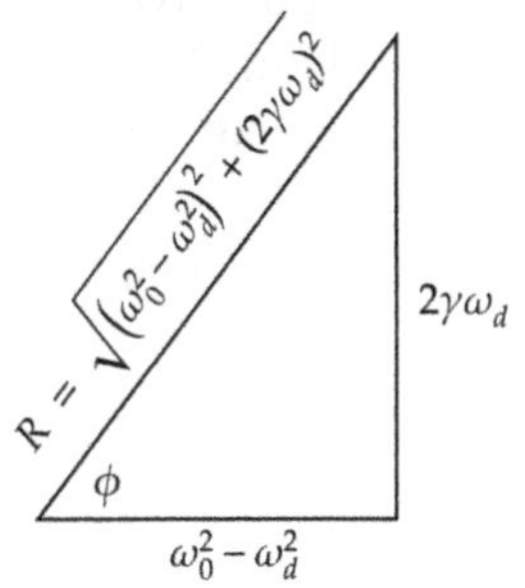

Figure 4.10. The triangle used in the derivation to find the equations relating ϕ and R.

Definition 4.3.2: The phase of a driven motion
The angle ϕ is defined as the **phase difference** between the driving impulse of the force and the *driven response* of the system to the force.

Through our mathematical manipulations of the particular solution we have obtained a simpler and more comprehensive solution to damped and driven motion in the form

$$x(t) = \frac{F}{R}\cos(\omega_d t + \phi) + e^{-\gamma t}[Ae^{\sqrt{\gamma^2-\omega_0^2}\,t} + Be^{-\sqrt{\gamma^2-\omega_0^2}\,t}]. \tag{4.38}$$

The above is our final solution of motion, let's make some observations on its form. Recall that earlier in the chapter we referred to the particular solution as its **stationary state solution**. This is because as $t \to \infty$, the exponential solution, called the **transition phase**, tends to zero.

Remark 4.10. Angular frequencies overview
At this point in the chapter, we have come across quite a few different types of angular frequencies. Below is a quick run through of how each frequency has been defined

- $\omega \to$ the general angular frequency of an oscillating body.
- $\omega_0 \to$ the **natural frequency** is the frequency at which the oscillator would naturally move if undergoing simple harmonic motion. It is intrinsic and defined by its structural properties.
- $\tilde{\omega} \to$ the **angular frequency of damped motion** occurs when the oscillator is undergoing *underdamped motion* and oscillating at a frequency which is dependent both on ω_0 and the damping rate γ.
- $\omega_d \to$ the **driving angular frequency** is the frequency at which the driving force moves the damped oscillator.

4.3.1 Driven oscillator curves

We can see that our solution depends upon two frequencies, the driving frequency and the natural frequency of the oscillator. We will observe the different responses our system can have based on the value that the ω_d will take compared with that of ω_0.

From equation (4.38), we know that the **amplitude** of the oscillator is:

$$\frac{F}{R} = \frac{F_d}{m\sqrt{(\omega_0^2 - \omega_d^2)^2 + (2\gamma\omega_d)^2}} \tag{4.39}$$

and the **phase difference** of the oscillator, is given by:

$$\tan \phi = \frac{2\gamma\omega_d}{\omega_0^2 - \omega_d^2}, \tag{4.40}$$

where the formula was obtained by looking at the triangle in figure 4.10. Looking at the equations given above we see that both the amplitude and the phase difference depend upon both the internal and external frequencies.

First let us look at the dependence of the amplitude from the angular frequency of the driving force. While the numerator has a linear dependence, the denominator is minimal when

$$\omega_d^2 = \omega_0^2 - 2\gamma^2. \tag{4.41}$$

When $\gamma \ll \omega_0$, the case of very weak damping, the above reduces to $\omega_d \approx \omega_0$. The maximum of the amplitude will occur very close to the case when the driving frequency is equal to the inner frequency of the oscillator.

Definition 4.3.3: Resonance
The maximal amplitude of oscillation occurs for a given frequency, close as we have seen to the natural frequency of the oscillator, called *resonance frequency*. Driven oscillators are one of the many physical systems exhibiting the phenomenon of *resonance*.

Figure 4.11 shows how the amplitude varies as a function of the driving angular velocity ω_d for two different intensities of damping.

Looking at the triangle with which we defined ϕ, see figure 4.10, we can see that the angle ϕ will be restricted to the interval $0 < \phi < \pi$. We can now study how the **phase angle** will vary based on the value of the driving frequency ω_d, by distinguishing between *three* main cases.

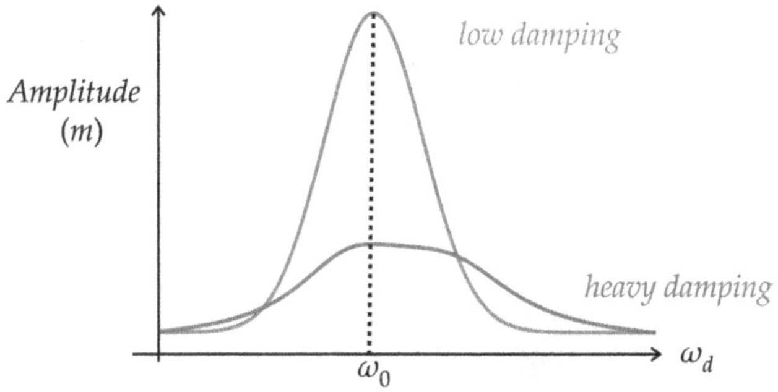

Figure 4.11. Shows the peak in amplitude at resonance, and the behaviour of the oscillator for different levels of damping.

1. $\omega_d \approx 0$.
   In particular when $\gamma\omega_d \ll \omega_0^2 - \omega_d^2$ we will have that $\phi \approx 0$
   This is because $\ddot{x}$ and $\dot{x}$ will both become very small, since they are proportional to ω_d^2 and ω_d, respectively. We will be left with $x \propto \cos\omega_d t$ meaning that the phase will be close to zero. Intuitively, if the driving force changes very slowly, it is also expected that the system will have time to follow it closely.

2. $\omega_d \approx \omega_0$
   This is the case of **resonance**, therefore by substituting the identity above into equation (4.40) we have that $\phi = \pi/2$.
   The positive sign of the phase is because the motion of the block lags the driving force by a quarter of a cycle at resonance.

3. $\omega_d \to \infty$.
   More precisely, if $\gamma\omega_d \ll \omega_d^2 - \omega_0^2$ we will have that $\phi \approx \pi$.
   This is because $\ddot{x}$ will dominate, since it is proportional to ω_d^2. We will be left with $\ddot{x} \propto \cos\omega_d t$, therefore $\ddot{x}$ will be in phase with the force. We now recall the property of a sinusoidal function, where the second derivative is 180° out of phase with x. From which we conclude that x will be 180° out of phase with the force.

Figure 4.12 shows a graph of how the phase difference changes based on the values of the driving frequency and the damping strength.

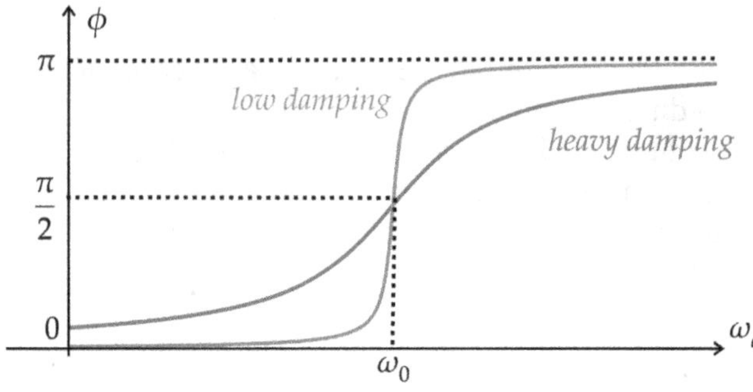

Figure 4.12. Variation of the phase shift between driving motion and oscillator response for high and low damping.

4.3.2 Quality factor

In figure 4.11 you can observe that the peak in amplitude at resonance has a change in sharpness based on the strength of the damping, indicated by the *damping rate* γ.

For larger values of γ we can see that the change in amplitude at resonance will not be very significant. We can define a **quality factor** to describe the sharpness of the peak at resonance.

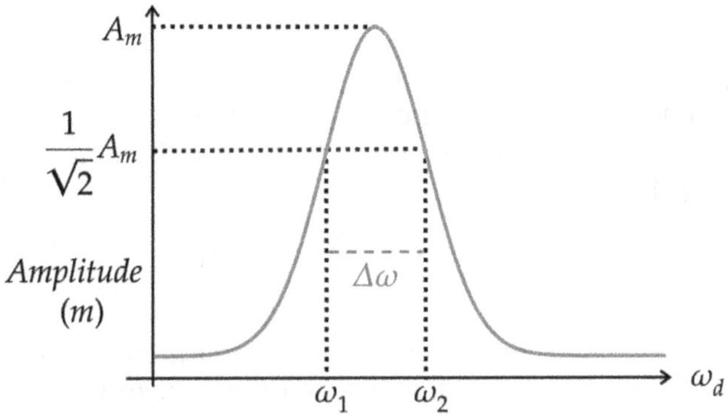

Figure 4.13. Illustrates the width of peak used to determine the quality factor.

We can define a measure called the **width of the peak** $\Delta\omega$ which, *when it exists*, is given by the difference of the two frequencies, ω_1 and ω_2 at which the amplitude has reduced to $1/\sqrt{2}$ its peak value A_m (figure 4.13). The width is defined as;

$$\Delta\omega \equiv |\omega_2 - \omega_1|. \tag{4.42}$$

Definition 4.3.4: Quality factor
Gives a measure of the width of the peak between the two points where the amplitude has dropped by a factor of $1/\sqrt{2}$ with respect to the maximal resonance amplitude. It is given by

$$Q = \frac{\omega_0}{\Delta\omega}. \tag{4.43}$$

The quality factor defined above has a second expression, which is used to determine its dependence on the damping intensity. We know the amplitude reaches $1/\sqrt{2}$ of its peak value when

$$\omega_d \approx \omega_0 \pm \frac{1}{2}\Delta\omega,$$

where:

$$\left|(\omega_0 \pm \frac{1}{2}\Delta\omega)^2 - \omega_0^2\right| \approx 2\gamma\omega_0$$
$$|\Delta\omega|\omega_0 \approx 2\gamma\omega_0$$
$$\Delta\omega = 2\gamma.$$

Therefore, the equivalent equation for the quality factor.

$$Q = \frac{\omega_0}{2\gamma}. \tag{4.44}$$

4.3.3 Work done by driving force

Let us calculate the amount of work done by a driving force during one full period of oscillation. We must consider that friction is not a conservative force, therefore a part of the energy will be lost in the form of heat due to friction. A period of forced oscillation is given by:

$$T_d = \frac{2\pi}{\omega_d}.$$

Let's now find the work as the *integral of the power with respect to time*. To do this we will need to change the work integral into a time integral

$$\begin{aligned} W &= \int F\,\mathrm{d}x \\ &= \int_0^T F\frac{\mathrm{d}x}{\mathrm{d}t}\,\mathrm{d}t. \end{aligned}$$

Now that we have changed the integration variable, we can find the work done over one period. We know the equation of our force to be $F_d \cos \omega_d t$. By substituting the force and the derivative to the stationary solution, equation (4.37), we have that:

$$W = -\frac{F_d^2}{mR}\int_0^T \cos(\omega_d t)\sin(\omega_d t + \phi)\,\mathrm{d}t.$$

Note that the minus sign derives from the derivative of the sinusoidal solution. We then substitute $\omega_d t = \theta$ so that $\mathrm{d}t = \mathrm{d}\theta/\omega_d$, which gives us

$$\begin{aligned} W &= -\frac{F_d^2}{mR}\int_0^{2\pi} \cos(\theta)\sin(\theta + \phi)\,\mathrm{d}\theta \\ &= -\frac{F_d^2}{mR}\int_0^{2\pi} \cos(\theta)[\sin\theta\cos\phi + \cos\theta\sin\phi]\,\mathrm{d}\theta. \end{aligned}$$

We can use the following integral relations:

$$\int_0^{2\pi} \sin\theta\cos\theta\,\mathrm{d}\theta = 0 \text{ and } \int_0^{2\pi} \cos^2\theta\,\mathrm{d}\theta = \pi.$$

Therefore, the work done by the driving force during one period of oscillation is

$$W = -\pi\frac{F_d^2}{mR}\sin\phi. \tag{4.45}$$

The average power dissipated by the driving force is related to the work by the relation

$$\bar{P} = \frac{W}{T_d} = \frac{\omega_d}{2\pi}W.$$

Substituting equation (4.45) above we obtain

$$\bar{P} = -\frac{F_d^2}{2\,mR}\omega_d \sin\phi.$$

We also know from equation (4.36) that

$$\sin\phi = \frac{2\gamma\omega_d}{R}.$$

Substituting this into the equation for average power,

$$\bar{P} = -\frac{F_d^2}{2\,mR}\omega_d \sin\phi$$
$$= -\frac{F_d^2}{\not{2}mR}\omega_d\frac{\not{2}\gamma\omega_d}{R}$$
$$\bar{P} = -\frac{\omega_d^2}{R^2}\frac{F_d^2}{m}\gamma.$$

We know the relationship $v_{\max} = \frac{F_d}{mR}\omega_d$ where v_m is the **maximum speed reached**. Therefore, $\frac{\omega_d}{R} = \frac{v_{\max} m}{F_d}$ substituting this identity into the equation for power above we get

$$\bar{P} = m\gamma v_{\max}^2 \tag{4.46}$$

We can express this in terms of the maximum value of the kinetic energy,

$$\bar{P} = 2K_{\max}\gamma \tag{4.47}$$

4.4 Coupled oscillators

Definition 4.4.1: Coupled oscillators

Coupled oscillations occur when two or more oscillating systems are connected, so that kinetic energy can be exchanged between them.

For example, if two pendulums are attached with a spring, and one is initially set in motion, it will start driving the stationary pendulum until all its kinetic energy is transferred to the second pendulum. The second pendulum will in turn start driving the first one, now stationary, and the process could be repeated indefinitely in an ideal situation with no friction.

Let us consider two masses attached to each other and to two walls by three identical springs, with the same elastic constant k (figure 4.14). Let's call $x_1(t)$ and $x_2(t)$ the displacements of the first and second masses with respect to their

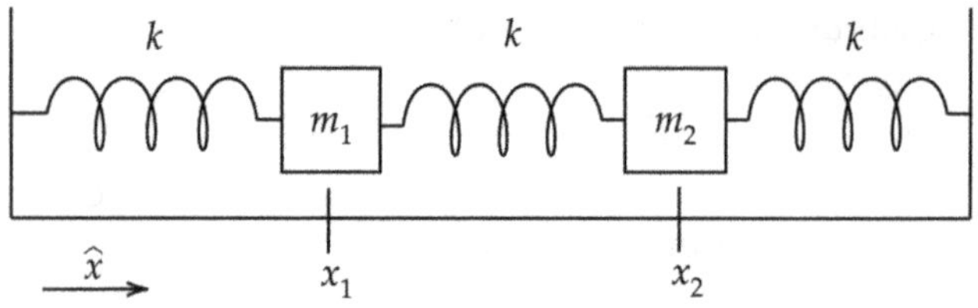

Figure 4.14. Illustrating the setup of the coupled oscillator, with two masses and three springs.

equilibrium position. In this position also the spring between the two masses is considered to be at rest.

By observing the image above, we can see that the middle spring will be stretched by a distance $x_2 - x_1$. Let's look at the forces acting on the masses, in the absence of friction:

$$m_1 \rightarrow F_1 = -kx_1 + k(x_2 - x_1)$$
$$m_2 \rightarrow F_2 = -kx_2 + k(x_2 - x_1).$$

To construct the equation of motion of the system, first we must consider the two masses individually:

$$\begin{cases} m_1\ddot{x}_1 = -kx_1 - k(x_2 - x_1) \\ m_2\ddot{x}_2 = -kx_2 + k(x_2 - x_1) \end{cases}.$$

By using $\omega^2 \equiv \frac{k}{m}$ and assuming the masses of the two objects are the same $m_1 = m_2 = m$, we can rewrite the equations of motion as

$$\begin{cases} \ddot{x}_1 + 2\omega^2 x_1 - \omega^2 x_2 = 0 \\ \ddot{x}_2 + 2\omega^2 x_2 - \omega^2 x_1 = 0 \end{cases}.$$

This pair of coupled equations cannot be immediately solved as both differential equations have a dependence on two independent variables, x_1 and x_2. By summing and subtracting the two equations above we can obtain two new differential equations, dependent on only one quantity each.

Summing the two equations above:

$$(\ddot{x}_1 + \ddot{x}_2) + \omega^2(x_1 + x_2) = 0, \tag{4.48}$$

and by subtracting we obtain

$$(\ddot{x}_1 - \ddot{x}_2) + 3\omega^2(x_1 - x_2) = 0. \tag{4.49}$$

We can see that if we consider $(x_1 + x_2)$ and $(x_1 - x_2)$ as the two new independent variables, the system will decouple into two independent differential equations which will be much easier to solve. The two new independent variables are called the **normal modes of the coupled oscillations**.

Definition 4.4.2: Normal modes

The motion of a coupled oscillator can be described by two modes. The sum of the positions for the two masses $(x_1 + x_2)$ represents the motion of the **centre of mass**, or the average motion of the oscillators, and $(x_1 - x_2)$ is the motion of the bodies' **relative separation**.

In case only one mode is excited, the other variable combination remains constant. When $(x_1 + x_2)$ is constant, the bodies must be going in opposite directions when their oscillations are in anti-phase, as seen in figure 4.15.

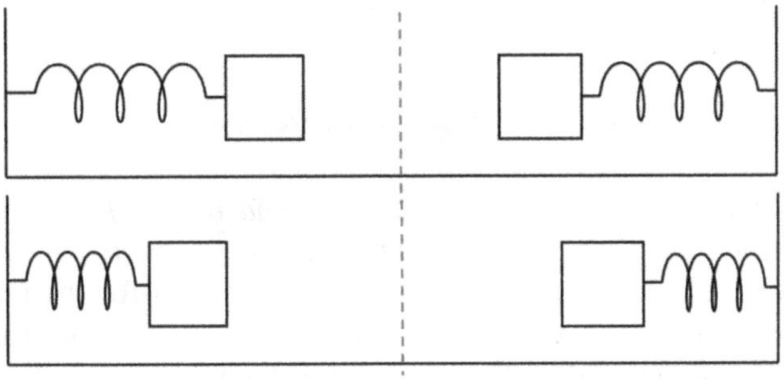

Figure 4.15. Clarification of the type of movement of the **first normal mode** of the coupled oscillator.

The second mode of oscillation happens when $(x_1 - x_2) = C$, where C is a constant value (figure 4.16). This will occur when the two masses are moving in the same direction with constant distance between them, and the oscillations of the two bodies are exactly in phase.

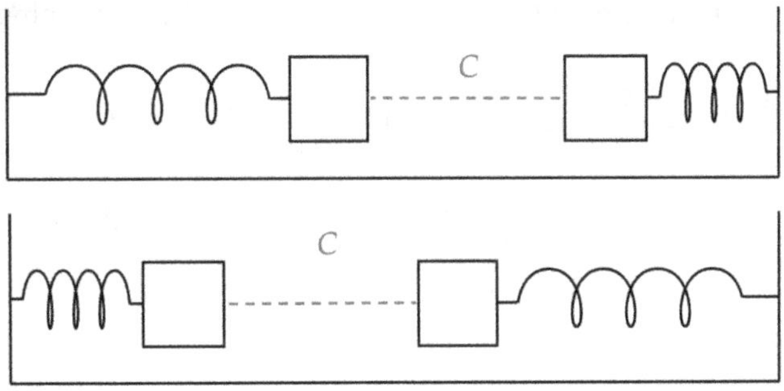

Figure 4.16. Clarification of the type of movement of the **second normal mode** of the coupled oscillator.

We can now look at how to solve our two simultaneous equations for the coupled oscillator.

We can start by renaming the independent variables, $(x_1 + x_2) = \alpha$ and $(x_1 - x_2) = \beta$

$$\begin{cases} \ddot{\alpha} + \omega^2 \alpha = 0 \\ \ddot{\beta} + 3\omega^2 \beta = 0 \end{cases}.$$

The above solutions look exactly like simple harmonic oscillators, whose solution to the equation of motion we already know. We therefore have that:

$$\alpha = (x_1 + x_2) = A_+ \cos(\omega t + \phi_+) \tag{4.50}$$

$$\beta = (x_1 - x_2) = A_- \cos(\sqrt{3}\,\omega t + \phi_-). \tag{4.51}$$

The factor of $\sqrt{3}$ is introduced for β the second mode of oscillation, as in the differential equation there is a factor of 3 in front of the angular velocity squared. The constants above $A_\pm$ depend upon the initial conditions set for the problem.

The variables defined above as α and β are the new coordinates that describe the movements of the modes, called the **normal coordinates.**

> Remark 4.11. The derivation above shows how the motion of a coupled oscillator can be more simply approximated by the sum of two simpler motions, described by the two normal coordinates. In some cases the oscillations can be described by one single mode, based on the initial conditions, but in general they will be described by a linear combination of the two modes.

Finally we derive the equations for the motions of the individual oscillators by summing and subtracting equations (4.50) and (4.51), thus obtaining:

$$\begin{cases} x_1(t) = \dfrac{A_+}{2}\cos(\omega t + \phi_+) + \dfrac{A_-}{2}\cos(\sqrt{3}\,\omega t + \phi_-) \\ x_2(t) = \dfrac{A_+}{2}\cos(\omega t + \phi_+) - \dfrac{A_-}{2}\cos(\sqrt{3}\,\omega t + \phi_-) \end{cases}. \tag{4.52}$$

Exercise 4.8. Consider figure 4.14. In this case, we will have two masses, m_1 and m_2, placed on a friction-less surface and attached to three ideal springs with identical spring constants, k.

(i) Initially the springs are held in their equilibrium positions, m_1 is kept at rest whilst m_2 is given a velocity $\boldsymbol{v} = v\hat{\boldsymbol{x}}$. Ignoring gravity, write the equations of motion for the two masses in terms of x_1 ad x_2 their relative displacements from the equilibrium positions.

(ii) Label the normal modes as $y_1 = x_1 + x_2$ and $y_2 = x_1 - x_2$. Show that the modes obey the equation of motion which describes simple harmonic motion, and that the ratio of the oscillation frequencies of y_1 and y_2 is $\frac{1}{\sqrt{3}}$.

(iii) The two external springs are fixed to stationary walls, use these boundary conditions to show that the solution to y_1 is

$$y_1 = v\sqrt{\frac{m}{k}} \sin\left(t\sqrt{\frac{k}{m}}\right).$$

(iv) Now consider the addition of air friction as a damping factor. The friction force opposes the motion of the two masses and has a magnitude of βv_r, where v_r is defined as the relative velocity between the two masses. Write down the equation of motion of the masses in terms of x_1 ad x_2 with the addition of the damping force.

(v) Using the equations of motion obtained above in terms of y_1 and y_2, consider a trial solution for y_2 of Ae^{qt}. Show that $x_1 = x_2$ as $t \to \infty$, and hence derive the maximal amplitude of oscillation of either of the two masses as $t \to \infty$.

Exercise 4.9. The extended pendulum

Let's consider an extended body pendulum, as seen in figure 4.17. The body has a total mass M, formed by a rod of length L and mass m_1 and a spherical ball of radius l and mass m_2. Clearly, the length of the whole extended pendulum will be $L + 2l$.

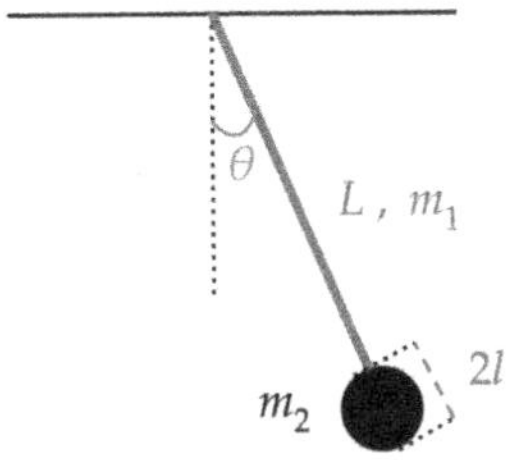

Figure 4.17. The form of the extended body pendulum described above.

The equation of motion takes the form

$$\left(m_2 + \frac{m_1}{3}\right)L^2\ddot{\theta} + \left(m_2 + \frac{m_1}{2}\right)gL\theta = 0.$$

Find the frequency of the oscillatory motion of the extended pendulum. Then, using the solution to the *simple pendulum exercise* earlier in this chapter, show that the frequency determined is that of a simple pendulum as $m_1 \to 0$.

Exercise 4.10. A mass m is vertically hanging from an ideal massless spring, attached to the ceiling at a point A. The mass is released from rest at a time $t = 0$ from a displacement x_0 from the equilibrium position, as seen in figure 4.18

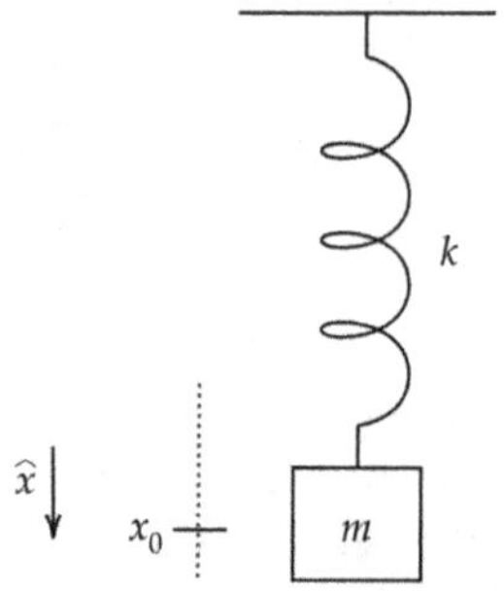

Figure 4.18. A mass hanging vertically from a massless spring.

(i) Write down the equation of motion which describes the moving mass in terms of the displacement x from the equilibrium position. Show that the mass will move with simple harmonic motion.
(ii) Verify that the equation of motion is satisfied by a solution of the form

$$x(t) = A\cos\omega t + B\sin\omega t.$$

Finding the values of the constants given the boundary conditions above.
(iii) We will now consider the presence of air friction, which will create a damping force of $-\beta\dot{x}$. Draw a diagram of how $x(t)$ from part b) changes as a function of time if $\frac{\beta}{2m} < \sqrt{\frac{k}{m}}$.

Exercise 4.11. A mass m can move in the (x, y)-plane, under a potential

$$V(x, y) = \frac{1}{2}k_x(x - x_0)^2 + \frac{1}{2}k_y(y - y_0)^2.$$

(i) Find the equilibrium position in 2D, and the forces acting on the mass in a generic position in the (x, y)-plane.
(ii) Calculate the equations of motion of the mass on the x and y planes.
(iii) At time $t = 0$, the mass is left at rest from position (a, b). Calculate the two-dimensional position of the mass as a function of time.

Mario Campanelli

Chapter 5

Angular momentum and central forces

5.0 Introduction

Up until now we have used the Cartesian coordinate system (x,y,z) to describe the motion of point-like objects. This system is the most commonly used when dealing with linear situations. However, it becomes much harder to use Cartesian coordinates when treating rotating bodies, spinning systems or angular accelerations. Fortunately, the position of particles can also be defined using **polar coordinates** (r,θ, z), or **spherical coordinates** (r,θ,ϕ). These coordinates greatly simplify the calculations when dealing with rotating systems.

In this chapter we will be introducing polar coordinates in two dimensions (so, at constant z), and demonstrating how to transform between the Cartesian and polar systems. This reference system will then be applied to describe the **angular momentum** of a rotating object, as well as used to characterize **central forces**.

5.1 Polar coordinates

Given a point in space, a dedicated reference frame can be defined using a unit radial vector $\hat{\boldsymbol{r}}$ describing the direction between the origin and that point, and the unit transverse vector $\hat{\boldsymbol{\theta}}$. Just like the Cartesian coordinates $\hat{x}$ and $\hat{y}$, $\hat{\boldsymbol{r}}$ and $\hat{\boldsymbol{\theta}}$ are orthogonal to each other. Figure 5.1 shows the relationship between the new and the original coordinate systems.

Figure 5.2 is an enlargement of the yellow circle in figure 5.1 and shows the relationship between the unit vectors from the two coordinate systems. We can use it to derive the equations for $\hat{\boldsymbol{r}}$ and $\hat{\boldsymbol{\theta}}$:

$$\hat{\boldsymbol{r}} = \cos\theta\hat{\boldsymbol{i}} + \sin\theta\hat{\boldsymbol{j}} \tag{5.1}$$

$$\hat{\boldsymbol{\theta}} = -\sin\theta\hat{\boldsymbol{i}} + \cos\theta\hat{\boldsymbol{j}}\,. \tag{5.2}$$

doi:10.1088/978-0-7503-2690-2ch5

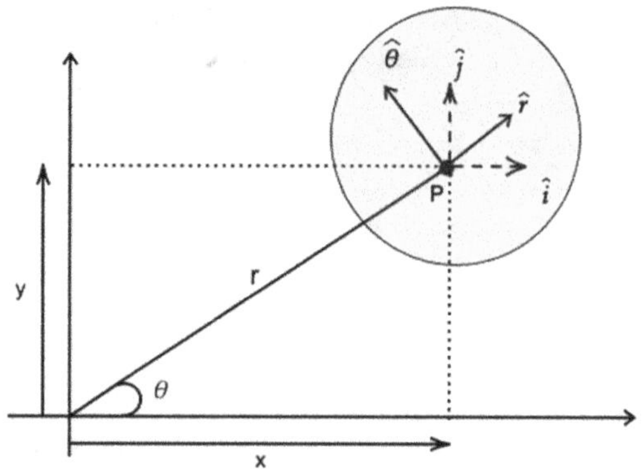

Figure 5.1. Relationship between local and Cartesian coordinates.

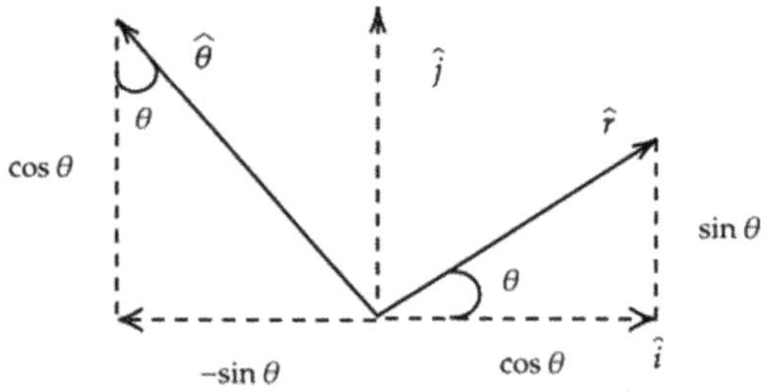

Figure 5.2. An enlargement of the unit vectors in figure 5.1.

As a particle moves, the directions of $\hat{\boldsymbol{r}}$ and $\hat{\boldsymbol{\theta}}$ vary, and in general the rate of change is not constant. Let's consider the velocity of the particle, i.e. the rate of change of position with time.

$$v = \frac{\mathrm{d}r}{\mathrm{d}t} = \frac{\mathrm{d}}{\mathrm{d}t}(r\hat{\boldsymbol{r}}) \tag{5.3}$$

$$v = \frac{\mathrm{d}r}{\mathrm{d}t}\hat{\boldsymbol{r}} + r\frac{\mathrm{d}\hat{\boldsymbol{r}}}{\mathrm{d}t}$$

$$v = \dot{r}\hat{\boldsymbol{r}} + r\dot{\hat{\boldsymbol{r}}}.$$

In equation (5.3) we have evaluated the differential $\frac{\mathrm{d}}{\mathrm{d}t}(r\hat{\boldsymbol{r}})$ and used the notation $\frac{\mathrm{d}r}{\mathrm{d}t} = \dot{r}$ to write the equation for the velocity of the particle. We can see that to find the velocity of the particle we need to find the time derivative of the unit radial vector $\hat{\boldsymbol{r}}$. As seen in equation (5.1):

$$\hat{\boldsymbol{r}} = \cos\theta\hat{\boldsymbol{i}} + \sin\theta\hat{\boldsymbol{j}}\,\dot{\hat{\boldsymbol{r}}} = -\sin\theta\dot{\theta}\hat{\boldsymbol{i}} + \cos\theta\dot{\theta}\hat{\boldsymbol{j}}$$

which can then be simplified to:

$$\dot{\hat{r}} = \dot{\theta}(-\sin\theta\hat{\boldsymbol{i}} + \cos\theta\hat{\boldsymbol{j}})$$
$$\dot{\hat{r}} = \dot{\theta}\hat{\boldsymbol{\theta}}.$$

Hence the velocity of the particle can be written as:

$$v = \dot{r} = \dot{r}\hat{r} + r\dot{\theta}\hat{\boldsymbol{\theta}}. \tag{5.4}$$

Similarly, the acceleration of the particle can also be considered as the rate of change of the velocity.

$$\boldsymbol{a} = \frac{\mathrm{d}v}{\mathrm{d}t}$$
$$\boldsymbol{a} = \frac{\mathrm{d}}{\mathrm{d}t}\left(\frac{\mathrm{d}r}{\mathrm{d}t}\hat{r} + r\frac{\mathrm{d}\theta}{\mathrm{d}t}\hat{\boldsymbol{\theta}}\right)$$
$$\boldsymbol{a} = \frac{\mathrm{d}^2r}{\mathrm{d}t^2}\hat{r} + \frac{\mathrm{d}r}{\mathrm{d}t}\frac{\mathrm{d}\hat{r}}{\mathrm{d}t} + \frac{\mathrm{d}r}{\mathrm{d}t}\frac{\mathrm{d}\theta}{\mathrm{d}t}\hat{\boldsymbol{\theta}} + r\frac{\mathrm{d}^2\theta}{\mathrm{d}t^2}\hat{\boldsymbol{\theta}} + \frac{\mathrm{d}\theta}{\mathrm{d}t}\frac{\mathrm{d}\hat{\boldsymbol{\theta}}}{\mathrm{d}t}$$

$$\boldsymbol{a} = \ddot{r}\hat{r} + \dot{r}\dot{\hat{r}} + \dot{r}\dot{\theta}\hat{\boldsymbol{\theta}} + r\ddot{\theta}\hat{\boldsymbol{\theta}} + r\dot{\theta}\dot{\hat{\boldsymbol{\theta}}}. \tag{5.5}$$

Equation (5.2) includes the derivative of the unit transverse vector. This can be expressed as follows:

$$\hat{\boldsymbol{\theta}} = -\sin\theta\hat{\boldsymbol{i}} + \cos\theta\hat{\boldsymbol{j}}$$
$$\frac{\mathrm{d}\hat{\boldsymbol{\theta}}}{\mathrm{d}t} = -\cos\theta\frac{\mathrm{d}\theta}{\mathrm{d}t}\hat{\boldsymbol{i}} - \sin\theta\frac{\mathrm{d}\theta}{\mathrm{d}t}\hat{\boldsymbol{j}}$$
$$\frac{\mathrm{d}\hat{\boldsymbol{\theta}}}{\mathrm{d}t} = -\frac{\mathrm{d}\theta}{\mathrm{d}t}(\cos\theta\hat{\boldsymbol{i}} + \sin\theta\hat{\boldsymbol{j}})$$

$$\frac{\mathrm{d}\hat{\boldsymbol{\theta}}}{\mathrm{d}t} = -\frac{\mathrm{d}\theta}{\mathrm{d}t}\hat{r} = -\dot{\theta}\hat{r}. \tag{5.6}$$

Substituting in $\dot{\hat{\boldsymbol{\theta}}} = -\dot{\theta}\hat{r}$ and $\dot{\hat{r}} = \dot{\theta}\hat{\boldsymbol{\theta}}$ in the acceleration equation:

$$\boldsymbol{a} = \ddot{r}\hat{r} + \dot{r}\dot{\hat{r}} + \dot{r}\dot{\theta}\hat{\boldsymbol{\theta}} + r\ddot{\theta}\hat{\boldsymbol{\theta}} + r\dot{\theta}\dot{\hat{\boldsymbol{\theta}}}$$
$$\boldsymbol{a} = \ddot{r}\hat{r} + \dot{r}\dot{\theta}\hat{\boldsymbol{\theta}} + \dot{r}\dot{\theta}\hat{\boldsymbol{\theta}} + r\ddot{\theta}\hat{\boldsymbol{\theta}} - r\dot{\theta}\dot{\theta}\hat{r}$$

$$\boldsymbol{a} = (\ddot{r} - r\dot{\theta}^2)\hat{r} + (2\dot{r}\dot{\theta} + r\ddot{\theta})\hat{\boldsymbol{\theta}}. \tag{5.7}$$

By performing the derivation above we have obtained the equation for the acceleration of an object in polar coordinates. As we can see, the acceleration can be divided into the radial $\hat{r}$ and transverse $\hat{\boldsymbol{\theta}}$ directions.

We can now use this equation to understand the motion of objects in real life situations such as circular motion, which is covered in the next subsection.

Example 5.1. A bead moves outwards along a friction-less spoke of a bicycle wheel at a constant speed u. The bead is at the centre of the wheel at $t = 0$. The bead has a constant angular speed ωt. Find the velocity and acceleration of the bead.
We are provided with the information

$$\dot{r} = u$$
$$\dot{\theta} = \omega t.$$

From this we can get the value for r,

$$r = ut.$$

Substituting these values into the equation (5.4), we get:

$$\boldsymbol{v} = \dot{r}\hat{\boldsymbol{r}} + r\dot{\theta}\hat{\boldsymbol{\theta}}$$
$$\Longrightarrow \boldsymbol{v} = u\hat{\boldsymbol{r}} + u\omega t^2\hat{\boldsymbol{\theta}}.$$

This gives us the velocity of the bead in polar coordinates. To find the acceleration of the bead we substitute the values into equation (5.7),

$$\boldsymbol{a} = (\ddot{r} - r\dot{\theta}^2)\hat{\boldsymbol{r}} + (2\dot{r}\dot{\theta} + r\ddot{\theta})\hat{\boldsymbol{\theta}}$$
$$\boldsymbol{a} = -u\omega^2 t^3\hat{\boldsymbol{r}} + (2u\omega t + u\omega t)\hat{\boldsymbol{\theta}}$$
$$\Longrightarrow \boldsymbol{a} = -u\omega^2 t^3\hat{\boldsymbol{r}} + 3u\omega t\hat{\boldsymbol{\theta}}.$$

Exercise 5.1. Express $\hat{\boldsymbol{r}}$ and $\hat{\boldsymbol{\theta}}$ in terms of the Cartesian unit vectors $\hat{\boldsymbol{i}}$ and $\hat{\boldsymbol{j}}$. Show that:

$$\frac{\mathrm{d}\hat{\boldsymbol{r}}}{\mathrm{d}t} = \dot{\theta}\hat{\boldsymbol{\theta}}$$

and hence demonstrate:

$$\boldsymbol{v} = \frac{\mathrm{d}\boldsymbol{r}}{\mathrm{d}t} = \dot{r}\hat{\boldsymbol{r}} + r\dot{\theta}\hat{\boldsymbol{\theta}}.$$

Finally, show that:

$$\boldsymbol{a} = \frac{\mathrm{d}\boldsymbol{v}}{\mathrm{d}t} = (\ddot{r} - r\dot{\theta}^2)\hat{\boldsymbol{r}} + (2\dot{r}\dot{\theta} + r\ddot{\theta})\hat{\boldsymbol{\theta}}.$$

◆

5.2 Circular motion

Definition 5.2.1: Uniform circular motion
A particle moving at constant velocity in a circular trajectory with a constant radius r is said to be undergoing uniform circular motion.

Since the radius is constant, $\dot{r}$ and $\ddot{r}$ will both be equal to zero. This makes the calculation for the velocity of the particle much simpler.

$$\boldsymbol{v} = \dot{r}\hat{\boldsymbol{r}} + r\dot{\theta}\hat{\boldsymbol{\theta}}$$

$$\boldsymbol{v} = r\dot{\theta}\hat{\boldsymbol{\theta}}. \tag{5.8}$$

Similarly, for the acceleration:

$$\boldsymbol{a} = (\ddot{r} - r\dot{\theta}^2)\hat{\boldsymbol{r}} + (2\dot{r}\dot{\theta} + r\ddot{\theta})\hat{\boldsymbol{\theta}}$$

$$\boldsymbol{a} = (-r\dot{\theta}^2)\hat{\boldsymbol{r}} + (r\ddot{\theta})\hat{\boldsymbol{\theta}}. \tag{5.9}$$

Remark 5.1. The angular velocity $\dot{\theta}$ is constant for uniform circular motion, the rate of change of angular velocity is then $\ddot{\theta} = 0$.

This allows us to further simplify the acceleration to:

$$\boldsymbol{a} = -r\dot{\theta}^2\hat{\boldsymbol{r}}. \tag{5.10}$$

The angular velocity of the particle $\dot{\theta}$ can be written as ω. From a previous definition we know that $v = \omega r$. This allows us to write the velocity and acceleration as seen below.

$$\boldsymbol{v} = r\omega\hat{\boldsymbol{\theta}} \tag{5.11}$$

$$\boldsymbol{a} = -r\omega^2\hat{\boldsymbol{r}} = -\frac{v^2}{r}\hat{\boldsymbol{r}}. \tag{5.12}$$

Remark 5.2. The negative sign in equation (5.12) tells us that the acceleration is directed towards the centre of the circle.

This acceleration is known as **centripetal acceleration**. The force which is responsible for this acceleration can be obtained from Newton's second law.

$$\boldsymbol{F} = m\boldsymbol{a}$$

$$\boldsymbol{F} = -m\frac{v^2}{r}\hat{\boldsymbol{r}}. \tag{5.13}$$

This is called the **centripetal force**. The direction of the centripetal force is perpendicular to the tangential velocity, so this force can only change the direction of the velocity vector, but since it cannot perform any work on the particle, no increase in the speed can happen. This ensures circular motion takes place.

> Remark 5.3. We know that a cross product between two vectors gives us a third vector perpendicular to the first two. Since $\boldsymbol{r}$ and $\boldsymbol{p}$ are both in the x–y plane, the angular momentum vector resulting from their cross product must always be pointing in the $\hat{z}$ direction, and the particle cannot leave the plane of motion.

Exercise 5.2. The movement of a particle in a plane is described by the distance r from a fixed origin and the angle θ which its position vector makes with a fixed axis. Sketch the directions of the unit vectors $\hat{\boldsymbol{r}}$ and $\hat{\boldsymbol{\theta}}$.
Write down expressions for the velocity $\boldsymbol{v}$ and acceleration $\boldsymbol{a}$ in this coordinate system.
Using the expressions above to determine the force $\boldsymbol{F}$ necessary to maintain a mass m in a uniform circular motion with angular velocity ω and radius r.

◆

5.3 Angular momentum

Consider a particle of mass m moving with a velocity $\boldsymbol{v}$ at a position $\boldsymbol{r}$ relative to the centre of motion.

> **Definition 5.3.1: Angular momentum**
> The angular momentum is defined as the cross product of the position vector with the momentum of the particle.
>
> $$\boldsymbol{L} = \boldsymbol{r} \times \boldsymbol{p} \tag{5.14}$$
>
> $$\boldsymbol{L} = m\boldsymbol{r} \times \boldsymbol{v}. \tag{5.15}$$

By definition, $\boldsymbol{L}$ is a vector as it is the cross product between two vectors. Therefore, it gives us the direction of rotation of the particle.

> **Definition 5.3.2: Torque**
> If we consider a force acting on the particle, the moment of force or torque about the centre is defined as:
>
> $$\boldsymbol{\tau} = \boldsymbol{r} \times \boldsymbol{F} \tag{5.16}$$
>
> $$\boldsymbol{\tau} = m\boldsymbol{r} \times \frac{\mathrm{d}\boldsymbol{v}}{\mathrm{d}t}. \tag{5.17}$$

In the equation above we used the fact that $\boldsymbol{F} = m\boldsymbol{a}$. To simplify the equation for the torque, we can perform the *time derivative* of the cross product of the radius with the velocity.

$$\begin{aligned}\frac{\mathrm{d}(\boldsymbol{r} \times \boldsymbol{v})}{\mathrm{d}t} &= \frac{\mathrm{d}\boldsymbol{r}}{\mathrm{d}t} \times \boldsymbol{v} + \boldsymbol{r} \times \frac{\mathrm{d}\boldsymbol{v}}{\mathrm{d}t} \\ &= \boldsymbol{v} \times \boldsymbol{v} + \boldsymbol{r} \times \frac{\mathrm{d}\boldsymbol{v}}{\mathrm{d}t} = \boldsymbol{r} \times \frac{\mathrm{d}\boldsymbol{v}}{\mathrm{d}t}\end{aligned} \tag{5.18}$$

This is because the cross product of a vector with itself is zero since $\sin 0 = 0$. Therefore, the torque can be written as:

$$\boldsymbol{\tau} = m\boldsymbol{r} \times \frac{\mathrm{d}\boldsymbol{v}}{\mathrm{d}t} \equiv \frac{\mathrm{d}(m\boldsymbol{r} \times \boldsymbol{v})}{\mathrm{d}t}$$

$$\boldsymbol{\tau} = \frac{\mathrm{d}\boldsymbol{L}}{\mathrm{d}t}. \tag{5.19}$$

Remark 5.4. We can see from the derivation above that the torque is the rate of change of the angular momentum with respect to time. This means that when the angular momentum is constant, there is no force being applied on the particle and $\boldsymbol{\tau} = 0$.

Remark 5.5. We can also obtain the absolute value of the torque using the cross product.

$$\boldsymbol{\tau} = \boldsymbol{r} \times \boldsymbol{F}$$
$$|\boldsymbol{\tau}| = |\boldsymbol{r}||\boldsymbol{F}| \sin\theta$$

$$|\boldsymbol{\tau}| = |\boldsymbol{r}_{\perp}||\boldsymbol{F}| = |\boldsymbol{F}_{\perp}||\boldsymbol{r}|. \tag{5.20}$$

Here the absolute value of the torque is given by the scalar product of the force with the perpendicular distance from the origin to the line of action of the force.

Angular momentum in polar coordinates

The angular momentum can be expressed in polar coordinates provided that both $\boldsymbol{p}$ and $\boldsymbol{r}$ are co-planar. Since the angular momentum is given by the cross product of the two vectors it is perpendicular to the plane in which they lie: for instance, if we take the motion to be in the x–y plane, then the angular momentum will be found in the z plane.

$$\boldsymbol{L}_z = m\boldsymbol{r} \times \boldsymbol{v} = m\boldsymbol{r} \times (\dot{r}\hat{\boldsymbol{r}} + r\dot{\theta}\hat{\boldsymbol{\theta}}).$$

Given that $\boldsymbol{r} \times \hat{\boldsymbol{r}} = 0$ the angular momentum in polar coordinates will be given by:

$$\boldsymbol{L}_z = mr^2\dot{\theta}. \tag{5.21}$$

We can also rewrite the above as

$$\boldsymbol{L}_z = mrv_\theta \tag{5.22}$$

and from this we obtain the known velocity relation:

$$v_\theta = r\omega = r\frac{\mathrm{d}\theta}{\mathrm{d}t}. \tag{5.23}$$

Exercise 5.3. A particle is in a position $\boldsymbol{r}$ relative to the origin of the coordinate system O. If the particle has mass m and is moving with a velocity $\boldsymbol{v}$ with a force $\boldsymbol{F}$ acting on it, state the angular momentum $\boldsymbol{L}$ of the particle about O and the torque $\boldsymbol{\tau}$ of the force.
Subsequently prove that the torque $\boldsymbol{\tau}$ is equivalent to the rate of change of the angular momentum $\boldsymbol{L}$ in time.

Exercise 5.4. Using the definition of angular momentum in equation (5.14) show that $|\boldsymbol{L}| = mr^2\dot{\theta}$ and define the direction of $\boldsymbol{L}$.
Hence show that if a particle is subject to a central force

$$2\dot{r}\dot{\theta} + r\ddot{\theta} = 0.$$

◆

5.4 Central forces

> **Definition 5.4.1: Central forces**
> A Central force is a force which has a direction given by its position vector with respect to the central point O of the system, and whose magnitude only depends upon the distance between the point of application and O.
> This means that central forces are either directed towards or away from the central point.

The magnitude of central forces is dependent on the distance from the centre, *i.e. the radius of the rotation.* Central forces can thus be written as:

$$\boldsymbol{F}(\boldsymbol{r}) = F(r)\hat{\boldsymbol{r}}. \tag{5.24}$$

If the force is directed away from the centre, it is repulsive and can be called a centrifugal force. If it is directed towards the centre of the circle it is attractive and

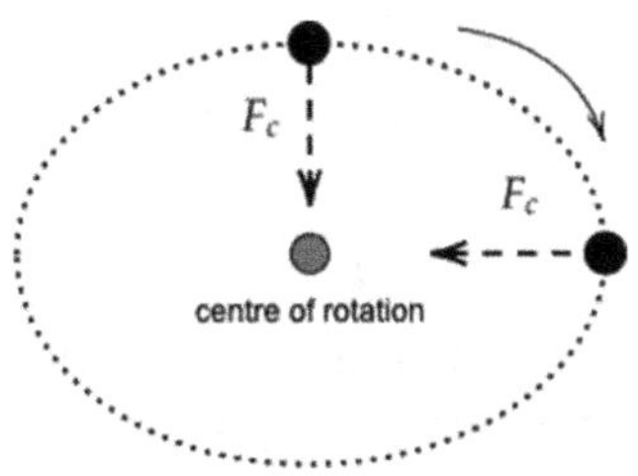

Figure 5.3. Direction of central force.

can be called a centripetal force. Figure 5.3 shows the change in direction of an attractive central force during circular motion.

Remark 5.6. Some common examples of central forces include:

(i) Electrostatic force between two point charges

$$\boldsymbol{F}(\boldsymbol{r}) = \frac{q_1 q_2}{2\pi\epsilon_0 r^2}\hat{\boldsymbol{r}}.$$

For like charges this force is repulsive, *i.e. centrifugal*, and for unlike charges it is attractive, *i.e. centripetal.*

(ii) Gravitational force between two spherical masses

$$\boldsymbol{F}(\boldsymbol{r}) = -G\frac{M_1 M_2}{r^2}\hat{\boldsymbol{r}}.$$

The presence of central forces on a physical system implies that there will be a form of rotational movement, hence it is appropriate to describe the system using polar coordinates. The **equation of motion** for central forces can be written in polar coordinate form as seen below:

$$\boldsymbol{F} = F(r)\hat{\boldsymbol{r}} = m\boldsymbol{a}.$$

Using equation (5.7) we can rewrite the above as

$$F(r)\hat{\boldsymbol{r}} = m[(\ddot{r} - r\dot{\theta}^2)\hat{\boldsymbol{r}} + (2\dot{r}\dot{\theta} + r\ddot{\theta})\hat{\boldsymbol{\theta}}]. \tag{5.25}$$

We can separate the equation of motion seen above into radial and transverse components. An important result comes from observing that central forces only act in the **radial direction**, thus the acceleration must only have a radial term. For movement under the effect of central forces the *transverse term* of the equation of motion must therefore be equal to zero, as seen below.

$$\begin{cases} m(\ddot{r} - r\dot{\theta}^2) = F(r) \\ m(2\dot{r}\dot{\theta} + r\ddot{\theta}) = 0 \end{cases}. \tag{5.26}$$

Angular momentum of central forces
We will now look at the angular momentum of central motion in polar coordinates. From its definition,

$$L = mr^2\dot{\theta}. \tag{5.27}$$

We will study how the angular momentum varies in time, to do this we take the *time derivative* of the angular momentum. By applying the product rule we obtain the following,

$$\begin{aligned}\frac{\mathrm{d}L}{\mathrm{d}t} &= \frac{\mathrm{d}}{\mathrm{d}t}(mr^2\dot{\theta})\\ \frac{\mathrm{d}L}{\mathrm{d}t} &= m(2r\dot{r}\dot{\theta} + r^2\ddot{\theta})\\ \frac{\mathrm{d}L}{\mathrm{d}t} &= mr(2\dot{r}\dot{\theta} + r\ddot{\theta}).\end{aligned} \tag{5.28}$$

We can see that the term in the brackets in equation (5.28) is equivalent to the transverse acceleration from equation (5.26). However, we know that the transverse acceleration must be *zero* for central motion, therefore we can conclude that:

$$\frac{\mathrm{d}L}{\mathrm{d}t} = 0. \tag{5.29}$$

Remark 5.7. We have therefore proven that the angular momentum is **constant** during motion under the effect of *any* central force.

Example 5.2. Consider a car of mass m moving in a circular path around a roundabout. The coefficient of friction is μ. What is the maximum speed the car can achieve without skidding on the surface?
In order for the car to move in a circle, it needs to have a centripetal force to change its direction.

$$\boldsymbol{F} = m\left(-\frac{v^2}{r}\right)\hat{\boldsymbol{r}}.$$

The negative direction signifies the direction of the force, towards the origin. We need to draw a free body diagram to clearly see the forces acting on the car.

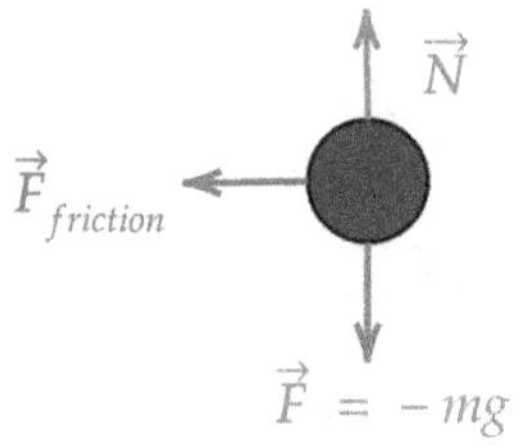

In the vertical direction, the normal reaction force balances the weight of the car. Hence:

$$N = mg.$$

The Centripetal force needed for the car to remain in a circle is provided by the friction force. We know from previous knowledge that friction is $\boldsymbol{F} = \mu N$

$$-\boldsymbol{F}_{friction} = m\left(-\frac{v^2}{r}\right)$$

$$m\left(\frac{v^2}{r}\right) = \mu mg$$

$$v_{max} = \sqrt{\mu g r}\,.$$

Exercise 5.5.

(i) Define what is meant by *central force*.

(ii) Prove that the angular momentum of a particle subject to a central force is constant.

◆

5.4.1 Potential energy for central forces

As we saw above, the magnitude of central forces only depend on the distance from the centre of motion as seen below.

$$\boldsymbol{F}(\boldsymbol{r}) = F(r)\hat{\boldsymbol{r}}.$$

A central force, whose magnitude only depends upon the distance from the origin, is **always** a *conservative force*. For this reason, we can always define a potential energy for a central force.

Definition 5.4.2: Potential energy of central forces

The potential energy arising from a central force is only dependent on the radial distance from the origin and can be defined as:

$$V(r) = -\int F(r)\mathrm{d}r. \tag{5.30}$$

So

$$F(r) = -\frac{\mathrm{d}V}{\mathrm{d}r}. \tag{5.31}$$

Exercise 5.6. An object of mass m moves in the x–y plane subject to a force associated to the potential energy

$$V(x, y) = C(x^2 + y^2)^2.$$

(i) Find the force $\boldsymbol{F}$ acting on the object in a general position (x, y).
(ii) Define a *central force*. Hence show that the force obtained above is a central force with centre at the origin. Subsequently deduce two dynamical quantities conserved during this motion.
(iii) The object is then launched from the point $(R, 0)$. State what the velocity vector should be for the object to move in an anticlockwise circular orbit in the x–y plane. Determine the period of such an orbit and the angular momentum $\boldsymbol{L}$.

Exercise 5.7. A particle of mass m is attached to a light in-extensible string. The free extremity of the string is passed through a small friction-less hole in the horizontal plane where the particle lies, as seen in the figure below.

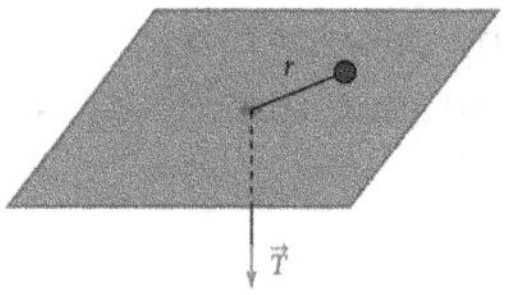

The particle is set in motion at a distance r_0 from the hole, the free extremity of the string is held fixed to make the particle move in a circular path with a constant angular velocity ω_0.

(i) State the angular momentum of the particle about the whole.
(ii) At a time $t = 0$ the string starts to be pulled down through the hole at a constant speed V, resulting in a decrease of the radius of the circular motion in time of $r(t) = (r_0 - Vt)$. What is the kinematic quantity which is conserved in the particle's resultant motion?
(iii) Given the above, determine the angular velocity of the particle ω at a time t and the velocity of the particle $\boldsymbol{v}$.
(iv) Using the radial equation of motion of the particle, find the value of the tension in the spring $\boldsymbol{T}$.

Exercise 5.8. A particle with mass m is rotating with a time dependent angular frequency $\dot{\theta}$ and with a radial velocity component which decreases at a constant rate α such that $\dot{r} = -\alpha$. The particle is in the radial position R_0 with $\dot{\theta} = \alpha$ at the time $t = 0$.
Using equations (5.4) and (5.7), and assuming that the angular momentum is conserved, show that at a general time t the angular velocity is

$$\dot{\theta}(t) = \frac{R_0^2 \alpha}{(R_0 - \alpha t)^2}$$

and that the angular component of the acceleration is zero.

Exercise 5.9. A Ladybird is initially at rest in the mid-point of a horizontal stick of length A. At a time $t = 0$ the stick starts to rotate around a vertical axis at one of its extremities with an angular velocity given by $\omega = \omega_0 t$, *where* ω_0 *is constant.*
Also in the instant $t = 0$, the ladybird starts to walk away from the axis of rotation with a constant speed v.
Using equations (5.4) and (5.7) determine the angular momentum vector $\boldsymbol{L}$ of the ladybird at a time t.

Exercise 5.10. Two beads of mass m are placed at the top of a friction-less hoop of mass M and radius R which is at rest in the vertical plane on top of a friction-less vertical support. The beads are given a tiny impulse and due to gravity they slide down from position (P), *seen in the image below*, in the clockwise and anticlockwise direction.

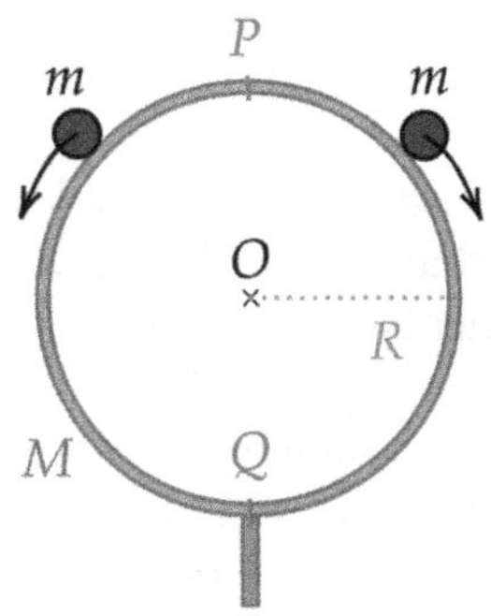

Determine the minimum value of $X = \frac{m}{M}$, X_{MIN} for which the loop will rise off the support before the beads will have reached the bottom position (Q).

Chapter 6

Centre of mass and collisions

6.0 Introduction

Until now, you've probably always solved problems in the laboratory frame of reference, which simply means that you used the given information/measurements as these were measured by instruments at rest in the lab. However, this is just one of the many ways we can solve problems. Nature has little to no interest in what we take measurements with respect to. Although most of the time using the lab frame might seem the most intuitive option, the maths can be easier if we calculate quantities with respect to something else, for example the centre of mass.

6.1 The centre of mass

We start then with the definition of the centre of mass.

Definition 6.1.1: Centre of mass

The *centre of mass* of a set of point-like objects is defined as the average of the positions of the individual masses, weighted by the magnitude of the masses

$$\boldsymbol{R}_{\mathrm{CM}} = \frac{\sum_i m_i \boldsymbol{r}_i}{M} \tag{6.1}$$

where $M = \sum_i m_i$ is the sum of all masses and $\boldsymbol{r}_i$ is their position.

Figure 6.1 is a schematic representation of the centre of mass of a system made of two bodies, mass m_1 and mass m_2.

doi:10.1088/978-0-7503-2690-2ch6

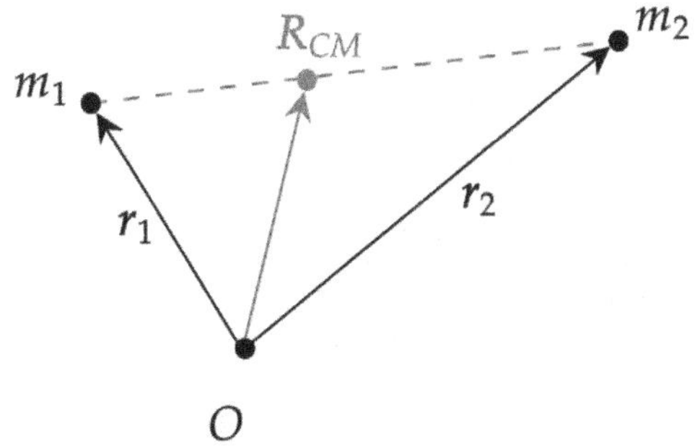

Figure 6.1. Centre of mass $\boldsymbol{R}_{\mathrm{CM}}$ of the system which comprehends two point-like bodies, m_1 and m_2, in positions $\boldsymbol{r}_1$ and $\boldsymbol{r}_2$, respectively.

We can also define the centre of mass velocity by taking the derivative of its position.

$$\boldsymbol{V}_{\mathrm{CM}} = \frac{\mathrm{d}\boldsymbol{R}_{\mathrm{CM}}}{\mathrm{d}t} = \frac{\sum_i m_i \frac{\mathrm{d}\boldsymbol{r}_i}{\mathrm{d}t}}{M} = \frac{\sum_i m_i \boldsymbol{v}_i}{M}. \tag{6.2}$$

In a system composed of N bodies, we can see the *centre of mass* as an **isolated system with the mass of all N bodies**. This becomes useful when calculating the momentum of the centre of mass,

$$\boldsymbol{P}_{\mathrm{CM}} = M\boldsymbol{V}_{\mathrm{CM}} = \frac{M}{M}\sum_i m_i \boldsymbol{v}_i = \sum_i \boldsymbol{p}_i \tag{6.3}$$

where we have used the result of equation (6.2). We can see the momentum of the centre of mass is simply the sum of the individual momenta of the bodies in the system.

If we now take the derivative of $\boldsymbol{P}_{\mathrm{CM}}$

$$\frac{\mathrm{d}\boldsymbol{P}_{\mathrm{CM}}}{\mathrm{d}t} = M\frac{\mathrm{d}^2\boldsymbol{R}_{\mathrm{CM}}}{\mathrm{d}t^2} = \boldsymbol{F}_{\mathrm{ext}} \tag{6.4}$$

we obtain a force (from Newton's second law) which we can call external force, $\boldsymbol{F}_{\mathrm{ext}}$. Equation (6.4) also tells us that the momentum of the centre of mass $\boldsymbol{P}_{\mathrm{CM}}$ is constant in absence of external forces $\boldsymbol{F}_{\mathrm{ext}}$ acting on the system, i.e.

$$\frac{\mathrm{d}\boldsymbol{P}_{\mathrm{CM}}}{\mathrm{d}t} = 0 \text{ or } \boldsymbol{V}_{\mathrm{CM}} = \text{constant} \iff \boldsymbol{F}_{\mathrm{ext}} = 0.$$

This is the conservation of momentum of a system of N bodies when no external forces act on the system. The fact that the momentum of the centre of mass is constant when no external forces are involved is one of the reasons why calculations using the centre of mass can be significantly easier and more straightforward.

Example 6.1. Centre of mass

Assume there is a system which includes two bodies. The first is of mass $m_1 = 1$ kg at position $\boldsymbol{r}_1 = (3\hat{\boldsymbol{i}}, -\hat{\boldsymbol{k}})$ while the second is of mass $m_2 = 3$ kg at position $\boldsymbol{r}_2 = (\hat{\boldsymbol{i}}, -2\hat{\boldsymbol{j}}, 3\hat{\boldsymbol{k}})$.

What is the position of the centre of mass, $\boldsymbol{R}_{\text{CM}}$?

We apply equation (6.1),

$$\begin{aligned}\boldsymbol{R}_{\text{CM}} &= \frac{\sum_i m_i \boldsymbol{r}_i}{M}\\ &= \frac{1\left(3\hat{\boldsymbol{i}} - \hat{\boldsymbol{k}}\right) + 3\left(\hat{\boldsymbol{i}} - 2\hat{\boldsymbol{j}} + 3\hat{\boldsymbol{k}}\right)}{1+3}\\ &= \frac{3+3}{4}\hat{\boldsymbol{i}} - \frac{6}{4}\hat{\boldsymbol{j}} + \frac{-1+9}{4}\hat{\boldsymbol{k}}\\ \Rightarrow \boldsymbol{R}_{\text{CM}} &= \frac{3}{2}\hat{\boldsymbol{i}} - \frac{3}{2}\hat{\boldsymbol{j}} + 2\hat{\boldsymbol{k}}.\end{aligned}$$

◆

6.1.1 The centre of mass frame

When no external forces are acting on the system it is often convenient to measure positions and velocities *relative to the centre of mass*, especially in collision problems.

> Remark 6.1. Using measurements of positions and velocities relative to the centre of mass is referred to as working in *the centre of mass frame*; the original position and velocities relative to an external laboratory observer, i.e. measurements usually given in a problem set, are said to be in *the laboratory frame.*

If we use the centre of mass as the origin, new positions in this frame can be written as

$$\boldsymbol{r}_i' = \boldsymbol{r}_i - \boldsymbol{R}_{\text{CM}} \tag{6.5}$$

(you can see how in figure 6.2 position vector $\boldsymbol{r}_1$ is a sum of $\boldsymbol{r}_1'$ and $\boldsymbol{R}_{\text{CM}}$, i.e. $\boldsymbol{r}_1 = \boldsymbol{r}_1' + \boldsymbol{R}_{\text{CM}}$).

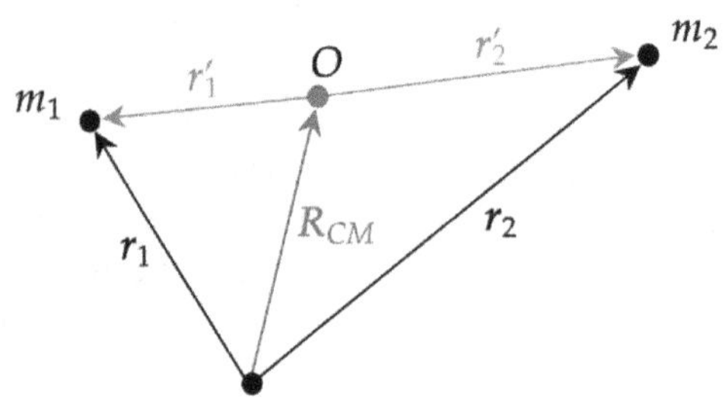

Figure 6.2. Representation of position vectors of two bodies, $\boldsymbol{r}_1'$ and $\boldsymbol{r}_2'$, in the centre of mass frame.

Remark 6.2. This transformation from the lab to the centre of mass frame is an example of a Galilean transformation (uniform velocity boost). Newton's laws are exactly the same in the CM frame, provided only internal forces act between the particles that depend only on their relative positions and velocities. As we saw in the previous section, if there is an external force acting on the system the velocity of the centre of mass will undergo some acceleration, i.e. $\mathrm{d}\boldsymbol{V}_{\mathrm{CM}}/\mathrm{d}t \neq 0$. In this case the centre of mass frame would become a non-inertial/accelerated frame of reference, which will be discussed in section 8.4.6; for now let's not worry about non-inertial frames since we will encounter only internal forces in this chapter.

Analogously, velocities are

$$\boldsymbol{v}_i' = \boldsymbol{v}_i - \boldsymbol{V}_{\mathrm{CM}}. \tag{6.6}$$

We would have obtained the same by differentiating equation (6.5) as follows

$$\begin{aligned}\boldsymbol{v}_i' &= \frac{\mathrm{d}\boldsymbol{r}_i'}{\mathrm{d}t}\\ &= \frac{\mathrm{d}}{\mathrm{d}t}(\boldsymbol{r}_i - \boldsymbol{R}_{\mathrm{CM}})\\ &= \frac{\mathrm{d}\boldsymbol{r}_i}{\mathrm{d}t} - \frac{\mathrm{d}\boldsymbol{R}_{\mathrm{CM}}}{\mathrm{d}t}\\ \Rightarrow \boldsymbol{v}_i' &= \boldsymbol{v}_i - \boldsymbol{V}_{\mathrm{CM}}.\end{aligned}$$

The corresponding momenta can be written as

$$\boldsymbol{p}_i' = m_i\boldsymbol{v}_i' = m_i(\boldsymbol{v}_i - \boldsymbol{V}_{\mathrm{CM}}). \tag{6.7}$$

Notice here that the sum of the momenta in the centre of mass frame is:

$$\begin{aligned}\sum_i \boldsymbol{p}_i' &= \sum_i m_i(\boldsymbol{v}_i - \boldsymbol{V}_{\mathrm{CM}})\\ &= \sum_i m_i\boldsymbol{v}_i - M\boldsymbol{V}_{\mathrm{CM}}\\ &= \boldsymbol{P}_{\mathrm{CM}} - \boldsymbol{P}_{\mathrm{CM}}\\ &= 0,\end{aligned}$$

i.e. the total momentum of the system can be assumed to be entirely *carried by the centre of mass*. This is extremely useful in two-body collision problems, where the velocities of the components in the centre of mass frame must always be in opposite directions such that $\sum_i \boldsymbol{p}_i' = 0$. In this frame, physical processes are generally described more symmetrically and therefore the result becomes more intuitive.

6.1.2 Relative displacement and reduced mass

We are now going to define a vector for the relative displacement between two bodies as well as a relative velocity vector. This becomes extremely useful when solving problems since, as we will shortly see, these vectors remain unchanged when switching from the lab frame to the centre of mass frame. This property also makes calculations easier when solving two-body collision problems.

$$\boldsymbol{r} = \boldsymbol{r}_1 - \boldsymbol{r}_2$$

and

$$\boldsymbol{v} = \boldsymbol{v}_1 - \boldsymbol{v}_2$$

remain the same in the centre of mass frame

$$\boldsymbol{r} = \boldsymbol{r}_1' - \boldsymbol{r}_2'$$

and

$$\boldsymbol{v} = \boldsymbol{v}_1' - \boldsymbol{v}_2'.$$

We can now use the fact that $\sum_i \boldsymbol{p}_i' = 0$ to write an expression for $\boldsymbol{v}_1'$ and $\boldsymbol{v}_2'$ in terms of the relative velocity:

$$\begin{aligned}
\boldsymbol{v}_1' &= \boldsymbol{v}_1 - \boldsymbol{V}_{\text{CM}} \\
&= \boldsymbol{v}_1 - \frac{m_1\boldsymbol{v}_1 + m_2\boldsymbol{v}_2}{m_1 + m_2} \\
&= \frac{(m_1 + m_2)\boldsymbol{v}_1 - m_1\boldsymbol{v}_1 - m_2\boldsymbol{v}_2}{m_1 + m_2} \\
&= \frac{m_2(\boldsymbol{v}_1 - \boldsymbol{v}_2)}{m_1 + m_2} \\
&= \frac{m_2\boldsymbol{v}}{m_1 + m_2}
\end{aligned}$$

such that:

$$\boldsymbol{v}_1' = \frac{m_2}{M}\boldsymbol{v} \tag{6.8}$$

and similarly for $\boldsymbol{v}_2'$:

$$\boldsymbol{v}_2' = \frac{-m_1}{M}\boldsymbol{v}. \tag{6.9}$$

We can use these two equations to find new expressions for $\boldsymbol{p}_1'$ and $\boldsymbol{p}_2'$:

$$\boldsymbol{p}_1' = m_1\boldsymbol{v}_1' = m_1\frac{m_2}{M}\boldsymbol{v} = \frac{m_1m_2}{m_1 + m_2}\boldsymbol{v} = \mu\boldsymbol{v} \tag{6.10}$$

$$\boldsymbol{p}_2' = m_2\boldsymbol{v}_2' = m_2\frac{-m_1}{M}\boldsymbol{v} = \frac{-m_1m_2}{m_1 + m_2}\boldsymbol{v} = -\mu\boldsymbol{v} \tag{6.11}$$

where we have defined the *reduced mass* μ as follows:

$$\mu = \frac{m_1 m_2}{M} = \frac{m_1 m_2}{m_1 + m_2} \quad \text{or} \quad \frac{1}{\mu} = \frac{1}{m_1} + \frac{1}{m_2}. \tag{6.12}$$

Example 6.2. Example about changing frames of reference

Consider two objects (measurements taken by an observer at rest with the laboratory):

- mass 1, $m_1 = 4$ kg, $\boldsymbol{v}_1 = 3\hat{\boldsymbol{i}}$ m s^{-1}
- mass 2, $m_2 = 1$ kg, $\boldsymbol{v}_2 = -6\hat{\boldsymbol{i}}$ m s^{-1}.

Find the total momentum of the system. Find velocities in the *centre of mass frame* and check that the sum of momenta in the centre of mass frame is zero.

We begin by calculating individual momenta in the lab frame:

$$\begin{aligned} \boldsymbol{p}_1 &= m_1\boldsymbol{v}_1 = 4 \cdot 3\hat{\boldsymbol{i}} \\ &= 12\hat{\boldsymbol{i}} \\ \boldsymbol{p}_2 &= m_2\boldsymbol{v}_2 = 1 \cdot (-6\hat{\boldsymbol{i}}) \\ &= -6\hat{\boldsymbol{i}} \end{aligned}$$

such that total momentum $\boldsymbol{p}_{\text{tot}}$ is

$$\boldsymbol{p}_{\text{tot}} = \boldsymbol{p}_1 + \boldsymbol{p}_2 = 6\hat{\boldsymbol{i}}.$$

Velocities in the centre of mass frame can be found using equation (6.6), but since we're dealing with a two-body system it is straightforward to find the relative velocity between the two bodies and then use equations (6.8) and (6.9):

$$\begin{aligned} \boldsymbol{v} &= \boldsymbol{v}_1 - \boldsymbol{v}_2 = [3 - (-6)]\hat{\boldsymbol{i}} \\ &= 9\hat{\boldsymbol{i}} \\ \boldsymbol{v}_1' &= \frac{m_2}{M}\boldsymbol{v} = \frac{5}{9}\hat{\boldsymbol{i}} \\ \boldsymbol{v}_2' &= -\frac{m_1}{M}\boldsymbol{v} = -\frac{36}{9}\hat{\boldsymbol{i}}. \end{aligned}$$

We can now check if the sum of the individual momenta in the centre of mass frame adds up to zero

$$\begin{aligned} \sum_{i=1}^{2} m_i\boldsymbol{v}_i' &= m_1\boldsymbol{v}_1' + m_2\boldsymbol{v}_2' \\ &= 4\left(\frac{5}{9}\hat{\boldsymbol{i}}\right) + 1\left(-\frac{36}{9}\hat{\boldsymbol{i}}\right) = 0. \end{aligned}$$

Calculations would have been even more trivial if we had used the reduced mass with equations (6.10) and (6.11)

$$\mu = \frac{m_1 m_2}{m_1 + m_2} = 4/5$$
$$\boldsymbol{p}_1' = 4/5 \cdot 9\hat{\boldsymbol{i}}$$
$$\boldsymbol{p}_2' = -4/5 \cdot 9\hat{\boldsymbol{i}}$$
$$\sum_{i=1}^{2} \boldsymbol{p}_i' = [36/5 - 36/5]\hat{\boldsymbol{i}} = 0.$$

◆

6.1.3 Kinetic energy in centre of mass frame

Kinetic energy, $K = \frac{1}{2}mv^2$ in the lab frame, can be written in terms of the centre of mass,

$$\begin{aligned} K &= \frac{1}{2}\sum_i m_i v_i^2 \\ &= \frac{1}{2}\sum_i m_i(v_i - V_{\mathrm{CM}} + V_{\mathrm{CM}})^2 \\ &= \frac{1}{2}\sum_i m_i(v_i' + V_{\mathrm{CM}})^2 \\ &= \frac{1}{2}\sum_i m_i(v_i'^2 + 2v_i' \cdot V_{\mathrm{CM}} + V_{\mathrm{CM}}^2) \\ &= \frac{1}{2}\sum_i m_i v_i'^2 + \sum_i m_i v_i' \cdot V_{\mathrm{CM}} + \frac{1}{2}\sum_i m_i V_{\mathrm{CM}}^2 \end{aligned}$$

but since we know that the sum of all individual momenta in the centre of mass frame is zero, i.e. $\sum_i m_i v_i' = 0$, see equation (6.7), we can obtain an expression which has only two terms

$$\begin{aligned} K &= \frac{1}{2}\sum_i m_i v_i'^2 + \frac{1}{2}\sum_i m_i V_{\mathrm{CM}}^2 \\ &= K_{\mathrm{rel}} + K_{\mathrm{CM}}. \end{aligned}$$

The two terms can be identified as:

- K_{CM}, the kinetic energy the system would have if the total mass M were moving at the centre of mass velocity V_{CM}
- K_{rel}, the kinetic energy coming from the motion of the component bodies relative to the centre of mass.

In the case of a system of two bodies, we can find an expression for K_{rel} in terms of reduced mass by direct evaluation:

$$
\begin{aligned}
K_{\text{rel}} &= \frac{1}{2}\sum_{i=1}^{2} m_i v_i'^2 \\
&= \frac{1}{2}m_1 v_1'^2 + \frac{1}{2}m_2 v_2'^2 \\
&= \frac{m_1}{2}\left(\frac{m_2}{M}v\right)^2 + \frac{m_2}{2}\left(\frac{-m_1}{M}v\right)^2 \\
&= \frac{v^2}{2}\left(\frac{m_1 m_2^2 + m_2 m_1^2}{M^2}\right) \\
&= \frac{v^2}{2}\left(\frac{(m_1 m_2)(m_2 + m_1)}{M^2}\right) \\
&= \frac{v^2}{2}\left(\frac{(m_1 m_2)(M)}{M^2}\right) \\
&= \frac{v^2}{2}\left(\frac{m_1 m_2}{M}\right)
\end{aligned}
$$

$$\Rightarrow K_{\text{rel}} = \frac{1}{2}\mu v^2 \tag{6.13}$$

such that kinetic energy can be written as

$$K = \frac{1}{2}\mu v^2 + K_{\text{CM}}. \tag{6.14}$$

In the case where there are no external forces we know that $V_{\text{CM}} = \text{constant}$, therefore the change in kinetic energy is

$$\Delta K = K_{\text{final}} - K_{\text{initial}} \tag{6.15}$$

$$= \frac{1}{2}\mu v^2 + K_{\text{CM}} - \left(\frac{1}{2}\mu \boldsymbol{u}^2 + K_{\text{CM}}\right) \tag{6.16}$$

$$= \frac{1}{2}\mu(v^2 - \boldsymbol{u}^2) \tag{6.17}$$

where $v = v_1 - v_2$ is the final relative velocity between the two bodies in the system, $\boldsymbol{u} = \boldsymbol{u}_1 - \boldsymbol{u}_2$ is the initial relative velocity and K_{CM} cancels out in the difference.

Remark 6.3. Equation (6.17) can be extremely useful when solving problems and can avoid many calculations. It tells us that in absence of external forces the change in kinetic energy is possible only due to a change in relative velocity between the two bodies.

Exercise 6.1. A system comprises three bodies, their positions are described by the following vectors:

- $r_1 = -\hat{i} + 3\hat{j}$
- $r_2 = -\hat{i} - 2\hat{j}$
- $r_3 = 3\hat{i}$.

The centre of mass is found at the origin of the axis, i.e. $\boldsymbol{R}_{\text{CM}} = 0\hat{i} + 0\hat{j}$.

What are the masses of the three bodies knowing that $m_1 + m_2 = 4$ kg?

Exercise 6.2. A projectile is shot with speed $\boldsymbol{u} = u_x\hat{i} + u_y\hat{j}$, take this launching point as the origin of your axis. At some distance x from the launching point, at its peak height, the projectile breaks up in two identical parts. No external forces are involved such that the horizontal component of the centre of mass velocity remains constant. One of the pieces falls back and reaches exactly the initial launching point. Where does the second piece land?
You can ignore air friction.

◆

6.2 Collisions

In classical mechanics all collisions can be categorized into two main types: *elastic* collisions and *inelastic* collisions. Independently of the nature of the collision, the main procedure to follow when solving collision problems consists of **writing down the conservation of momenta (always satisfied) and mechanical energy (only satisfied for elastic collisions, otherwise restitution coefficients can be given)**, and then figure out whether to use the lab or centre of mass frame to calculate the variable of our interest.

6.2.1 Elastic and inelastic collisions

Definition 6.2.1: Elastic collision

The total kinetic energy of the system is conserved after an elastic collision.

$$\Delta K = K_{\text{final}} - K_{\text{initial}} = 0. \tag{6.18}$$

Notice how we can rewrite the two terms in the above definition as

$$K_{\text{initial}} = \frac{1}{2}m_1u_1^2 + \frac{1}{2}m_2u_2^2 = \frac{1}{2}\mu\boldsymbol{u}^2 + \frac{1}{2}MV^2_{\text{CM}_{\text{initial}}} \tag{6.19}$$

$$K_{\text{final}} = \frac{1}{2}m_1 v_1^2 + \frac{1}{2}m_2 v_2^2 = \frac{1}{2}\mu v^2 + \frac{1}{2}MV_{\text{CM}_{\text{final}}}^2. \tag{6.20}$$

In the case where there are no external forces, knowing that V_{CM} remains constant and therefore the total momentum remains constant, we can use equation (6.17) and see that

$$\Delta K = \frac{1}{2}\mu(v^2 - \boldsymbol{u}^2) = 0 \Leftrightarrow |\boldsymbol{u}| = |v|$$
$$\Rightarrow \boldsymbol{u} = \pm\, \boldsymbol{v}$$

the final relative velocity is equal in magnitude to the initial relative velocity and has either the same or the opposite direction. Problems involving elastic collisions can be greatly simplified knowing that the relative velocity can only change its direction after the collision.

Definition 6.2.2: Inelastic collision
The total kinetic energy of the system is not conserved after an inelastic collision

$$\Delta K = K_{\text{final}} - K_{\text{initial}} \neq 0 \tag{6.21}$$

$$\Rightarrow K_{\text{final}} < K_{\text{initial}} \tag{6.22}$$

$$\Rightarrow |v| < |\boldsymbol{u}|. \tag{6.23}$$

In this case, the final relative velocity is smaller than the initial relative velocity since some energy has been lost or dissipated. To fully solve problems involving inelastic collisions, the amount or fraction of lost energy must be stated. Remember that momentum is always conserved, so even for a fully inelastic collision energy is only lost in the centre of mass system, while the system will continue to move in the lab frame.

Remark 6.4. Whilst some kinetic energy is lost in an inelastic collision, the total energy of the system is still conserved in both types of collision. In the elastic case the kinetic energy remains in that form, in the inelastic case some kinetic energy goes into the form of heat (which at microscopic level turns out to be kinetic energy as well).

6.2.2 Coefficient of restitution

Definition 6.2.3: Coefficient of restitution

The fraction of momentum lost for an inelastic collision in the centre of mass system is called *coefficient of restitution*. If we throw a ball to a fixed wall, or to the ground, assuming that the wall (or the Earth) is infinitely heavier than the ball, we can consider that the centre of mass of the ball + wall system does not change. In this case we have

$$\boldsymbol{v}_f \cdot \hat{\boldsymbol{n}} = e(\boldsymbol{v}_i \cdot \hat{\boldsymbol{n}}) \tag{6.24}$$

where $\hat{\boldsymbol{n}}$ is the normal to the surface of impact, $0 \leqslant e < 1$, $\boldsymbol{v}_i$ is the initial velocity and $\boldsymbol{v}_f$ is the final velocity.

Also notice that:

- In the one dimensional case it simply reduces to:

$$|\boldsymbol{v}_f| = e|\boldsymbol{v}_i|$$

- for $e = 1$ the problem reduces to an elastic collision case
- for $e = 0$ we have the super inelastic case where $\boldsymbol{v}_1 \cdot \hat{\boldsymbol{n}} = e(\boldsymbol{u}_1 \cdot \hat{\boldsymbol{n}}) = 0$.

6.2.3 Single-body collision with a rigid wall

Consider the physical situation displayed in figure 6.3 below.

The wall is considered to be infinitely heavy, so the centre of mass of the wall + ball system coincides with the lab frame. If the wall can be considered to be smooth, i.e. no friction, then the impulse given to the particle by the wall is perpendicular to the wall (Newton's third law) and there is no component parallel to the wall's surface. As a consequence, the tangential component of velocity is unaltered:

$$v_i \sin\beta = v_f \sin\alpha.$$

On the other hand, the normal component will be decreased by the coefficient of restitution

$$\boldsymbol{v}_f \cdot \hat{\boldsymbol{n}} = e\boldsymbol{v}_i \cdot \hat{\boldsymbol{n}} \tag{6.25}$$

$$v_f \cos\beta = e v_i \cos\alpha. \tag{6.26}$$

Dividing the two previous equations we obtain

$$\tan\beta = \frac{1}{e}\tan\alpha \tag{6.27}$$

and the outgoing angle will be larger than the incoming one, since the perpendicular component is reduced while the tangent one will stay the same.

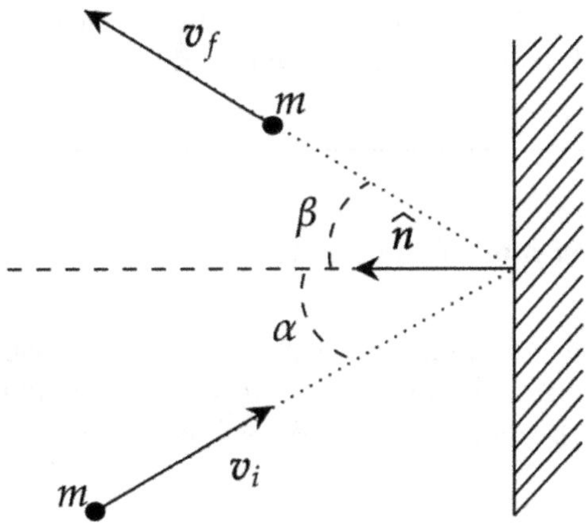

Figure 6.3. Schematic representation of collision of a particle with a wall, angle of incidence α and angle of reflection β.

If the collision is perfectly elastic, $e = 1$, the normal component of the velocity remains unchanged and there is no loss of kinetic energy,

$$v_i \cdot \hat{\boldsymbol{n}} = v_f \cdot \hat{\boldsymbol{n}} \tag{6.28}$$

$$v_i \cos\alpha = v_f \cos\beta \tag{6.29}$$

and

$$\Delta K = 0$$
$$\Rightarrow K_{\text{initial}} = K_{\text{final}}$$
$$\Rightarrow \frac{1}{2}mv_i^2 = \frac{1}{2}mv_f^2.$$

Therefore, in the elastic case this is true only if:

$$v_i = v_f, \tag{6.30}$$

which also implies, due to equation (6.29), that:

$$\alpha = \beta. \tag{6.31}$$

This means the normal component of the velocity is reversed and the angle α of collision is the same as the angle β.

Remark 6.5. 3D case

Notice how the same procedure can be followed in the 3D case. In this case the wall is a surface and there are four angles of collision to consider. Two angles describe the direction of the particle before the collision and the other two will describe the direction after the collision when the particle bounces off. In an exercise, if this is stated at the beginning, it can be helpful and save some time since you avoid writing everything in vector notation.

6.2.4 Collision between two bodies of finite mass

Head-on collision

Consider the following situation, as shown in figure 6.4, in one dimension: a particle of mass m_1 is moving with initial velocity u_1 towards a second particle, with mass m_2 initially at rest, i.e. with velocity $u_2 = 0$. We will consider this situation in the centre of mass frame.

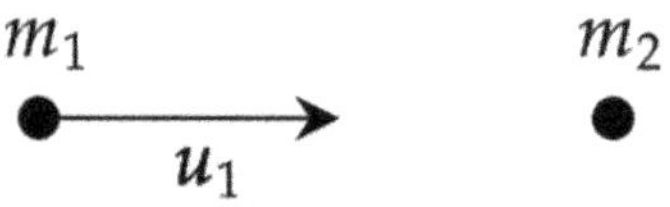

Figure 6.4. Initial situation of collision between two bodies, m_1 with velocity $\boldsymbol{u}_1$ moving towards mass m_2 which is at rest.

> Remark 6.6. Since we are working in one dimension we can just use magnitudes of the velocities rather than their vector form. This is fine as long as we **stick with the same convention**. In this case we are going to use that positive quantities go towards the right direction, therefore $u_1 > 0$.

We start by writing down the velocity of the centre of mass using equation (6.2)

$$V_{\mathrm{CM}} = \frac{\sum_i m_i u_i}{M} = \frac{m_1 u_1}{M}$$

as well as the initial velocities in the centre of mass frame

$$\begin{aligned} u_1' &= u_1 - V_{\mathrm{CM}} \\ &= \left(1 - \frac{m_1}{M}\right)u_1 = \frac{m_2}{M}u_1 \\ u_2' &= u_2 - V_{\mathrm{CM}} \\ &= -\frac{m_1}{M}u_1 = -\frac{m_1}{m_2}u_1' \end{aligned}$$

(in the last passage we used the fact that $u_1 = u_1' M/m_2$ from above) and the relative velocity between the two particles

$$u = u_1 - u_2 = u_1.$$

Then we find the **initial relative kinetic energy** using the formula derived in section 6.1.3:

$$\begin{aligned} K_{\text{initial}} &= K_1 + K_2 \\ &= \frac{1}{2}m_1 u_1'^2 + \frac{1}{2}m_2 u_2'^2 \\ &= \frac{1}{2}\mu u_1^2. \end{aligned}$$

Notice we have used the fact that the relative velocity $u = u_1$.

Now we let the final relative velocity be v, then we can write

$$v_1' = \frac{m_2}{M}v, \qquad v_2' = \frac{-m_1}{M}v \tag{6.32}$$

while **final relative kinetic energy is**

$$\begin{aligned} K_{\text{final}} &= K_1 + K_2 \\ &= \frac{1}{2}m_1 v_1'^2 + \frac{1}{2}m_2 v_2'^2 \\ &= \frac{1}{2}\mu v^2. \end{aligned}$$

At this point the problem has been described in mathematical terms quite generally; in order to make more accurate predictions we need to know whether the collision is elastic or inelastic:

- **Elastic case**

  In an elastic collision kinetic energy is conserved, hence:

$$K_{\text{initial}} = K_{\text{final}} \tag{6.33}$$

$$\frac{1}{2}\mu u_1^2 = \frac{1}{2}\mu v^2 \tag{6.34}$$

$$\text{i.e. } \Delta K = 0 \iff v = \pm u_1. \tag{6.35}$$

  The two allowed values for v are a result of our initial convention for the calculation of the relative velocity, i.e. $u = u_1 - u_2$, where u is positive if the distance between the two particles decreases, while it is negative if the distance increases. Hence, we can interpret $v = u_1 = u$ as the situation before the collision, while $v = -u_1$ is as after the collision.

Remark 6.7. Notice how if we had defined the relative velocity as $u = u_2 - u_1$ we would have interpreted the two solutions in the opposite way, again nature doesn't care how we describe a physical situation, what matters is, once we decide what convention to use, to stick with those conventions and definitions.

We can now substitute the final relative velocity into equation (6.32)

$$v = -u = -u_1$$
$$\Rightarrow v_1' = \frac{m_2}{M}(-u_1) = \frac{-m_2}{M}u_1$$
$$\Rightarrow v_2' = \frac{-m_1}{M}(-u_1) = \frac{m_1}{M}u_1$$

and then translate back to laboratory frame v_1

$$v_1 = v_1' + V_{CM}$$
$$= \frac{-m_2}{M}u_1 + \frac{m_1}{M}u_1$$
$$\Rightarrow v_1 = \frac{m_1 - m_2}{M}u_1$$

same for v_2

$$v_2 = v_2' + V_{CM}$$
$$= \frac{m_1}{M}u_1 + \frac{m_1}{M}u_1$$
$$\Rightarrow v_2 = 2\frac{m_1}{M}u_1.$$

At this point it is interesting to analyse three special cases (figure 6.5):

(A) if $m_1 = m_2$:

$$v_1 = 0, \qquad v_2 = u_1$$

such that all momentum is transferred to the second particle after the collision.

(B) if $m_2 \gg m_1$:

$$v_1 \approx -u_1, \qquad v_2 \approx 0$$

and the problem becomes collision with a rigid wall, as in section 6.2.3 with angle to the normal equal to zero (since in this example we are limited to one dimension)

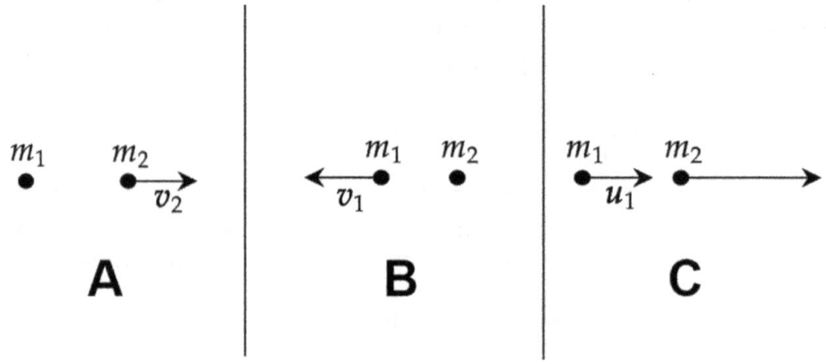

Figure 6.5. Schematic representation of the three cases of the situation after the collision.

(C) if $m_1 \gg m_2$:

$$v_1 \approx u_1, \qquad v_2 \approx 2u_1.$$

Notice how in all three special cases relative velocity $v = v_1 - v_2$ is always equal to $-u_1$.

> Remark 6.8. In a head-on 1D elastic collision the relative velocity of the two bodies after the collision is always equal in magnitude and opposite in direction to the initial relative velocity between the two.

- **Inelastic case**

  For inelastic collisions we need to take into account the coefficient of restitution e. Remember that it refers to quantities in the centre of mass system, and this is why for generic inelastic collisions we always need to express the system in the centre of mass frame (a possible exception are fully inelastic cases).

$$v_1' = -eu_1', \qquad v_2' = -eu_2'.$$

Hence, if we go back to the lab frame:

$$\begin{aligned} v_1 &= v_1' + V_{CM} \\ &= \frac{-em_2}{M}u_1 + \frac{m_1}{M}u_1 \\ \Rightarrow v_1 &= \frac{m_1 - em_2}{M}u_1. \end{aligned}$$

It is the same for v_2:

$$\begin{aligned} v_2 &= v_2' + V_{CM} \\ &= \frac{em_1}{M}u_1 + \frac{m_1}{M}u_1 \\ \Rightarrow v_2 &= \frac{(1 + e)m_1}{M}u_1. \end{aligned}$$

Special case if $e = 0$.

This scenario is more intuitively described in the lab frame of reference. In the super inelastic case the two bodies will stick together immediately after the collision and continue to move along the same direction together.

Then, using conservation of momentum, we can describe the system after the collision as follows

$$p_{tot} = m_1u_1 = (m_1 + m_2)v.$$

Where $(m_1 + m_2)$ is the mass of the new object after the collision and v the velocity of the new object.

Example 6.3. Two-body collision in 1D
Following the example in section 6.1.2, the two objects now undergo an elastic collision. Here is the data given in the previous example:

- mass 1, $m_1 = 4$ kg, $u_1 = 3$ m s^{-1}
- mass 2, $m_2 = 1$ kg, $u_2 = -6$ m s^{-1}.

Our goal is to use the centre of mass frame to find the two resulting final velocities and express these in the lab frame.

Since no external forces are present we already know that V_{CM} **is constant**. Also, because the collision is elastic we know that kinetic energy is conserved. We also know that in this case, because of equation (6.17) the magnitudes of relative velocities before and after the collision are the same but opposite in direction, i.e. (since from before $\boldsymbol{u} = 9\hat{\boldsymbol{i}}$ m s^{-1})

$$\begin{aligned} \boldsymbol{v} &= -\boldsymbol{u} \\ v &= -9 \text{ m s}^{-1}. \end{aligned}$$

At this point we already have enough information to find the final velocities in the centre of mass frame.

$$\begin{aligned} v_1' &= \frac{m_2}{M} v = -\frac{9}{5} \text{ m s}^{-1} \\ v_2' &= \frac{-m_1}{M} v = \frac{36}{5} \text{ m s}^{-1}. \end{aligned}$$

We now go back to the lab frame, we can do this easily using the relation between velocity vectors and V_{CM}

$$\begin{aligned} V_{\text{CM}} &= \frac{p_{\text{tot}}}{M} = \frac{6}{5} \text{ m s}^{-1} \\ v_1 &= v_1' + V_{\text{CM}} \\ &= \frac{6}{5} - \frac{9}{5} = -\frac{3}{5} \text{ m s}^{-1} \\ v_2 &= v_2' + V_{\text{CM}} \\ &= \frac{36}{5} + \frac{6}{5} = \frac{42}{5} \text{ m s}^{-1}. \end{aligned}$$

Check momentum has been conserved:

$$p_{\text{tot}} = m_1 v_1 + m_2 v_2 = \frac{42}{5} - \frac{12}{5} = \frac{30}{5} = 6 \text{ kg m s}^{-1}.$$

Exercise 6.3. Repeat example 1.3 without using the centre of mass frame. [*Hint:* you will have to solve a system of two equations to find the two velocities.]

◆

Glancing collision—general case

We have seen that in two-body collision problems, in the centre of mass frame of reference, the two objects initially approach with relative speed u and after the collision recede with relative speed v. Notice again we're are using magnitudes (speed) rather than vectors (velocity). We have also found that these two speeds are equal if collision is elastic,

$$\Delta K = 0$$
$$\frac{1}{2}\mu u^2 = \frac{1}{2}\mu v^2$$
$$\Rightarrow v \pm u$$

or smaller if inelastic because of the coefficient of restitution

$$\Rightarrow v = eu.$$

We can also inspect the change in **relative** kinetic energy

$$\frac{\mu v^2}{2} = e^2\frac{\mu u^2}{2} \tag{6.36}$$

and see that it reduces by a factor of e^2. Notice that this is the **relative kinetic energy**, K_{rel}, the **kinetic energy of the centre of mass**, K_{CM}, remains unchanged since there is no external force applied to the two-body system that would change the velocity of the centre of mass.

In general, as discussed in section 6.1.2, the final velocities of the two bodies in the centre of mass frame

$$v_1' = \frac{m_2}{M}v, \qquad v_2' = \frac{-m_1}{M}v$$

lie on two circles, of radii $\frac{m_2}{M}eu$ and $\frac{m_1}{M}eu$, respectively. By circles we mean that if we drew all the possible relative velocity outcomes (i.e. all possible directions for the final velocities), we would get two circles, these are shown in figure 6.6.

Notice also the two special cases:

- if $m_1 = m_2 = m$: the two circles coincide, both of radius $\frac{m}{M}eu$
- if $m_1 = m_2 = m$ and $e = 1$: the two circles coincide and the radius reduces to $\frac{m}{M}u = \frac{m}{2m}u = \frac{u}{2}$.

Now consider all possible velocity outcomes in lab frame; to switch back we just need to add the centre of mass velocity to the velocities in the centre of mass frame i.e.

$$v_1 = v_1' + V_{\text{CM}}, \qquad v_2 = v_2' + V_{\text{CM}}. \tag{6.37}$$

Here V_{CM} can be seen as the vector describing the displacement of the centre of the circles, since this is constant when no external forces acting on the system are present.

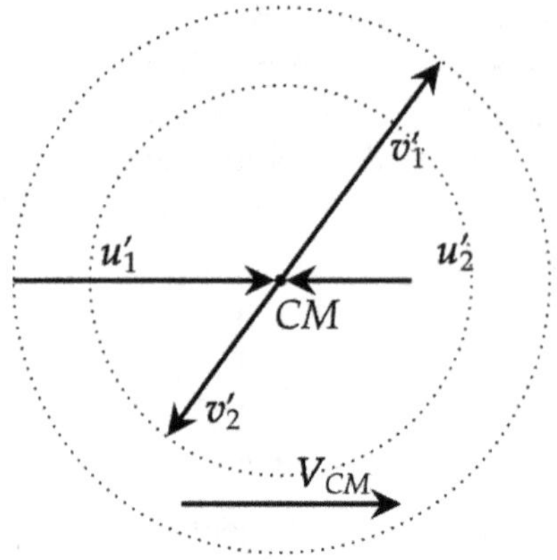

Figure 6.6. Representation of all velocity outcomes after collision of two bodies viewed from the centre of mass frame.

We can look at two interesting cases here:

- if conditions for the second special case above are met ($m_1 = m_2 = m$ and $e = 1$) and the second body (target) is initially at rest (as in section 6.2.4) then the velocity of the centre of mass is $u/2$ and the circle in the lab frame passes through the origin, while the first body (projectile) remains stationary after the collision.
- If $m_1 = m_2 = m$ but $e \leqslant 1$ the radius of the circle is reduced and it no longer passes through the origin.

Glancing collision of two finite-radius balls—elastic case

We now see how to relate the angles in the lab frame and the centre of mass frame, in the case where one of the balls is initially at rest (figure 6.7). We shall assume balls are smooth, so that the impulse on each ball can only be along the line of their centres. We are also going to assume that the collision is elastic.

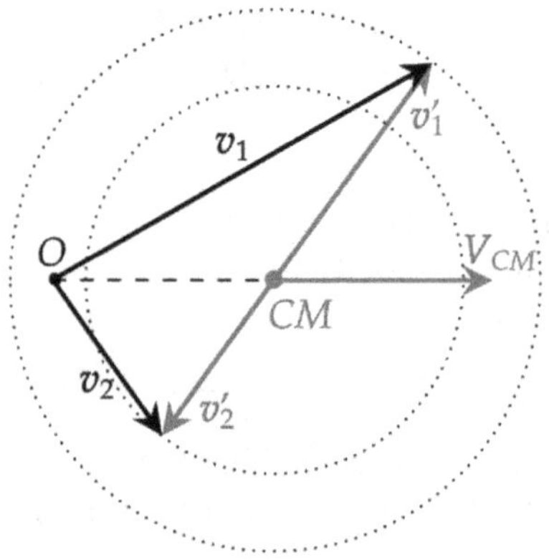

Figure 6.7. General case of glancing collision, Lab and centre of mass frame relation.

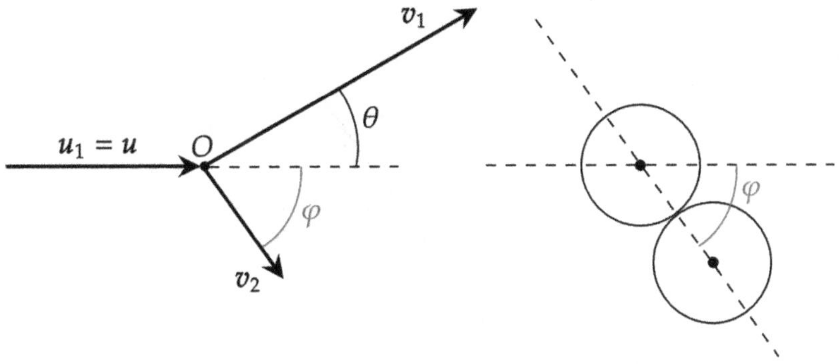

Figure 6.8. Left: collision as viewed from the laboratory frame of reference. Right: centre lines of the two colliding balls.

Consider the initial situation we studied for head-on collision. Ball with mass m_1 approaching with velocity $\boldsymbol{u}_1$ ball of mass m_2 which is instead stationary. The impulse received by ball 2 is

$$\boldsymbol{I}_2 = m_2 \boldsymbol{v}_2 \tag{6.38}$$

which is along the line of the centres (you can see this clearly from the lab view of the collision in figure 6.8). Notice from equation (6.38) that $\boldsymbol{v}_2$ is along the same direction as $\boldsymbol{I}_2$, i.e. along the centre line of ball 2 in figure 6.8.

The impulse on ball one, due to Newton's third law, will be

$$\boldsymbol{I}_1 = -\boldsymbol{I}_2 = m_1 \boldsymbol{v}_1 - m_1 \boldsymbol{u}_1 \tag{6.39}$$

such that

$$\boldsymbol{I}_2 = m_1 \boldsymbol{u}_1 - m_1 \boldsymbol{v}_1. \tag{6.40}$$

Notice how ball 1 emerges from the collision at an angle θ as shown in figure 6.8. Now let us change the frame of reference and look at the angles from the centre of mass point of view. In this frame of reference the final velocities need to be in opposite directions such that the deflection angle, θ', of both balls is the same. This is shown in figure 6.9.

As seen previously in section 6.2.4, relative velocity is initially $\boldsymbol{u} = \boldsymbol{u}_1$ and final relative velocity has the same magnitude as the initial relative velocity, i.e. $|\boldsymbol{u}| = |\boldsymbol{v}| \Rightarrow v = \pm u_1$. We can then notice that $m_1\boldsymbol{u}_1'$ and $m_1\boldsymbol{v}_1'$ have the same magnitudes since

$$\boldsymbol{u}_1' = \frac{m_2}{M}\boldsymbol{u}, \qquad \boldsymbol{v}_1' = \frac{m_2}{M}\boldsymbol{v}.$$

The centre of mass velocity instead is $\boldsymbol{V}_{\mathrm{CM}} = \frac{m_1}{M}\boldsymbol{u}$ (again have a look at section 6.2.4 if you're unfamiliar with these formulas).

From this it follows that

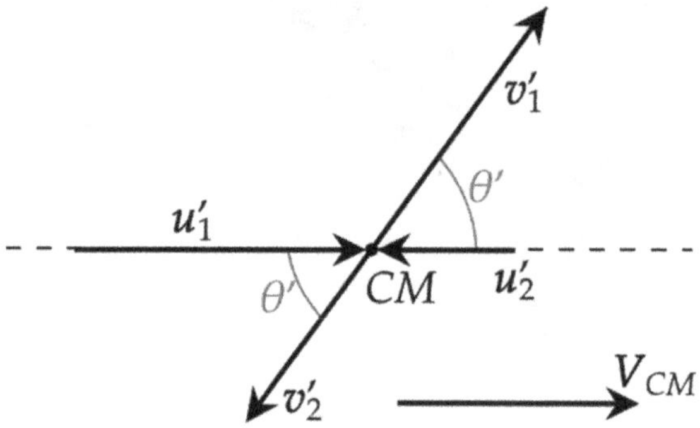

Figure 6.9. Representation of the velocities and respective angles from the centre of mass frame of reference.

$$\theta' = \pi - 2\varphi, \qquad \varphi = \frac{1}{2}(\pi - \theta'). \tag{6.41}$$

This relation should be clear also from figure 6.10.

Now take the motion on the x–y plane with the original velocity, $\boldsymbol{u}_1$, along the positive x direction. Final velocities, in the centre of mass frame, can be written in their vector form as follows:

$$\boldsymbol{v}_1' = \frac{m_2 u}{M}(\cos\theta'\hat{\boldsymbol{i}} + \sin\theta'\hat{\boldsymbol{j}}) \tag{6.42}$$

$$\boldsymbol{v}_2' = \frac{m_1 u}{M}(-\cos\theta'\hat{\boldsymbol{i}} - \sin\theta'\hat{\boldsymbol{j}}). \tag{6.43}$$

Switching now to lab frame we have final velocities:

$$\boldsymbol{v}_1 = \boldsymbol{v}_1' + \boldsymbol{V}_{\mathrm{CM}} = \frac{m_2 u}{M}(\cos\theta'\hat{\boldsymbol{i}} + \sin\theta'\hat{\boldsymbol{j}}) + \frac{m_1}{M}u\hat{\boldsymbol{i}} \tag{6.44}$$

$$= \frac{m_2 u}{M}\left[\left(\cos\theta' + \frac{m_1}{m_2}\right)\hat{\boldsymbol{i}} + \sin\theta'\hat{\boldsymbol{j}}\right] \tag{6.45}$$

$$\boldsymbol{v}_2 = \boldsymbol{v}_2' + \boldsymbol{V}_{\mathrm{CM}} = \frac{m_1 u}{M}(-\cos\theta'\hat{\boldsymbol{i}} - \sin\theta'\hat{\boldsymbol{j}}) + \frac{m_1}{M}u\hat{\boldsymbol{i}} \tag{6.46}$$

$$= \frac{m_1 u}{M}[(1 - \cos\theta'\hat{\boldsymbol{i}}) - \sin\theta'\hat{\boldsymbol{j}}]. \tag{6.47}$$

The **angle of deflection**, θ, is therefore, from trigonometry,

$$\tan\theta = \frac{\boldsymbol{v}_1 \cdot \hat{\boldsymbol{j}}}{\boldsymbol{v}_1 \cdot \hat{\boldsymbol{i}}} = \frac{\sin\theta'}{\cos\theta' + m_1/m_2}$$
$$= \frac{\sin(2\varphi)}{-\cos(2\varphi) + m_1/m_2}.$$

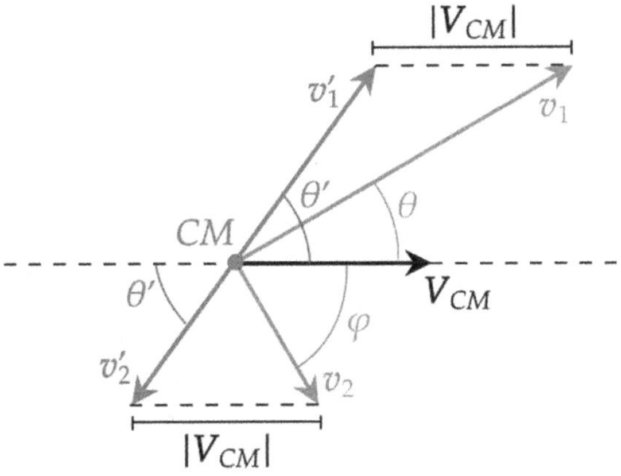

Figure 6.10. Overlap of Lab and centre of mass frames.

Now recall from the beginning of section 6.2.4

$$V_{CM} = \frac{m_1}{M}\boldsymbol{u} = \frac{m_1}{m_2}\boldsymbol{u}_1'$$
$$\Rightarrow V_{CM} = \frac{m_1}{m_2}u_1' = \frac{m_1}{m_2}v_1'$$

and let's take a look at momentum conservation. The initial total momentum is just

$$\boldsymbol{p}_{\text{tot}} = m_1\boldsymbol{u}_1 \tag{6.48}$$

since the second ball is at rest, i.e. $\boldsymbol{u}_2 = 0$. We also know from section 6.1 that the total momentum is carried by the centre of mass

$$\begin{aligned}\boldsymbol{p}_{\text{tot}} &= M\boldsymbol{V}_{CM}\\ &= m_1\boldsymbol{V}_{CM} + m_2\boldsymbol{V}_{CM}\\ &= m_1\boldsymbol{V}_{CM} + m_2\left(\frac{m_1}{M}\boldsymbol{u}\right)\\ &= m_1\boldsymbol{V}_{CM} + \mu\boldsymbol{u}\\ &= m_1\boldsymbol{V}_{CM} + m_1\boldsymbol{u}_1'.\end{aligned}$$

Particularly we're interested in the following two results:

$$m_1\boldsymbol{u}_1 = m_1\boldsymbol{V}_{CM} + \mu\boldsymbol{u} \tag{6.49}$$

$$m_1\boldsymbol{u}_1 = m_1\boldsymbol{V}_{CM} + m_1\boldsymbol{u}_1'. \tag{6.50}$$

These can be represented graphically as in figure 6.11.

Hence, the following cases:

- If $m_2 > m_1 \rightarrow v_1' > V$: the left-hand vertex of the triangle in figure 6.11 lies inside the circle swept out by the top vertex as θ' varies from 0 to π. Hence the maximum value of θ is also π.

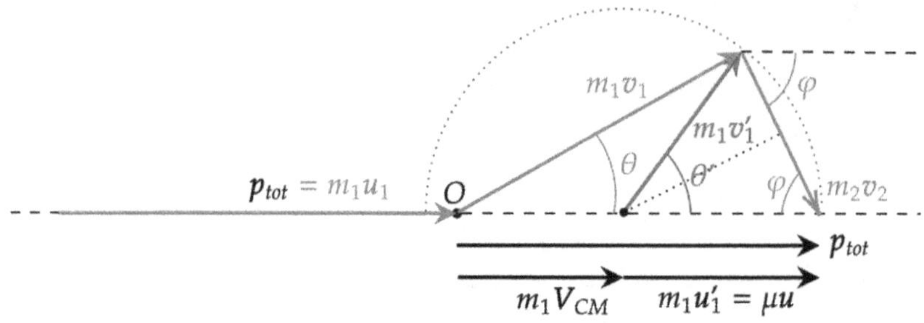

Figure 6.11. Schematic representation of the conservation of momentum of an elastic glancing collision.

- If $m_2 < m_1 \rightarrow v_1' < V$: the far-left vertex lies outside the circle. In this case the maximum deflection angle θ of ball 1 is when vectors v_1 and v_1' are perpendicular between each other. We can call this $\theta_{\max}$ and is given by

$$\sin\theta_{\max} = \frac{v_1'}{V} = \frac{m_2}{m_1}. \tag{6.51}$$

The final speed of particle 2 in the lab frame can be found from the geometry of the isosceles triangle on the right-hand side of figure 6.11 (remember $m_1v_1' = m_1\boldsymbol{u}_1'$):

$$\begin{aligned}\frac{1}{2}m_2v_2 &= \mu u \sin\left(\frac{\theta'}{2}\right)\\ \Rightarrow v_2 &= 2\frac{m_1}{M}u\sin\left(\frac{\theta'}{2}\right)\\ &= 2\frac{m_1}{M}u\cos(\varphi)\end{aligned}$$

where in the last passage we subbed in equation (6.41).

We can now investigate the fraction of original kinetic energy transferred to ball 2

$$\frac{K_{\text{transf}}}{K_{\text{initial}}} = \frac{m_2v_2^2}{m_1u^2} \tag{6.52}$$

$$= \frac{4m_1m_2}{M^2}\sin^2(\theta'/2) \tag{6.53}$$

$$= \frac{4m_1m_2}{M^2}\cos^2\varphi \tag{6.54}$$

where we have a maximum when $\cos\varphi = 1$ or $\sin(\theta'/2) = 1$, i.e. when

$$\varphi = 0, \qquad \theta' = \pi \tag{6.55}$$

which is the head-on collision case treated previously. In fact, if we plug in these angles into equations (6.45) and (6.47)

$$v_1 = \frac{m_2 u}{M}\left[\left(\cos\pi + \frac{m_1}{m_2}\right)\hat{\boldsymbol{i}} + \sin\pi\hat{\boldsymbol{j}}\right] \tag{6.56}$$

$$= \frac{m_2 u}{M}\left[\left(\frac{m_1}{m_2} - 1\right)\hat{\boldsymbol{i}} + 0\hat{\boldsymbol{j}}\right] \tag{6.57}$$

$$v_1 = \frac{m_1 u}{M}[(1 - \cos\pi\hat{\boldsymbol{i}}) - \sin\pi\hat{\boldsymbol{j}}] \tag{6.58}$$

$$= \frac{m_1 u}{M}[(2\hat{\boldsymbol{i}}) - 0\hat{\boldsymbol{j}}] \tag{6.59}$$

these are indeed the same results obtained in section 6.2.4 for the elastic case. Finally, in the case where $m_1 = m_2$ we have

$$\boldsymbol{u} = v_1 + v_2 \tag{6.60}$$

therefore, conservation of kinetic energy

$$\boldsymbol{K}_{\text{tot}} = \frac{1}{2}m_1\boldsymbol{u}_1^2 = \frac{1}{2}m_1(v_1 + v_2)^2 \tag{6.61}$$

$$= \frac{1}{2}m_1\left(v_1^2 + 2v_1 \cdot v_2 + v_2^2\right) \tag{6.62}$$

$$= \frac{1}{2}m_1\left(v_1^2 + v_2^2\right) \tag{6.63}$$

since v_1 and v_2 form a right angle (in the lab frame of course). Hence,

$$\theta + \varphi = \pi/2 \tag{6.64}$$

and also

$$\theta' = \pi - 2(\pi/2 - \theta) = 2\theta. \tag{6.65}$$

Exercise 6.4. What is the definition of the reduced mass μ in a two-body system (masses in the system are m_1 and m_2)?
Explain why the velocities in the *centre of mass frame* in a two-body system obey the following relation

$$m_1 v_1' + m_2 v_2' = 0.$$

Knowing that relative velocity is defined as $v = v_1 - v_2$, show that the individual momentum of the first body in the two-body system can be defined as

$$p_1' = \mu v.$$

Exercise 6.5. For a system of two particles with masses m_1 and m_2, show how the kinetic energy as viewed in the *centre of mass frame* can be expressed in terms of the reduced mass of the two particles and their relative velocity $v = v_1 - v_2$.
Hence state the theorem that describes how the relative velocity of the two particles changes during an elastic collision between them.

Exercise 6.6. In an isolated two-body system of reduced mass μ, two particles undergo a glancing inelastic collision with coefficient of restitution e. Explain why in the *centre of mass frame* the kinetic energy of the centre of mass, K_{CM} as defined in the book, remains unaffected.
What part of the kinetic energy is reduced and by what factor?

Exercise 6.7. Cooking chicken
You are looking for renewable solutions to cook a full chicken of 2000 g. The cooking temperature for chicken needs to be at least 343.15 K, once this temperature is reached the chicken can be considered to be cooked. From an initial temperature of 293.15 K the chicken requires around 500 kJ to reach the cooking temperature that kills all the bacteria. You only have available a bowling ball of mass $m = 7.27$ kg to do this.

The collision is inelastic, coefficient of restitution $e = 0.01$, also assume that the chicken is held fixed on some support such that it cannot move after the collision.
Do you think this can be feasible? At what velocity does the bowling ball bounce back?

Exercise 6.8. A ball of mass $m = 1$ kg is allowed to fall freely from rest through a height $h = 10$ m above ground. At the end of its fall, it strikes a thin platform with mass $M = 10$ kg, which is held at a distance of 10 cm above the ground by a spring with elastic constant k. Both the length of the spring at the point where it is maximally compressed, and the thickness of the platform can be neglected.

(i) Calculate the velocity of the ball at the moment it hits the platform.

(ii) Suppose that after the collision the ball remains attached to the platform. Calculate the velocity of the whole system (ball + platform) immediately after the collision.

(iii) Calculate the elastic constant of the spring such that the system (ball + platform) reaches the ground with zero speed. [*Hint:* use conservation of gravitational energy, not forgetting that at the time of the collision both ball and platform still have non-zero gravitation potential with respect to the ground.]

Exercise 6.9. Two identical tennis balls, each of mass m, collide in the (x, y)-plane, without any external force acting on the system. The initial velocity of the first ball is described by the vector $\boldsymbol{u}_1 = 2\hat{\boldsymbol{i}} + \hat{\boldsymbol{j}}$, that of the second ball by the vector $\boldsymbol{u}_2 = -2\hat{\boldsymbol{i}} + \hat{\boldsymbol{j}}$.

(i) Derive an expression for the momentum of each ball in their centre of mass system.

(ii) The two balls collide and the collision is inelastic with coefficient of restitution, e (defined as the ratio of the final and initial moment of the balls in the centre of mass system). Assuming the directions of both velocities in the centre of mass frame are reversed after the collision, derive an expression for the final velocities of the balls in the laboratory system.

IOP Publishing

Classical Mechanics

A professor-student collaboration

Mario Campanelli

Chapter 7

Orbits

7.0 Introduction: a historical note

Orbits were first accurately described by Johannes Kepler, whose results are captured in his three laws of planetary motion (figure 7.1(a)). From that starting point we have come very far in understanding the motion of objects in the Universe. Thanks to this progress we were able to send Sputnik 1 into low orbit in 1957 (figure 7.1(b)). Now, we have over a thousand satellites orbiting Earth and are sending spacecraft all over the Solar System.

In this chapter, we will study central forces and how astronomical objects are affected by them. We will see different types of orbital motion and trajectories, looking at this from the point of view of the effective potential to shine light on some

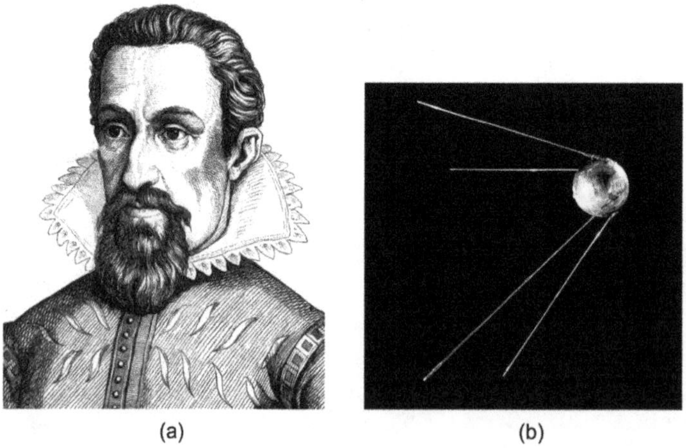

Figure 7.1. (a) Portrait of Johannes Kepler, courtesy of the Smithsonian Libraries https://library.si.edu/imagegallery/72831. (b) Sputnik 1. Courtesy of paukrus. CC BY-SA 2.0.

doi:10.1088/978-0-7503-2690-2ch7

important results otherwise ignored. Moreover, we shall look at how two objects interact in space and how a rocket advances.

7.1 Orbital forces

For orbits we consider the motion of a central force, a force whose magnitude at any point, other than the origin, depends only on the distance from that point to the origin. The direction of this force is parallel to the line connecting that point to the origin. The radial equation is given by:

$$F(r) = m(\ddot{r} - r\dot{\theta}^2) \tag{7.1}$$

and the angular momentum is:

$$L = mr^2\dot{\theta} = \text{constant.} \tag{7.2}$$

Such that the angular frequency is given by:

$$\dot{\theta} = \frac{L}{mr^2}.$$

The radial equation of motion becomes

$$F(r) = m\left(\ddot{r} - r\left(\frac{L}{mr^2}\right)^2\right) \tag{7.3}$$

$$F(r) = m\ddot{r} - \frac{L^2}{mr^3} \tag{7.4}$$

$$m\ddot{r} = F(r) + \frac{L^2}{mr^3}. \tag{7.5}$$

Notice that we have the same equation as the equation of motion for a particle moving in one dimension under an effective force

$$F_{eff} = F(r) + F_C \tag{7.6}$$

where F_C is the **centrifugal force**.

Definition 7.1.1: Centrifugal force

The centrifugal force is an inertial or **fictitious force** that acts on all bodies in a rotating frame of reference. It is positive and directed radially outwards (figure 7.2).

$$F_C = \frac{L^2}{mr^3}. \tag{7.7}$$

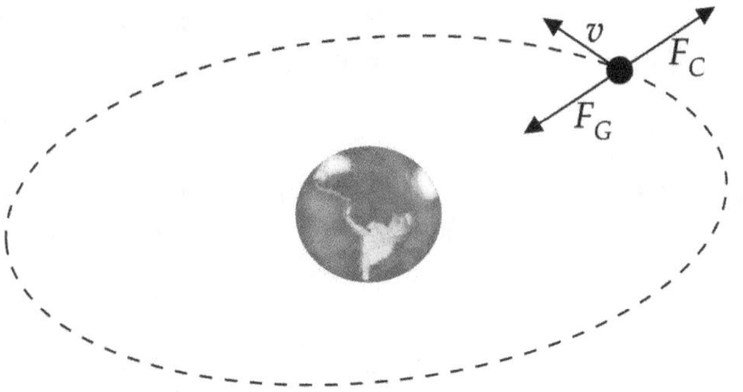

Figure 7.2. Centrifugal force.

In order for an object to be able to carry out a stable orbit, there needs to be a balancing central force, $F(r)$, which has the opposite direction to the centrifugal force. This centripetal component (radially inwards) comes in the from of the **gravitational force**.

7.1.1 Potentials

Definition 7.1.2: Centrifugal potential
The centrifugal force has an associated potential, called the centrifugal potential, such that:

$$V_C = \frac{1}{2}\frac{L^2}{mr^2}. \tag{7.8}$$

It is obtained by integrating the centrifugal force with respect to the distance, r.

$$\begin{aligned} V_C &= -\int F_C \, \mathrm{d}r \\ &= -\int \frac{L^2}{mr^3} \, \mathrm{d}r \\ &= \frac{1}{2}\frac{L^2}{mr^2} \end{aligned}$$

Therefore, we can say the body is moving in an effective potential

$$V_{eff} = V(r) + V_C(r) \tag{7.9}$$

Remark 7.1. For minimum effective potential we have a stable orbit. This is because the closer V_{eff} is to zero, $V(r)$ and $V_C(r)$ being opposite in direction, the closer the centrifugal and centripetal potentials are to perfect balance.

7.2 Circular motion approximation

Even for elliptical motion where the radius is not constant, if the force is central then the angular momentum is conserved because of the fundamental theorem of angular momentum.

Remark 7.2. Angular momentum is conserved even in elliptical orbits as there are no external forces and there is no external torque about the central mass, since the force coming from it, F_G and position vector are always at a 180° angle. This is considering the global system as, if the satellite loses linear momentum, the Earth will gain it and vice versa. This is applicable to any two orbiting objects.

$$L = mr^2\dot{\theta}.$$

Back to equation (7.4)

$$F(r) = m\ddot{r} - \frac{L^2}{mr^3}.$$

Let's consider the case where we have an **attractive** (hence the minus sign) power law:

$$F(r) = -Kr^n. \tag{7.10}$$

Definition 7.2.1. Power law
A power law is a functional relationship between two quantities, where one quantity varies proportionally as a power of the other.

We consider a circular orbit of radius r_0, the distance where the attractive and centrifugal forces are equal in magnitude and opposite in direction. Now, we obtain the angular frequency, ω, and the period of the oscillation, τ, by equating F_C and $F(r)$.

By equating F_C and $F(r)$ we can obtain the angular frequency ($\omega = \dot{\theta}$):

$$\omega = \sqrt{\frac{kr_0^{n-1}}{m}} \tag{7.11}$$

And the period would be:

$$\tau = \frac{2\pi}{\omega} = 2\pi\left(\frac{m}{Kr_0^{n-1}}\right)^{\frac{1}{2}}. \tag{7.12}$$

By taking $r = r_0 + x$ where x is very small, we approximate both sides of our equation by making a Taylor expansion around the equilibrium distance r_0, to obtain

$$m\ddot{x} = -x\left[nKr_0^{n-1} + 3\frac{L^2}{mr_0^4}\right] = -xKr_0^{n-1}[n+3]$$

where we have used the fact that $Kr_0^{n-1} = \frac{L^2}{mr_0^4}$, as the forces affecting the orbit need to be in equilibrium. As long as $n > -3$ the equation above describes a simple harmonic motion, with angular frequency Ω and period T. For $n < -3$ the orbit is unstable and r diverges.

> Remark 7.3. The reason why we took a Taylor expansion will become more clear when we discuss orbits seen from the point of view of an effective potential. It is related to the fact that the equilibrium position occurs at the minimum of the potential.

By applying the same method used above to obtain the angular frequency of the quasi-circular motion, equating (7.2) and F_C

$$\Omega = \left(\frac{Kr_0^{n-1}[n+3]}{m}\right)^{\frac{1}{2}} = (n+3)^{\frac{1}{2}}\omega \tag{7.13}$$

$$T = \frac{2\pi}{\Omega} = \frac{\tau}{(n+3)^{\frac{1}{2}}}. \tag{7.14}$$

So we can see that the radius of the orbit is oscillating about r_0, and the angular frequency of the oscillation and of the rotation are connected by a very simple relation that depends on the power law.

There are two important cases where we have a rational relationship between τ and T:

- **Hooke's law** ($n = 1$). In this case, we have that $T = \tau/2$, so the orbit closes in on itself with two oscillations per revolution.
- **Inverse square law** ($n = -2$). We have $T = \tau$ and the orbit closes in on itself with one oscillation per revolution. This is the case we encounter in elliptical orbits.

7.3 Motion under the inverse square law of force

Law 7.3.1: Inverse square law

Newton proposed that the inverse square law applied to gravitational motion. However, we can generalise it to any attractive or repulsive forces that behave in such a way:

$$\boldsymbol{F} = \frac{K}{r^2}\hat{\boldsymbol{r}} \tag{7.15}$$

K being negative for attractive forces and positive for repulsive forces.

By applying the relationship between the force and the potential to the equation seen above, we obtain the potential energy

$$V = \frac{K}{r} + C. \tag{7.16}$$

As the distance between any two objects considered increases ($r \to \infty$) we take the potential energy to be zero. Thus, $C = 0$.

Below are two examples of two forces which can be described using the inverse square law seen above.

Definition 7.3.1. Gravitational force

For the **gravitational force** we have $K = -GMm$, so it's always attractive. *Where the gravitational constant is given by* $G = 6.67 \times 10^{-11}$ N m^2 kg^{-2}.

Definition 7.3.2: Electrostatic force

For the **electrostatic force** $K = \frac{q_1 q_2}{4\pi\epsilon_0}$, where q_1 and q_2 are the charges which can be either positive of negative.

Now, we want to obtain the possible shapes of the orbit from the equation of motion. The difference with the previous case is that this time we will not do the quasi-circular approximation, so the treatment will be valid also for orbits, like those of comets, where the distance from the central mass varies a lot, and the Taylor expansion would not be valid.

The equation of motion of the body under a central force in polar coordinates is:

$$m[(\ddot{r} - r\dot{\theta}^2)\hat{\boldsymbol{r}} + (2\dot{r}\dot{\theta} + r\ddot{\theta})\hat{\boldsymbol{\theta}}] = \frac{K}{r^2}\hat{\boldsymbol{r}}. \tag{7.17}$$

We can separate the radial and transverse components, as follows:

$$m(\ddot{r} - r\dot{\theta}^2) = \frac{K}{r^2} \tag{7.18}$$

$$(2\dot{r}\dot{\theta} + r\ddot{\theta}) = 0. \tag{7.19}$$

Exercise 7.1. Calculate $\frac{\mathrm{d}}{\mathrm{d}t}(r^2\dot{\theta})$ and show it is equal to 0.

◆

From the exercise seen above, excluding the case where $r \neq 0$, we have shown that $r^2\dot{\theta}$ is constant. This means the angular momentum $L = mr^2\dot{\theta}$ also remains constant. We try now to determine the shape of the orbit, without considering its time dependence. In order to do this, we need to express quantities depending on time with respect to constants and distances. For instance, the angular velocity can be expressed as a function of the angular momentum:

$$\dot{\theta} = \frac{L}{mr^2}$$

and substitute this identity into the radial equation:

$$m\left(\ddot{r} - \frac{L^2}{m^2r^3}\right) = \frac{K}{r^2}. \tag{7.20}$$

To determine the shape of the orbit we will solve for r as a function of θ.

To make calculations easier, we can define the variable $u = \frac{1}{r}$ which will make it easier to determine the form of our results. Firstly we must express the time derivatives of r in terms of u and θ:

$$\begin{aligned}\dot{r} &= \frac{\mathrm{d}r}{\mathrm{d}t} = \frac{\mathrm{d}}{\mathrm{d}t}\left(\frac{1}{u}\right) = -\frac{1}{u^2}\frac{\mathrm{d}u}{\mathrm{d}t} = -\frac{1}{u^2}\frac{\mathrm{d}u}{\mathrm{d}\theta}\frac{\mathrm{d}\theta}{\mathrm{d}t} \\ &= -\frac{1}{u^2}\frac{\mathrm{d}u}{\mathrm{d}\theta}\left(\frac{L}{mr^2}\right) = -\frac{L}{m}\frac{\mathrm{d}u}{\mathrm{d}\theta}.\end{aligned}$$

Differentiating,

$$\ddot{r} = \frac{\mathrm{d}\dot{r}}{\mathrm{d}t} = \frac{\mathrm{d}}{\mathrm{d}t}\left(-\frac{L}{m}\frac{\mathrm{d}u}{\mathrm{d}\theta}\right).$$

Applying the chain rule

$$\frac{\mathrm{d}}{\mathrm{d}\theta}\left(-\frac{L}{m}\frac{\mathrm{d}u}{\mathrm{d}\theta}\right)\frac{\mathrm{d}\theta}{\mathrm{d}t}.$$

Finally, substituting $\dot{\theta}$

$$\ddot{r} = -\frac{L^2}{m^2}u^2\frac{\mathrm{d}^2u}{\mathrm{d}\theta^2}.$$

We can substitute these identities to obtain a time-independent radial equation

$$m\left(-\frac{L^2}{m^2}u^2\frac{\mathrm{d}^2u}{\mathrm{d}\theta^2} - u^3\frac{L^2}{m^2}\right) = Ku^2.$$

Since $u > 0$, we can multiply this expression by $-\frac{m}{L^2u^2}$

$$\frac{\mathrm{d}^2u}{\mathrm{d}\theta^2} + u = -\frac{mK}{L^2}.$$

To allow us to more effectively see the form of the equation we will change variables to $y = u + \frac{mK}{L^2}$

$$\frac{\mathrm{d}^2y}{\mathrm{d}\theta^2} + y = 0.$$

Now we can clearly see this is the equation for simple harmonic motion with solution $y = A\cos(\theta - \theta_0)$. Looking back at our equation in terms of u we can deduce a general solution in the form seen below.

$$u = A\cos(\theta - \theta_0) - \frac{mK}{L^2} = \frac{1}{r}. \tag{7.21}$$

ASIDE: Conic sections

The name conic section comes from the fact that if a cone is sliced by a plane, the intersection will be a curve which depends on the relative angle of the plane and the cone.

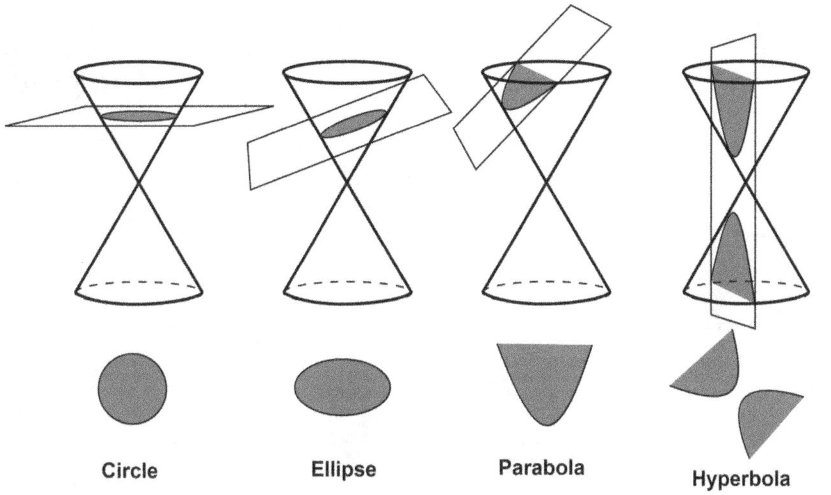

When the plane is perpendicular to the cone axis, the intersection is a circle; when the plane's angle is between the cone's axis and the cone's side, the intersection is an ellipse; when the plane is exactly parallel to the side of the cone the intersection is a parabola, and for even larger angles it is a hyperbola. The curves obtained this way can also be defined using a straight line, called directrix, and a point, called focus. Each point of these curves is characterized by a constant ratio, called eccentricity, between these two distances which will remain constant for a given curve. This can be seen more clearly in the image below.

- For an **ellipse**, the ratio is less than 1 (a circle is just a special case of the ellipse with eccentricity 0).
- For a **parabola**, the ratio is 1, so the two distances are equal.
- For a **hyperbola**, the ratio is greater than 1.

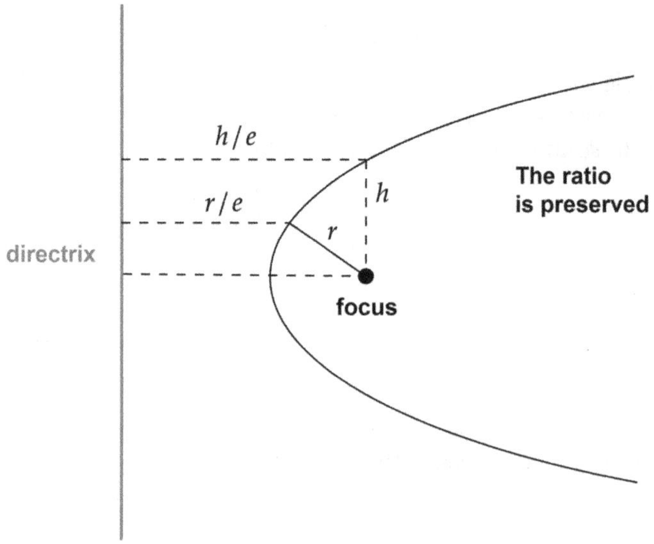

Let us calculate the equation of a conic curve from its definition in terms of constant distance between focus and directrix, shown in the above picture. Applying trigonometry we have that

$$r\cos\theta + \frac{r}{e} = \frac{h}{e}$$

where h is the value of r when $\theta = \pi/2$. So,

$$r(1 + e\cos\theta) = h \tag{7.22}$$

and in terms of u as before

$$u = \frac{1}{r} = \frac{1}{h}(1 + e\cos\theta). \tag{7.23}$$

Now, we choose $A > 0$ (from equation (7.21)) so that $\theta = 0$ corresponds to the minimum distance, and take $\theta_0 = 0$ to simplify the calculations. Then, our expression,

$$u = A\cos\theta - \frac{mK}{L^2}, \tag{7.24}$$

has the same form as the general equation of a conic section.

7.3.1 Trajectories

A **trajectory** is the curve that a body describes in space. In this section we will be looking at how attractive and repulsive forces affect trajectories (figure 7.3).

Attractive force
For an **attractive force**, $K < 0$, the trajectory curves towards the centre of the force. However, it can have three different shapes depending on the eccentricity.

- If $e > 1$ the trajectory is a **hyperbola.**
- If $e = 1$, the trajectory is a **parabola.**
- If $e < 1$, the trajectory is an **ellipse.**

Repulsive force
For a **repulsive force**, $K > 0$ the trajectory curves away from the centre of the force. So our equation has the form

$$u = \frac{1}{h}(e\cos\theta - 1). \tag{7.25}$$

This is an equation in the form of the second branch of a hyperbolic orbit.

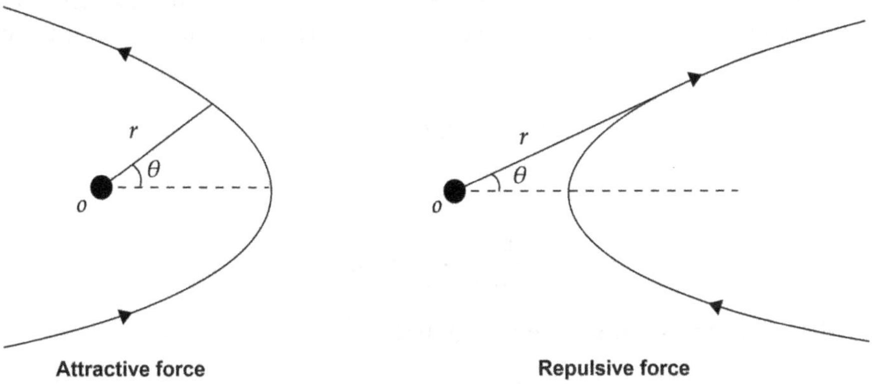

Figure 7.3. Trajectories under attractive and repulsive forces.

7.4 Orbits under an attractive force: elliptical orbits and Kepler's laws

7.4.1 Eccentricity

The objective of this section is to compare the general form of the attractive case for a conic section with the solution for the motion obtained by considering the energy. This will allow us to derive an expression for the eccentricity in terms of the energy and angular momentum.

Solution for the motion in terms of energy
Let's look at some descriptive properties of the orbit starting from a given total energy E and angular momentum L. Since we are dealing with the **attractive case**, recall that we have $K < 0$.

Since the gravitational force is a *conservative force* the total energy of the particle is constant throughout the motion. The energy equation is

$$E = \frac{1}{2}mv^2 + V \tag{7.26}$$

$$E = \frac{1}{2}m(\dot{r}^2 + r^2\dot{\theta}) + \frac{K}{r}. \tag{7.27}$$

Using the expressions for $\dot{r} = -\frac{L}{m}\frac{du}{d\theta}$ and $\dot{\theta} = \frac{L}{mr^2} = \frac{L}{m}u^2$ that we have previously derived:

$$\begin{aligned} E &= \frac{1}{2}m\left[\left(\frac{L}{m}\frac{du}{d\theta}\right)^2 + \frac{1}{u^2}\left(\frac{L}{m}u^2\right)^2\right] + Ku \\ &= \frac{1}{2}m\left[\frac{L^2}{m^2}\left(\frac{du}{d\theta}\right)^2 + \frac{L^2}{m^2}u^2\right] + Ku. \end{aligned}$$

We can see that the first term is always positive, so for $K < 0$ the energy can be positive, negative or zero. Recall from equation (7.21) that the equation of the orbit is

$$u = A\cos\theta - \frac{mK}{L^2}.$$

Differentiating with respect to θ

$$\frac{du}{d\theta} = -A\sin\theta.$$

So, the total expression for the energy becomes

$$E = \frac{1}{2}m\left[\frac{L^2}{m^2}(-A\sin\theta)^2 + \frac{L^2}{m^2}\left(A\cos\theta - \frac{mK}{L^2}\right)^2\right] + K\left(A\cos\theta - \frac{mK}{L^2}\right)$$

As the energy is conserved throughout the orbit, we can set $\theta = \frac{\pi}{2}$ just to make the algebra a bit simpler. Thus,

$$\begin{aligned} E &= \frac{1}{2}m\left[\frac{L^2A^2}{m^2} + \frac{L^2}{m^2}\frac{m^2K^2}{L^4}\right] - \frac{mK^2}{L^2} \\ &= \frac{A^2L^2}{2m} - \frac{mK^2}{2L^2}. \end{aligned}$$

Solving for A, we get

$$A = \frac{m|K|}{L^2}\sqrt{1 + \frac{2EL^2}{mK^2}} \tag{7.28}$$

where A is just a constant that is related to the eccentricity.

Comparing

The solution for the motion is

$$u = A\cos\theta - \frac{mK}{L^2}$$

and the general form for a conic section is:

$$\begin{aligned} u &= \frac{1}{h}(1 + e\cos\theta) \\ &= \frac{e}{h}\cos\theta + \frac{1}{h}. \end{aligned}$$

Comparing both, we can see

$$e = Ah \tag{7.29}$$

$$\frac{1}{h} = -\frac{mK}{L^2} \tag{7.30}$$

$$e = -\frac{AL^2}{mK} = \sqrt{1 + \frac{2EL^2}{mK^2}}. \tag{7.31}$$

Notice, h is entirely determined by the angular momentum. We can now connect the energy, eccentricity and the type of curve from the treatment of the conic section:

- If $E > 0$, then $e > 1$ and the trajectory is a **hyperbola**.
- If $E = 0$, then $e = 1$ and the trajectory is a **parabola**.
- If $E < 0$, then $e < 1$ and the trajectory is an **ellipse**.

Remark 7.4. Repulsive force

The result for a repulsive force is similar. As $K > 0$, E will always be positive and the trajectory will be hyperbolic which is also consistent with our previous treatment.

7.4.2 Understanding orbits from the effective potential

From equation (7.9) we have that the effective potential is given by:

$$V_{eff} = V + V_C V_{eff} = -\frac{|K|}{r} + \frac{L^2}{2mr^2}. \tag{7.32}$$

Depending on the total energy, E, we have different orbits:

- If $E < 0$ then the energy line crosses V_{eff} at $r = r_{\min}$ and $r = r_{\max}$. This is the case for an **ellipse**. In figure 7.4 we can see the effective potential graph for this case. *When the energy line, E, is tangent to the potential energy at minimum PE we are in the case of a circular orbit.*
- If $E > 0$ then $r_{\max}$ is infinite and we are in the presence of a **hyperbola**.
- If $E = 0$ then $r_{\max}$ is also infinite, but the particle will have zero energy at infinity. This is the case of a **parabola**.

Exercise 7.2. Draw a diagram similar to figure 7.4 for $E > 0$ (hyperbola) and $E = 0$ (parabola).

◆

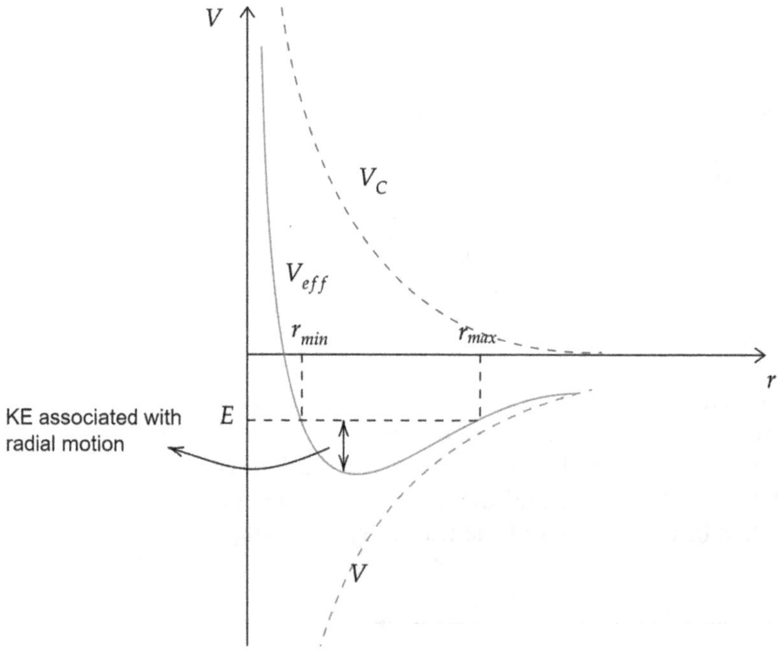

Figure 7.4. Effective potential graph for an ellipse.

7.4.3 Elliptical orbit

Let's consider an elliptical orbit and try to find the value of the elements that compose the equation of the orbit. Just by knowing the velocity at the position of closest approach, r_a, we can determine the semi-major and semi-minor axes.

First, the angular momentum and the energy are given by:

$$L = mr_a v_a \tag{7.33}$$

$$E = \frac{1}{2}mv_a^2 - \frac{|K|}{r_a}. \tag{7.34}$$

Then, from the general equation for an elliptical orbit:

$$r(1 + e\cos\theta) = h$$

we obtain for $r_a(\theta = 0)$ and $r_b(\theta = \pi)$.

$$r_a(1 + e) = h$$
$$r_b(1 - e) = h.$$

As can be seen from figure 7.5, we have:

$$2a = r_a + r_b$$

Thus, substituting r_a and r_b

$$2a = \frac{h}{1+e} + \frac{h}{1-e} = \frac{2h}{1-e^2}$$

such that

$$a = \frac{h}{1-e^2}. \tag{7.35}$$

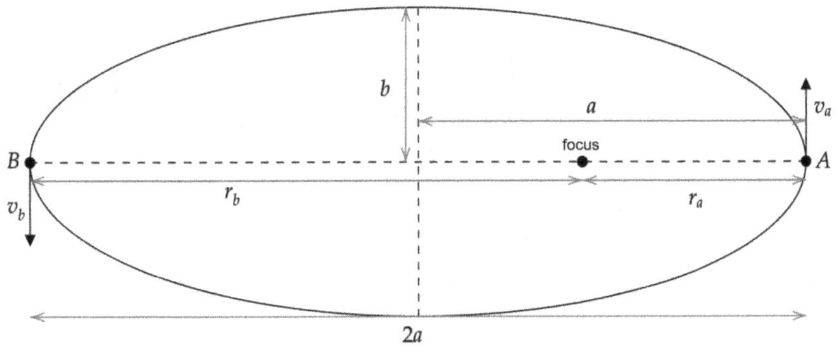

Figure 7.5. Elliptical orbit. Point A is the perihelion if the Sun is at the focus, and perigee if the Earth is at the focus. Similarly point B is the aphelion (Sun at the focus) or apogee (Earth at the focus).

The distance between the focus and the centre of the ellipse is:

$$a - r_a = \frac{h}{1 - e^2} - \frac{h}{1 + e}$$

we obtain (you can work through the calculations as extra practice, but you should find them straightforward) :

$$a - r_a = ea.$$

So, the ellipse crosses the y-axis when:

$$\cos\theta = -\frac{ea}{r} \tag{7.36}$$

as we can see figure 7.6 (*remember* $\cos(180 - \theta) = -\cos\theta$).

Then, substituting

$$r(1 + e\cos\theta) = r - ae^2 = h$$

$$\Rightarrow r = h + ae^2 = h\left(1 + \frac{e^2}{1 - e^2}\right) = a.$$

Thus, (applying Pythagoras theorem):

$$b^2 = r^2 - a^2e^2 = a^2(1 - e^2) = \frac{h^2}{1 - e^2}.$$

Hence

$$b = \frac{h}{\sqrt{1 - e^2}} \tag{7.37}$$

and we have that the general equation of an ellipse in Cartesian coordinates is:

$$\frac{x^2}{a^2} + \frac{y^2}{b^2} = 1. \tag{7.38}$$

Now, let's define a in terms of the energy. We can substitute the expressions for h and e we derived when calculating the eccentricity. Then,

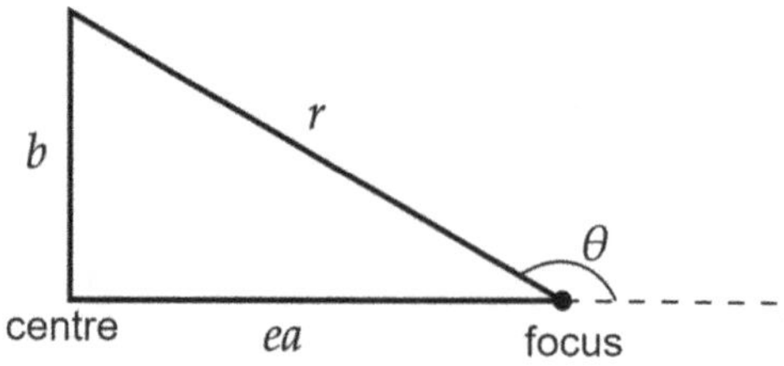

Figure 7.6. Intersection with y-axis.

$$a = \frac{-\frac{L^2}{mK}}{-\frac{2EL^2}{mK^2}} = \left| \frac{K}{2E} \right|. \tag{7.39}$$

We can see the elliptical orbit is only dependent on the energy.

7.4.4 Kepler's laws

Law 7.4.1: Kepler's first law
The planets move in elliptical orbits with the Sun at the focus.

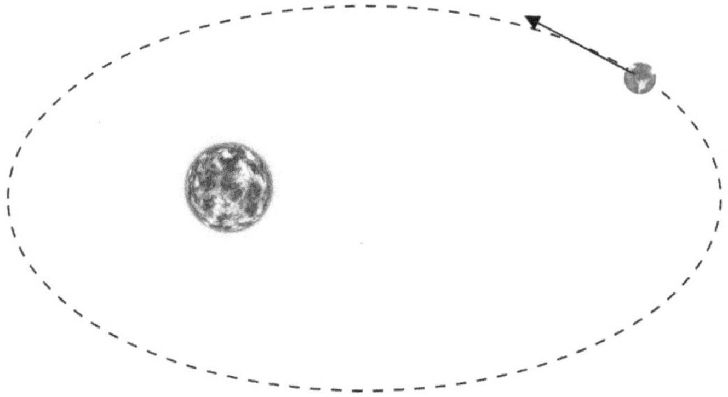

We have looked at this previously.

Law 7.4.2. Kepler's second law
The radius vector to a planet sweeps out an area at a rate that is independent of its position in the orbit.

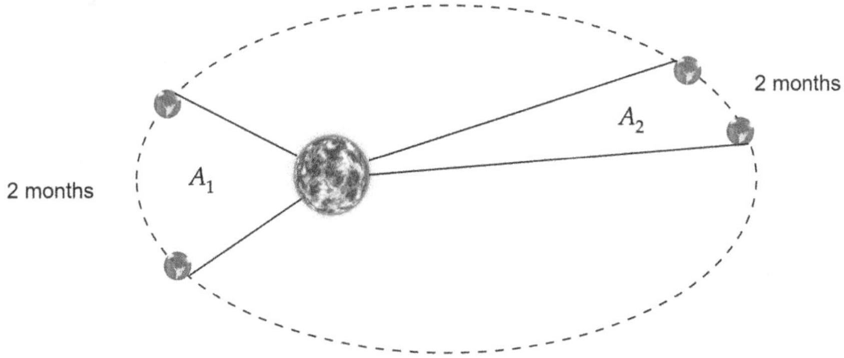

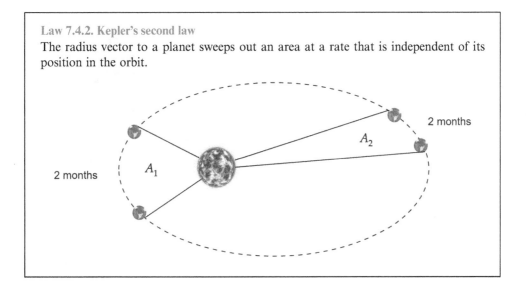

Proof. This can easily be proven taking into account conservation of angular momentum. The area swept out by the radius vector during a short period of time is

$$\mathrm{d}A = \frac{r(r\mathrm{d}\theta)}{2}$$

as can be seen in:

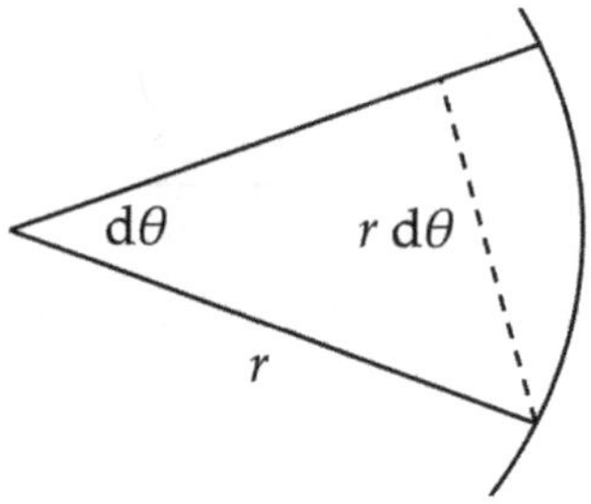

Then, differentiating:

$$\frac{\mathrm{d}A}{\mathrm{d}t} = \frac{1}{2}r^2\dot{\theta}\frac{L}{2m}$$

and substituting $L = mr^2\dot{\theta}$:

$$\frac{\mathrm{d}A}{\mathrm{d}t} = \frac{L}{2m} \tag{7.40}$$

which is constant as angular momentum is conserved for a central force. ◆

Law 7.4.3. Kepler's third law

The square of the period of an orbit, T, is proportional to the cube of the semi-major axis length, a.

$$T^2 \propto a^3. \tag{7.41}$$

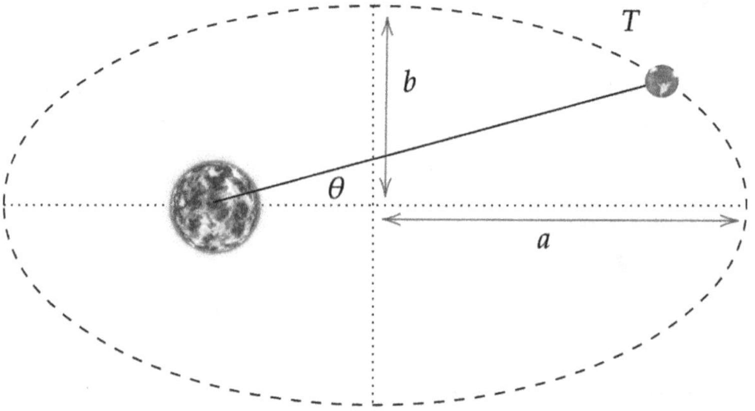

Proof. Integrating equation (7.4.4) over the time of a whole orbit gives:

$$A = \frac{LT}{2m}. \tag{7.42}$$

But the area of an ellipse is $A = \pi ab$, where a and b are the semi-major and semi-minor axes, respectively. We have that the semi-minor axis is equal to:

$$b = a\sqrt{1 - e^2}. \tag{7.43}$$

Then, squaring equation (7.42)

$$A^2 = \frac{L^2T^2}{4m^2} = \pi^2a^4(1 - e^2).$$

So,

$$\pi^2a^4 = \frac{L^2}{m(1 - e^2)}\frac{T^2}{4m}$$

using the fact that $L^2 = mKh$ from our derivation of eccentricity. We get

$$\pi^2a^4 = Ka\frac{T^2}{4m}$$

$$T^2 = \frac{4\pi^2ma^3}{K}.$$

For a gravitational force where $K = GMm$,

$$T^2 = \frac{4\pi^2a^3}{GM}. \tag{7.44}$$

◆

Exercise 7.3. TIP
State and derive Kepler's laws. *It is standard to ask to derive any of the laws, therefore you should become very familiar with them.*

◆

7.5 Orbits with positive energy: unbound orbits

Now we consider the case of positive-energy orbits ($E > 0$ and $e > 1$). The distance from the centre of force can now diverge to infinity. Let us take the conic equation for the **attractive** case

$$\frac{h}{r} = 1 + e\cos\theta. \tag{7.45}$$

And we have that r can go to infinity for a value of theta θ_∞:

$$1 + e\cos\theta_\infty = 0 \Rightarrow \cos\theta_\infty = -\frac{1}{e}$$

$$1 - e\cos\theta_\infty = 0 \Rightarrow \cos\theta_\infty = \frac{1}{e}$$

for both attractive and repulsive forces, respectively.

Definition 7.5.1. Impact parameter
The **impact parameter**, b represents the distance of closest approach to the centre of force if the object did not deviate from its initial trajectory, and continued as a straight line.

We can express the angular momentum and energy at the moment of collision in terms of the speed at infinity v_∞ and the impact parameter b.

$$L = mv_\infty b$$

$$E = \frac{1}{2}mv_\infty^2.$$

Then, the eccentricity is given by

$$e = \sqrt{1 + \frac{2EL^2}{mK^2}} = \sqrt{1 + \frac{m^2v_\infty^4b^2}{K^2}}. \tag{7.46}$$

Note that b replaces the semi-minor axis for positive-energy orbits. In Cartesian coordinates, the equation for a fully symmetric hyperbola around the origin is:

$$\frac{x^2}{a^2} - \frac{y^2}{b^2} = 1. \tag{7.47}$$

For very large values of x and y we can neglect the unity, obtaining the asymptotic lines of the hyperbola:

$$y = \pm\frac{b}{a}x. \tag{7.48}$$

However, for our particular case we have a shift of ae along the x-axis:

$$\frac{(x - ae)^2}{a^2} - \frac{y^2}{b^2} = 1. \tag{7.49}$$

Thanks to this shift it is acceptable to have solutions for $x = 0$.

We will now calculate the angle θ_∞. This result can be calculated **mathematically**, in a very straightforward manner by recalling that the gradient of the asymptotic lines is related to θ_∞ as seen below.

$$m = \tan\theta_\infty = \pm\frac{b}{a}.$$

Since $a = \frac{K}{2E}$, we have:

$$\tan\theta_\infty = \frac{2Eb}{K}. \tag{7.50}$$

Alternatively, we can reach the same solution through a **physical perspective**. Let's start by considering the x-component of the force on the body.

$$\frac{K}{r^2}\cos\theta = \frac{mK\dot{\theta}}{L}\cos\theta.$$

Then, we can apply Newton's law and simplify the mass of the orbiting body

$$\frac{\mathrm{d}v_x}{\mathrm{d}t} = \frac{K\dot{\theta}}{L}\cos\theta = \frac{K}{L}\frac{\mathrm{d}}{\mathrm{d}t}(\sin\theta)$$

by integrating

$$v_x = \frac{K}{L}\sin\theta + C.$$

But $v_x = 0$ at $\theta = 0$, so $C = 0$. At infinity, we have $v_x = v_\infty\cos\theta$, so

$$\tan\theta_\infty = \frac{v_\infty L}{K} = \frac{mv_\infty^2 b}{K}. \tag{7.51}$$

Then, using our expression for energy, we have:

$$\tan\theta_\infty = \frac{2Eb}{K} \tag{7.52}$$

as expected.

This is valid for both attractive and repulsive forces. For **attractive** forces we have that $\tan\theta_\infty < 0$ and thus $|\theta_\infty| > \pi/2$. However, for **repulsive** forces, we have $|\theta_\infty| < \pi/2$.

7.6 Reduced mass and the two-body problem

When considering a planet orbiting the Sun we have assumed that the Sun is fixed at the focus. However, the mass of the Sun, M, is finite and therefore the Sun and planet both move with respect to their centre of mass. This is *always* the case for any two astronomical objects.

> Remark 7.5. The considerations seen below can be applied to a two-body system of masses which are affected by *central forces* and **no external forces**.

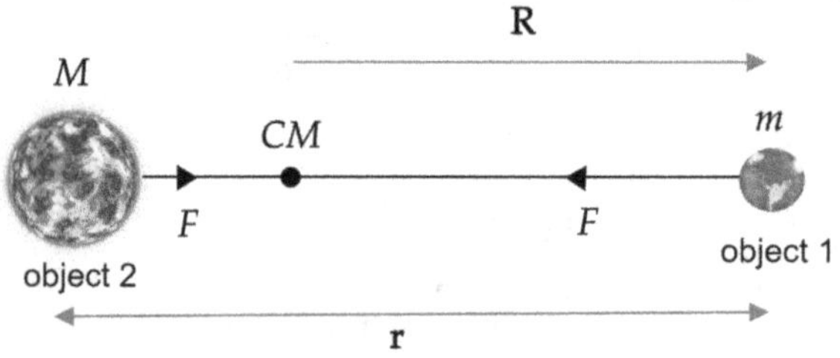

Figure 7.7. Two-body problem.

Then, as can be seen in figure 7.7, the position of object 1 (in this case Earth) relative to the centre of mass is:

$$\boldsymbol{r}_1' = \boldsymbol{r}_1 - \boldsymbol{R} = \frac{m_2}{m_1 + m_2}\boldsymbol{r}$$

where $\boldsymbol{r} = \boldsymbol{r}_1 - \boldsymbol{r}_2$ is the relative position of the two bodies. Then, the momentum of the first object with respect to the centre of mass will become:

$$\boldsymbol{p}_1' = m_1\boldsymbol{r}_1' = \mu\dot{\boldsymbol{r}} = \mu\boldsymbol{v}$$

where μ is the reduced mass and $\boldsymbol{v}$ is the relative velocity of object one with respect to the centre of mass.

With the same reasoning we can express the kinetic energy of the relative motion, the angular momentum and the force with respect to the centre of mass:

$$K_{rel} = \frac{1}{2}\mu v^2 \tag{7.53}$$

$$\boldsymbol{L} = \mu\boldsymbol{r} \times \boldsymbol{v} \tag{7.54}$$

$$\boldsymbol{F}_{12} = \frac{\mathrm{d}\boldsymbol{p}}{\mathrm{d}t} = \mu\frac{\mathrm{d}\boldsymbol{v}}{\mathrm{d}t} = \mu\frac{\mathrm{d}^2\boldsymbol{r}}{\mathrm{d}t^2}. \tag{7.55}$$

Hence, the real system in which both bodies orbit about a common centre of mass is equivalent to a body of reduced mass, μ orbiting at a distance $\boldsymbol{r}$ from the centre of mass, which is fixed. Moreover, all dynamical properties are conserved as there are no external forces.

Exercise 7.4. Applying the method detailed above, find the reduced mass of the Earth and Sun, and the consequent change in the period of the Earth's orbit from what we would have calculated assuming the Sun was fixed.

Example 7.1. Binary stars

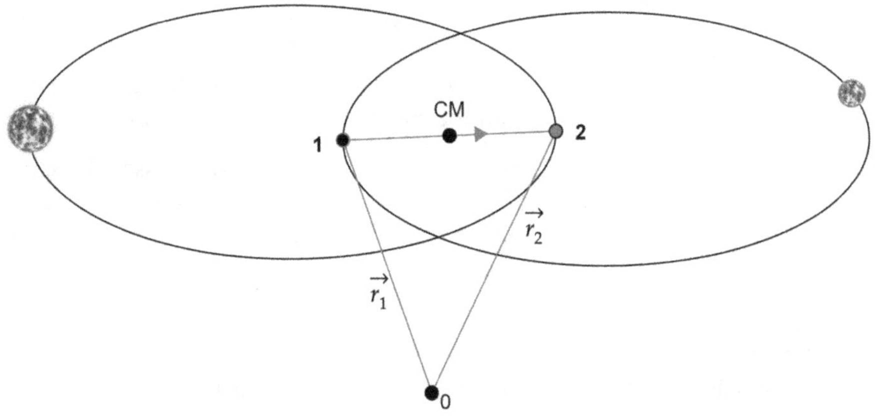

We define $\boldsymbol{R} = \boldsymbol{r}_1 - \boldsymbol{r}_2$ and $R = |\boldsymbol{r}_1 - \boldsymbol{r}_2|$.

We can apply Newton's second law considering that the only force between the Stars is the *gravitational force*. The force on Star 1 exerted by Star 2 is along $-\hat{\boldsymbol{R}}$.

$$M_1\ddot{\boldsymbol{r}}_1 = -G\frac{M_1M_2}{R^2}\hat{\boldsymbol{R}}.$$

And the force exerted on Star 2 by Star 1 is

$$M_2\ddot{\boldsymbol{r}}_2 = G\frac{M_1M_2}{R^2}\hat{\boldsymbol{R}}.$$

Then subtracting the two equations above:

$$\ddot{\boldsymbol{r}}_1 - \ddot{\boldsymbol{r}}_2 = -\frac{G}{R^2}(M_1 + M_2)\hat{\boldsymbol{R}} = \ddot{\boldsymbol{R}}.$$

Multiplying the expression above by the reduced mass $\mu = \frac{M_1M_2}{M_1 + M_2}$,

$$\mu\ddot{\boldsymbol{R}} = -G\frac{M_1M_2}{R^2}\hat{\boldsymbol{R}}.$$

We can also express this as

$$\ddot{\boldsymbol{R}} = -\frac{GM}{R^2}\hat{\boldsymbol{R}} = -\frac{GM}{R^3}\boldsymbol{R}$$

where $M = M_1 + M_2$.

Now, we express $\ddot{\boldsymbol{R}}$ in polar coordinates:

$$\ddot{\boldsymbol{R}} = (\ddot{R} - R\dot{\theta}^2)\,\hat{\boldsymbol{r}} + (R\ddot{\theta})\,\hat{\boldsymbol{\theta}}. \tag{7.56}$$

Taking the radial component,

$$\mu(\ddot{R} - R\dot{\theta}^2) = -G\frac{M_1M_2}{R^2} \tag{7.57}$$

$$\ddot{R} - R\dot{\theta}^2 = -G\frac{M_1M_2}{\mu}\frac{1}{R^2} = -\frac{GM}{R^2}. \tag{7.58}$$

Therefore, we can see that both orbits are elliptical and have the same period and eccentricity. Recall equation (7.18) and the subsequent treatment in section 7.3.

Exercise 7.5. A binary star system consists of two stars of masses M_1 and M_2 orbiting their common centre of mass. If the orbits of both stars are observed to be circular and to have period T, find an expression for the distance R between the stars.

◆

7.7 Variable mass problems

Previously, we only considered systems where the mass was constant. However, in reality mass is not always conserved, making the study of momentum conservation quite tricky. We shall demonstrate how to treat such cases through the following example.

Example 7.2. Rocket motion
In the case of a rocket, most of its mass gets ejected to propel it forward through an accelerating force. Since the rocket is not subject to external forces, the total momentum must be conserved. We shall assume the rocket moves horizontally, so that there is no gravitational force acting in its direction of motion.

Let's consider the rocket to have a time dependent variable mass $m(t)$, with a constant rate of change. The variation of the mass of the rocket as time passes will be of $\mathrm{d}m$, so that after an infinitesimal amount of time the mass of the rocket will be $m + \mathrm{d}m$. Therefore, the mass emitted by the rocket must be $-\mathrm{d}m$ in the negative x-direction. The relative speed between the rocket and the expelled mass is u.

At time t the rocket has an initial mass m and velocity v, as observed from an external reference frame.

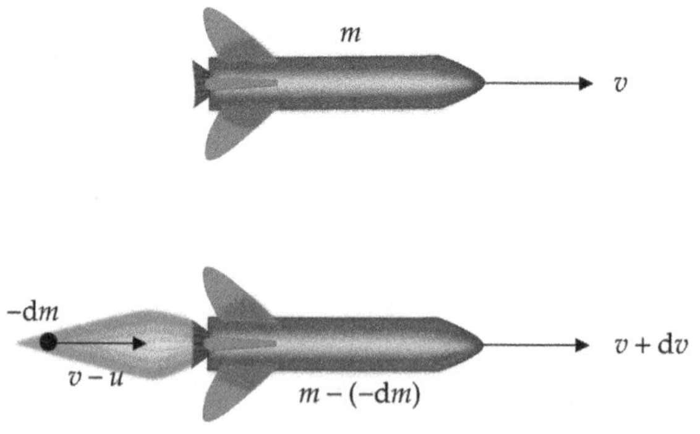

Since the overall momentum is conserved, its value must be the same at a time t as it is at $t + \mathrm{d}t$. So, we have:

$$mv = (m + \mathrm{d}m)(v + \mathrm{d}v) + (-\mathrm{d}m)(v - u).$$

The term, $dmdv$ obtained from the multiplication of the first parenthesis is a second order infinitesimal and it can therefore be neglected, obtaining:

$$mdv = -udm.$$

Applying an integral on both sides:

$$\int_{v_1}^{v_2} dv = -\int_{m_1}^{m_2} u\frac{dm}{m}$$

where m_1 and v_1 are, respectively, the mass of the rocket and velocity at a time t_1 and m_2 and v_2 are the mass and velocity at a time t_2.
Integrating:

$$v_2 = v_1 + u \ln \frac{m_1}{m_2} \tag{7.59}$$

which explains why rockets take off very slowly. Just before taking off, the rocket is at rest $v_1 = 0$. The second term of the equation is very small at the beginning, so the rocket advances slowly. In addition, real rockets take off vertically, so the additional gravitational acceleration from the Earth must also be taken into account.

Exercise 7.6. A duck floats at rest on Lake Geneva. At time $t = 0$ the duck is startled and starts flying upward with speed v_0, at constant angle α with respect to the horizontal. So scared is the duck that at time t_1 it simultaneously loses a feather of mass m_f, and emits some organic droppings of mass m_d, both initially at the same speed as the duck. The feather will feel the action of air resistance with force $\boldsymbol{F} = -\beta \boldsymbol{v}$, where $\boldsymbol{v}$ is the velocity of the feather, while the effect of air on the droppings is negligible.

1. Describe the forces acting on the feather and on the droppings at time t_1, separately in the horizontal and vertical directions.
2. Derive the terminal velocity v_T of the feather in air. For this part only assume that as soon as it is lost, the feather instantaneously loses its horizontal velocity, and falls vertically downwards with its constant terminal velocity. Derive an expression for the time it takes for the feather to fall back in the water. Also find the horizontal and vertical distances of the droppings from the position where the duck was at rest, as a function of time.
3. Now consider the general case where the initial motion of the feather cannot be neglected. Show that the velocity of the feather in the horizontal direction is

$$v_x(t) = v_0 \cos \alpha \exp\left(-\frac{\beta}{m}t\right),$$

and the one in the vertical direction

$$v_y(t) = (v_0 \sin \alpha + v_T)\exp\left(-\frac{\beta}{m}t\right) - v_T.$$

4. In the general case, will the maximal height be higher for the feather or for the droppings, and why?

IOP Publishing

Classical Mechanics
A professor-student collaboration
Mario Campanelli

Chapter 8

Rigid bodies

8.0 Introduction

So far we have always dealt with the motion of point-like objects and particles. However, we know that reality is more complex than this and we must learn to describe objects which have an extension in space and a particular shape.

We begin this chapter about motion of rigid bodies in 2D and 3D, by giving some basic definitions and presenting three theorems that we will use in this and in the next chapter. We also show how to calculate the centre of mass for a general rigid body with a given density.

To study the motion of a rigid body in space, we start by considering the simpler case of a planar object in the x–y plane. We will consider first a purely rotational motion about the z-axis, and then a more general motion always in the x–y plane. We will also introduce two useful theorems valid under these conditions. Finally, we will consider non-planar objects, but still undergoing a motion in the x–y plane.

In the final section, instead, we will move on to the treatment of a non-planar object undergoing a general motion in 3D space.

8.1 Preliminaries

Let's start by giving a definition of what is meant by *rigid body*.

Definition 8.1.1: Rigid body
A rigid body is a continuous system of points in space. It is a three-dimensional body which will have a defined shape, extended in space and constant in time. If an external force is applied to a rigid body, there is no relative displacement of component parts, i.e. it cannot be deformed.

doi:10.1088/978-0-7503-2690-2ch8

In this section, we look at three theorems that will be needed later on. However, we do not cover their proofs.

> **Theorem 8.1.1** Consider a rigid body undergoing arbitrary motion. Pick any point P in the body. Then, at any instant, the motion of the body can be written as the sum of the translational motion of P and a rotation around some axis (which may change with time) through P.

> **Remark 8.1.** In other words, this theorem is telling us that a general motion of a rigid body can always be decomposed into the sum of a translation and a rotation.

In the previous chapters, we treated the angular velocity only as a scalar quantity associated with a rotating particle. This is because, in 2D, a particle can rotate only in a plane, and in two ways: counterclockwise or clockwise. So, when defining the angular velocity, we just need to specify a scalar quantity for its modulus, and which way it is rotating. However, in a 3D space, a particle can rotate about any axis and therefore we need a vector in order to uniquely identify its rotation. Since in this chapter we will also consider the motion of a rigid body in 3D space, we need to extend our treatment by introducing the **angular velocity vector**.

> **Definition 8.1.2: Angular velocity vector**
> In the three-dimensional plane, angular velocity is not a scalar property of an object, but a **vector perpendicular** to the **plane of rotation** passing through the centre of rotation.
> The direction in which the angular velocity vector points can be found using the right-hand rule (figure 8.1).

> **Remark 8.2.** Throughout this chapter, we'll assume that, unless otherwise stated, the axis of rotation passes through the origin of the coordinate system.

We can use the angular velocity vector to obtain the velocity of any point in the rotating body. This is stated in the following theorem.

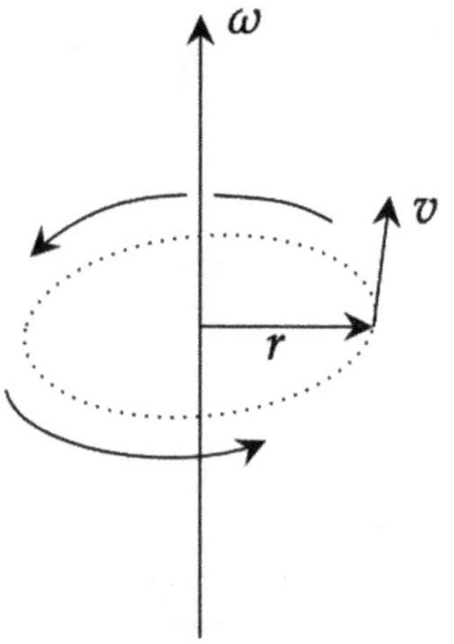

Figure 8.1. Schematic representation of the angular velocity vector along the z-axis.

Theorem 8.1.2: Velocity
Given an object rotating with angular velocity $\boldsymbol{\omega}$, the velocity $\boldsymbol{v}$ of a point at position $\boldsymbol{r}$ is given by

$$\boldsymbol{v} = \boldsymbol{\omega} \times \boldsymbol{r}. \tag{8.1}$$

Remark 8.3. Consider a rotating rigid body. To find the speed of a particular point, we need to look at the component of the position vector $\boldsymbol{r}$, called $r_\perp$, that is perpendicular to the angular velocity vector $\boldsymbol{\omega}$.
From figure 8.2, it is clear that $r_\perp = |\boldsymbol{r}| \sin\theta$. Thus, the speed is given by

$$v = |\boldsymbol{\omega}| r_\perp = |\boldsymbol{\omega}||\boldsymbol{r}| \sin\theta = |\boldsymbol{\omega} \times r|.$$

The final theorem we need to introduce tells us how the sum of angular velocities in different coordinate systems works.

Theorem 8.1.3 Let coordinate systems S_1, S_2, and S_3 have a common origin. Let S_1 rotate with angular velocity $\boldsymbol{\omega}_{1,2}$ with respect to S_2, and let S_2 rotate with angular velocity $\boldsymbol{\omega}_{2,3}$ with respect to S_3. Then S_1 rotates (instantaneously) with angular velocity

$$\boldsymbol{\omega}_{1,3} = \boldsymbol{\omega}_{1,2} + \boldsymbol{\omega}_{2,3} \tag{8.2}$$

with respect to S_3.

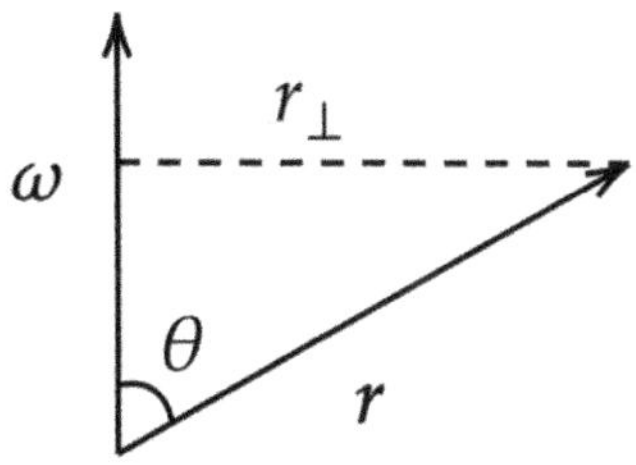

Figure 8.2. The projection of the position vector perpendicular to the angular velocity.

8.2 Centre of mass

Since a rigid body is made up of many particles, it's easier to speak about densities rather than considering each single particle. Table 8.1 summarises the different definitions of density we will encounter. In the formulae, valid for uniform densities, the parameter M is the mass, while L, A and V are, respectively, the length, the area and the volume.

In general, for a rigid body with volumetric mass density ρ, an infinitesimal element of mass $\mathrm{d}m$ is given by

$$\mathrm{d}m = \rho \, \mathrm{d}x \, \mathrm{d}y \, \mathrm{d}z. \tag{8.3}$$

In chapter 6, we stated that for a system of particles the position vector of the centre of mass is given by

$$\boldsymbol{R}_{\mathrm{CM}} = \frac{\sum_i \boldsymbol{r}_i \, m_i}{\sum_i m_i} = \frac{\sum_i \boldsymbol{r}_i \, m_i}{M},$$

where M is the total mass, and $\boldsymbol{r}_i$ and m_i are, respectively, the position vector of a particle and its mass.

As a rigid body is a continuous system of points in space (see definition 8.1.1), the position vector of the centre of mass $\boldsymbol{R}_{\mathrm{CM}}$ is now given by an integral, i.e.

$$\boldsymbol{R}_{\mathrm{CM}} = \frac{\int \boldsymbol{r} \, \mathrm{d}m}{\int \mathrm{d}m} = \frac{\int \boldsymbol{r} \, \mathrm{d}m}{M}, \tag{8.4}$$

Table 8.1. The different densities we will encounter.

Density	Symbol	Formula
Volumetric mass density	ρ	M/V
Area density	σ	M/A
Linear density	λ	M/L

where $\boldsymbol{r}$ is the position vector of a point in the rigid body and the integral guarantees we are considering every point in the body.

Therefore, the components of the position vector are given by

$$x_{\mathrm{CM}} = \frac{\int x\,\mathrm{d}m}{\int \mathrm{d}m};\quad y_{\mathrm{CM}} = \frac{\int y\,\mathrm{d}m}{\int \mathrm{d}m};\quad z_{\mathrm{CM}} = \frac{\int z\,\mathrm{d}m}{\int \mathrm{d}m};$$

and $\mathrm{d}m$ usually needs to be expressed using equation (8.3).

Example 8.1. Centre of mass of a thin rod

Case 1. Consider a rod going from $x = 0$ to $x = L$ with constant density λ. In this situation, we can neglect the y and z components, because the system can be approximated as one-dimensional.

The infinitesimal element of mass is therefore given by $\mathrm{d}m = \lambda\,\mathrm{d}x$. Let's simply apply equation (8.4) to find the position vector of the centre of mass. Of course, since we are in 1D, we consider only the x component, so the equation becomes

$$x_{\mathrm{CM}} = \frac{\int_0^M x\,\mathrm{d}m}{\int_0^M \mathrm{d}m} = \frac{\int_0^L x\,\lambda\,\mathrm{d}x}{\int_0^L \lambda\,\mathrm{d}x} = \frac{\cancel{\lambda}\int_0^L x\,\mathrm{d}x}{\cancel{\lambda}\int_0^L \mathrm{d}x} = \frac{\frac{1}{2}x^2|_0^L}{x|_0^L} = \frac{\frac{1}{2}L^{\cancel{2}}}{\cancel{L}} = \frac{1}{2}L,$$

as expected. *Note that we can take the linear density λ out of the integral because it is constant.*

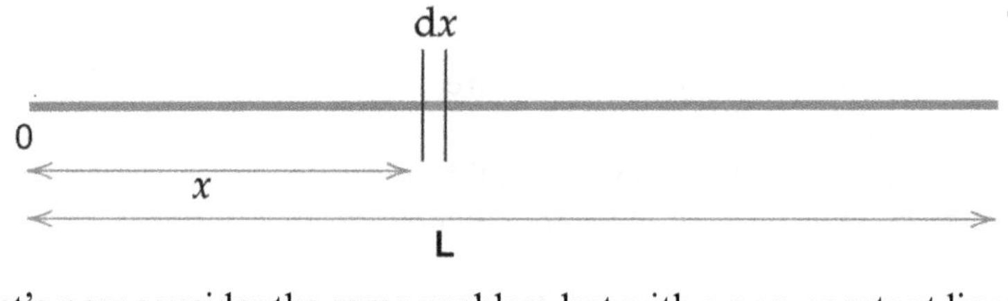

Case 2. Let's now consider the same problem but with a non-constant linear density, for instance $\lambda(x) = e^{-x}$. Again, we can neglect the y and z components.

The infinitesimal element of mass is now $\mathrm{d}m = e^{-x}\,\mathrm{d}x$. Since the density depends on the position x, we cannot take it out of the integral as before. We have

$$x_{\mathrm{CM}} = \frac{\int_0^L x\,e^{-x}\,\mathrm{d}x}{\int_0^L e^{-x}\mathrm{d}x}.$$

We can solve the integral in the numerator integrating by parts: integrate e^{-x} and then differentiate x.

$$x_{\rm CM} = \frac{[-e^{-x}\,x]_0^L - \int_0^L(-e^{-x})\mathrm{d}x}{-e^{-x}|_0^L} = \frac{-e^{-L}\,L - e^{-L} + 1}{-e^{-L} + 1} = \frac{-e^{-L}(L+1) + 1}{-e^{-L} + 1}.$$

Exercise 8.1. A stationary rod of length L, with non-uniform density, is lying on the x-axis with one extremity at the origin. The mass density is given by $\lambda = \lambda_0 \frac{x}{L}$.

Given this information, derive an expression for the total mass of the rod and the x position of its centre of mass.

◆

8.3 Flat object in x–y plane

Consider a flat rigid body that has an extension only in the x–y plane. Let this body undergo an arbitrary motion, i.e. a combination of translation and rotation, in that same plane. Let's find its angular momentum relative to the origin of the coordinate system.

From chapter 5, we know that the **angular momentum** of a particle is given by $\boldsymbol{L} = \boldsymbol{r} \times \boldsymbol{p}$, where $\boldsymbol{r}$ and $\boldsymbol{p}$ are, respectively, the position vector and the momentum of the particle.

For a system of particles of mass m_i, the angular momentum is simply given by the sum of the angular momenta of the particles, i.e.

$$\boldsymbol{L} = \sum_i \boldsymbol{L}_i = \sum_i \boldsymbol{r}_i \times \boldsymbol{p}_i. \tag{8.5}$$

However, we are considering a rigid body, which has a *continuous* distribution of mass, therefore we need to replace the sum with an integral:

$$\boldsymbol{L} = \int (\boldsymbol{r} \times \boldsymbol{p})\,\mathrm{d}x\,\mathrm{d}y. \tag{8.6}$$

> Remark 8.4. Remember that for now we are considering $\boldsymbol{r}$ and $\boldsymbol{p}$ to be in the x–y plane at all times, thus the angular momentum will always point in the $\hat{z}$ direction. See chapter 5, remark 5.3.

8.3.1 Rotation about the z-axis

Suppose the two-dimensional body is pivoted at the origin and rotates with angular frequency ω around the z-axis in the counterclockwise direction (as viewed from above). This setup is shown in figure 8.3.

An infinitesimal mass element $\mathrm{d}m$ is at a perpendicular distance $r = \sqrt{x^2 + y^2}$ from the pivot. Since we are considering a rigid body, all mass elements must have

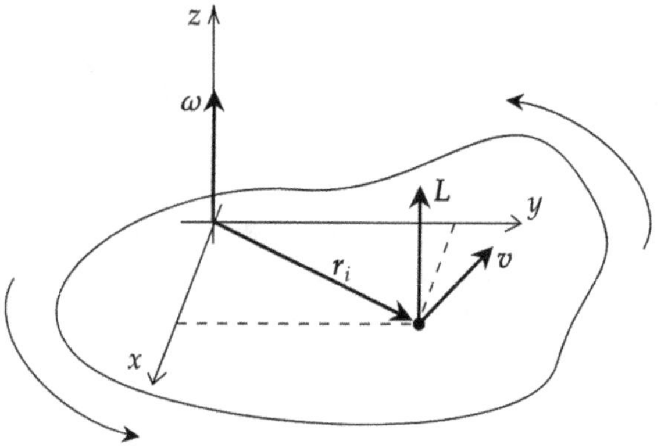

Figure 8.3. Rotating flat body in the x–y plane.

the same angular velocity. So the speed with which this element is travelling in a circle around the origin is $v = \omega r$. Let's now calculate the angular momentum relative to the origin for this infinitesimal element. Since it has a mass dm, its momentum will be $\boldsymbol{p} = v\,\mathrm{d}m$.

It follows that

$$\boldsymbol{L} = \int \boldsymbol{r} \times \boldsymbol{p} = \int r\,(v\,\mathrm{d}m)\,\hat{z} = \int r^2\,\omega\,\mathrm{d}m\,\hat{z} = \int (x^2 + y^2)\,\omega\,\hat{z}\,\mathrm{d}m,$$

where the integral is calculated over the whole body. Note that the z direction comes from remark 8.4.

Definition 8.3.1: Moment of inertia

We define the moment of inertia around the z-axis as

$$I_z \equiv \int r^2\,\mathrm{d}m = \int (x^2 + y^2)\,\mathrm{d}m. \tag{8.7}$$

This quantity represents how the system will initially resist the rotation and how, once in movement, it will want to remain in this movement. It is analogous of mass for rotations, but it also accounts for the distance of the masses from the rotational axis. Our experience tells us that, for a given mass, it is easier to start a rotation close to the rotational axis rather than far away from it, and this definition formalises that.

As usual, if the rigid body were made up of a discrete collection of point masses m_i, we would need to replace the integral with a sum in this definition. The moment of inertia around the z-axis would then be given by

$$I_z = \sum_i m_i r_i^2 = \sum_i m_i \left(x_i^2 + y_i^2\right)^2. \tag{8.8}$$

Following from the continuous definition, we have that

$$L_z = I_z\omega \quad \text{and} \quad L_x = L_y = 0. \tag{8.9}$$

Now that we have defined this quantity we can consider the kinetic energy. An infinitesimal element of mass has energy

$$\mathrm{d}K = \frac{1}{2}\,\mathrm{d}m\, v^2 = \frac{1}{2}\,\mathrm{d}m\,(r\omega)^2.$$

Therefore, the kinetic energy of the whole rigid body is given by

$$K = \int \frac{1}{2}\, r^2\, \omega^2\, \mathrm{d}m = \frac{1}{2}\,\omega^2 \int r^2\, \mathrm{d}m = \frac{1}{2}\, I_z\, \omega^2. \tag{8.10}$$

Exercise 8.2. A planar rigid body is rotating with angular velocity ω about a fixed axis a, with a moment of inertia given by I_a.

Use this information to determine the angular momentum of the object about the axis a and also its rotational kinetic energy.

Under which conditions is the angular momentum of the body about the axis conserved?

◆

8.3.2 General motion in the x–y plane

In the last subsection, we only considered a *rotation* in the x–y plane. We now need to study a rigid body undergoing both a translation and a rotation.

Remark 8.5. Centre of mass coordinates

We will now make use of the centre of mass coordinates (whose origin coincides with the centre of mass of the system) to write the angular momentum $\boldsymbol{L}$ and the kinetic energy K. The motion of a point in the rigid body relative to the origin is thus split into two parts: the motion of the point relative to the centre of mass, and the motion of the centre of mass relative to the origin of the coordinate system.

Working with different coordinate systems can be confusing, as the same variable can be expressed in many systems. In general, quantities in the centre of mass system are indicated with a prime, while the quantities indicating the relative movement between the centre of mass and the original system are indicated by capital letters. Here's a table of the variables we will use:

Symbol	Meaning
$\boldsymbol{r}'$	Position vector of the point relative to the centre of mass
$\boldsymbol{R}$	Position vector of the centre of mass relative to the origin
$\boldsymbol{r}$	Position vector of the point relative to the origin
$\boldsymbol{v}'$	Velocity of the point relative to the centre of mass
$\boldsymbol{V}$	Velocity of the centre of mass relative to the origin
$\boldsymbol{v}$	Velocity of the point relative to the origin
ω'	Angular velocity of the point relative to the centre of mass

From figure 8.4, it is clear that $\boldsymbol{r} = \boldsymbol{R} + \boldsymbol{r}'$. Differentiating this equation, we also see that $\boldsymbol{v} = \boldsymbol{V} + \boldsymbol{v}'$. If the body is rotating with angular speed ω' about the centre of mass (remaining in the x–y plane at all times), then a point at distance r will have speed $v' = \omega' r'$ relative to the centre of mass.

Let M be the total mass of the rigid body. Again, we want to calculate the angular momentum of the point relative to the origin:

$$\boldsymbol{L} = \int \boldsymbol{r}\times\boldsymbol{v}\ \mathrm{d}m = \int (\boldsymbol{R}+\boldsymbol{r}')\times(\boldsymbol{V}+\boldsymbol{v}')\ \mathrm{d}m = \int (\boldsymbol{R}\times\boldsymbol{V} + \cancelto{0}{\boldsymbol{r}'\times\boldsymbol{V}} + \cancelto{0}{\boldsymbol{R}\times\boldsymbol{v}'} + \boldsymbol{r}'\times\boldsymbol{v}')\ \mathrm{d}m.$$

The previous integral is divided into four parts. However,

$$\int \boldsymbol{r}' \times \boldsymbol{V}\ \mathrm{d}m = \int \boldsymbol{R} \times \boldsymbol{v}'\ \mathrm{d}m = \boldsymbol{0}.$$

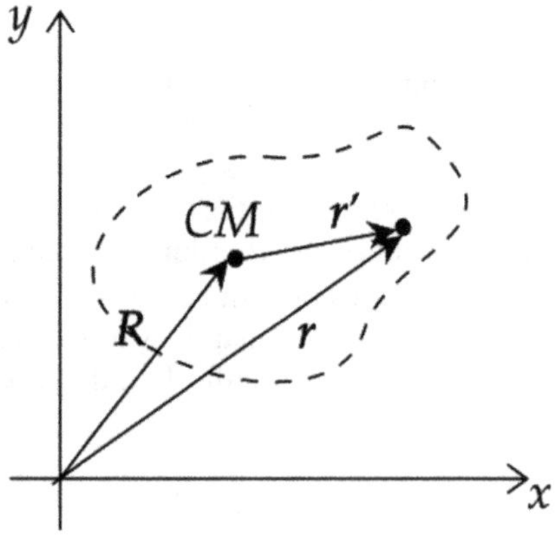

Figure 8.4. Schematic representation of position vectors relative to the centre of mass and relative to the origin of the axis.

Remark 8.6. This happens because the definition of the centre of mass is $\int \boldsymbol{r}' \, \mathrm{d}m = \boldsymbol{0}$ (meaning that the position of the centre of mass in the centre of mass frame is zero). It follows that $\int \boldsymbol{v}' \, \mathrm{d}m = \frac{\mathrm{d}(\int \boldsymbol{r}' \, \mathrm{d}m)}{\mathrm{d}t} = \boldsymbol{0}$. Since $\boldsymbol{R}$ and $\boldsymbol{V}$ are constant quantities, they can be taken out of the integrals, and therefore those integrals evaluate to zero.

We are left with the sum of two integrals:

$$\boldsymbol{L} = \int \boldsymbol{R} \times \boldsymbol{V} \, \mathrm{d}m + \int \boldsymbol{r}' \times \boldsymbol{v}' \, \mathrm{d}m$$
$$= M \, \boldsymbol{R} \times \boldsymbol{V} + \left(\int r'^2 \omega' \, \mathrm{d}m \right) \hat{z} = \boldsymbol{R} \times \boldsymbol{P} + \left(I_z^{\mathrm{CM}} \omega' \right) \hat{z},$$

where we have used definition 8.3.1, and 'CM' stands for *centre of mass*.

Hence, we have reached an important result:

Theorem 8.3.1: Angular momentum relative to the origin

The angular momentum relative to the origin of a rigid body undergoing a general motion is given by the sum of the angular momentum of the centre of mass relative to the origin and the angular momentum of the rigid body relative to the centre of mass. That is,

$$\boldsymbol{L} = \boldsymbol{R} \times \boldsymbol{P} + \left(I_z^{\mathrm{CM}} \omega' \right) \hat{z}. \tag{8.11}$$

Remark 8.7. Special case

If the centre of mass travels in a circle around the origin with angular speed Ω, then $V = \Omega R$ and $\boldsymbol{L} = (MR^2\Omega + I_z^{\mathrm{CM}} \omega')\hat{z}$.

Let's again look at the kinetic energy:

$$K = \int \frac{1}{2} v^2 \, \mathrm{d}m = \frac{1}{2} \int \left| \boldsymbol{V} + \boldsymbol{v}' \right|^2 \mathrm{d}m = \frac{1}{2} \int \left(V^2 + \cancelto{0}{2\boldsymbol{V} \cdot \boldsymbol{v}'} + v'^2 \right) \mathrm{d}m.$$

Analogously to the angular momentum derivation (see remark 8.6), we have that $\int \boldsymbol{V} \cdot \boldsymbol{v}' \, \mathrm{d}m = \boldsymbol{V} \cdot \int \boldsymbol{v}' \, \mathrm{d}m = 0$. Therefore, the previous equation simplifies to

$$K = \frac{1}{2}\int V^2\,\mathrm{d}m + \frac{1}{2}\int v'^2\,\mathrm{d}m$$
$$= \frac{1}{2}MV^2 + \frac{1}{2}\int r'^2\omega'^2\,\mathrm{d}m = \frac{1}{2}MV^2 + \frac{1}{2}I_z^{\mathrm{CM}}\omega'^2.$$

Theorem 8.3.2: Kinetic energy relative to the origin

The kinetic energy relative to the origin of a rigid body undergoing a general motion is given by the sum of the kinetic energy of the centre of mass relative to the origin and the kinetic energy of the rigid body relative to the centre of mass. That is,

$$K = \frac{1}{2}MV^2 + \frac{1}{2}I_z^{\mathrm{CM}}\omega'^2. \tag{8.12}$$

We now introduce two very useful theorems to calculate moments of inertia.

8.3.3 Theorem of parallel axis

The following theorem allows us to calculate the moment of inertia around an axis parallel to the axis passing through the centre of mass (figure 8.5).

Consider a rigid body (planar or non-planar) of total mass M, and any axis passing through its centre of mass. Further, consider a parallel axis, passing through a point A and at a distance a from the first axis.

When calculating the moment of inertia about an axis, only the perpendicular distance of a point mass from the axis matters. Therefore, we can squash any non-planar body onto a plane perpendicular to the axis, say the x–y plane. The rotation axis will therefore be the z-axis (in the centre of mass frame).

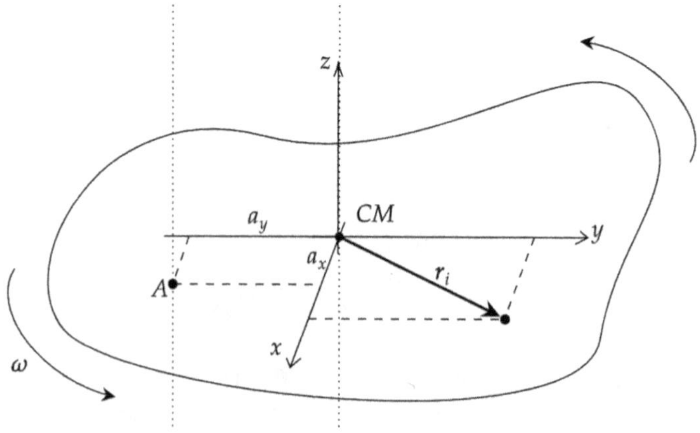

Figure 8.5. Schematic representation of a body lying on the x–y plane with centre of mass CM and axis of rotation going through point A.

The position vector $\boldsymbol{r}$ of a point P relative to the centre of mass is given by $r = \sqrt{x^2 + y^2}$. The distance a between the two axes can be written as $a = \sqrt{a_x^2 + a_y^2}$. Therefore, using Pythagoras' theorem, the total distance R between point P and the axis passing through point A is

$$R^2 = (x + a_x)^2 + (y + a_y)^2.$$

We know that the moment of inertia around the centre of mass is given by

$$I_{\mathrm{CM}} = \int r^2 \, dm = \int (x^2 + y^2) \, dm.$$

Thus, the moment of inertia around the new rotation axis is

$$\begin{aligned} I_A &= \int R^2 \, dm = \int \left[(x + a_x)^2 + (y + a_y)^2\right] dm \\ &= \int \left[(x^2 + y^2) + (a_x^2 + a_y^2) + 2xa_x + 2ya_y\right] dm \\ &= \int (x^2 + y^2) \, dm + \int (a_x^2 + a_y^2) \, dm + 2a_x \cancelto{0}{\int x \, dm} + 2a_y \cancelto{0}{\int y \, dm} \\ &= I_{\mathrm{CM}} + Ma^2, \end{aligned}$$

where $\int x \, dm = \int y \, dm = 0$ because of the definition of the centre of mass (see remark 8.6).

Theorem 8.3.3: Theorem of parallel axis

Consider an axis passing through the centre of mass of a rigid body of mass M, and let the moment of inertia calculated around it be I_{CM}. Then, the moment of inertia calculated around a parallel axis, passing through a point A and at a distance a from the first, is given by

$$I_A = I_{\mathrm{CM}} + Ma^2. \tag{8.13}$$

This theorem is valid

(i) for *any* arbitrary non-planar object;

(ii) *only* for an axis parallel to another axis passing through the centre of mass (no other point does the trick).

8.3.4 Theorem of perpendicular axis

Consider a flat object in the x–y plane. The moments of inertia of the body around the x- and the y-axes are defined analogously to the moment of inertia of the rigid body around the z-axis (see definition 8.3.1). Therefore, the moments of inertia of the body are given by

$$I_x = \int (y^2 + z^2)\, \mathrm{d}m, \quad I_y = \int (x^2 + z^2)\, \mathrm{d}m, \quad I_z = \int (x^2 + y^2)\, \mathrm{d}m.$$

However, since the flat object always remains in the x–y plane, we have that $z = 0$. Therefore, $I_x = \int y^2\, \mathrm{d}m$ and $I_y = \int x^2\, \mathrm{d}m$. The moment of inertia around the z-axis can thus be expressed as

$$I_z = \int (x^2 + y^2)\, \mathrm{d}m = \int x^2\, \mathrm{d}m + \int y^2\, \mathrm{d}m = I_y + I_x.$$

Theorem 8.3.4: Theorem of perpendicular axis
Given a flat rigid body in the x–y plane, its moment of inertia around the z-axis can be expressed as the sum of its moments of inertia around the x-axis and the y-axis. In formulae,

$$I_z = I_x + I_y. \tag{8.14}$$

This theorem is valid only for planar objects.

Exercise 8.3. State the theorems of *parallel axis* and of *perpendicular axis*, giving the conditions under which each one is valid.

Knowing that the moment of inertia of a uniform rod of mass M and length L through the axis perpendicular to the length and passing through one end of the rod is $\frac{1}{3}ML^2$, determine the moment of inertia of a uniform *square plate* of mass M and length L about the two axes given below:

(i) Passing through one side of the square plate.
(ii) Through one corner and perpendicular to the plane containing the square.

◆

8.3.5 Non-planar objects moving around an axis

We are now going to consider non-planar objects in the simplified situation where they can only rotate around an axis parallel to the z-axis. As a result, we will only be concerned about L_z, not L_x or L_y. In fact, in this case nearly all the results we have so far derived will hold.

This happens because we can slice an object that has an extension also in the z direction into planar objects which extend only in the x–y plane. It doesn't matter that these flat bodies will have different z values, equations (8.7) and (8.9) are still valid. As the non-planar object is given by the sum of all these slices, these equations must be valid also for the entire rigid body. Similarly, the kinetic energy of the whole body is still given by equation (8.10).

Moreover, also equations (8.11) and (8.12) still hold.
Finally, as already mentioned, the **parallel-axis theorem** is still *valid* while the **perpendicular-axis theorem** *isn't.*

> Remark 8.8. In the next section, we'll see that if the conditions given above, namely
> (i) the non-planar object rotates around an axis parallel to the z-axis,
> (ii) we are concerned only with L_z, not L_x or L_y,
>
> are not met, then we need to modify our treatment.

Example 8.2. Wheel rolling on a plane (without skidding)

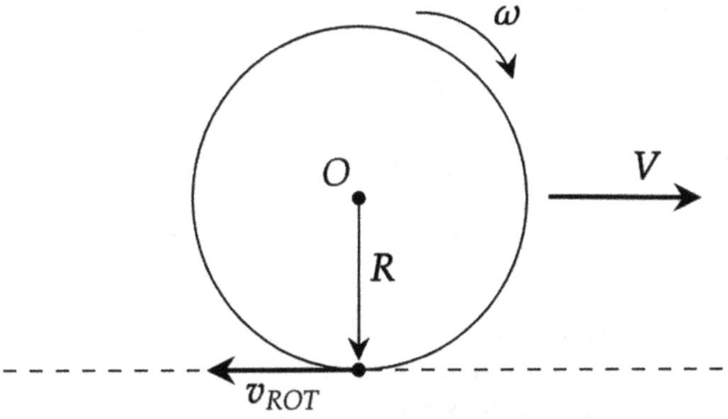

A rolling wheel can be seen as undergoing both a translational and rotational motion. The translational (or *linear*) part moves the centre of mass, while the radial part gives a rotation to the body around its centre of mass.
Consider a wheel rolling without skidding and with speed V. Additionally, the wheel has radius R and is rotating with angular speed ω. Since the wheel is not skidding, the instantaneous velocity v_A at the point of contact A at the bottom of the wheel must be zero, i.e.

$$v_A = v_{\text{LIN}} - v_{\text{ROT}} = V - R\omega = 0.$$

Therefore, it follows that there is no skidding if $V = R\omega$.

> **Definition 8.3.2: Pure roll**
> A rigid body which is rolling without slipping or skidding is said to be undergoing a **pure roll** motion. The mathematical condition for pure roll is that the linear and rotational speeds are equal.
> In formulae,
>
> $$V_{\mathrm{LIN}} = V_{\mathrm{ROT}}. \tag{8.15}$$

> Remark 8.9. Note that for a wheel of radius R rotating with angular velocity ω it is always true that $V_{\mathrm{ROT}} = \omega R$. However, if it is undergoing a pure roll motion, then also $V_{\mathrm{LIN}} = \omega R$.

Let's compute the kinetic energy of the wheel:

$$\begin{aligned} K = K_{\mathrm{LIN}} + K_{\mathrm{ROT}} &= \frac{1}{2}MV^2 + \frac{1}{2}I_0\omega^2 \\ &= \frac{1}{2}M(\omega R)^2 + \frac{1}{2}I_0\omega^2 = \frac{1}{2}\omega^2(I_0 + MR^2) = \frac{1}{2}\omega^2 I_A, \end{aligned}$$

where I_0 and I_A are the moments of inertia, respectively, around an axis through the centre of mass and around an axis through the point A. Note that both axes are perpendicular to the plane of the wheel.

Thus, the wheel can be considered as momentarily rotating about the point of contact A with the surface with angular velocity ω.

As we'll soon see in the following subsection, the moment of inertia of the wheel, which can be approximated as a ring, around an axis passing through the centre of mass and perpendicular to the plane of the wheel is $I_0 = MR^2$. Therefore, we can express the kinetic energy as

$$\frac{1}{2}\omega^2(I_0 + MR^2) = \frac{1}{2}\omega^2(MR^2 + MR^2) = M\omega^2R^2 = MV^2.$$

Exercise 8.4. What condition is satisfied when a rotating rigid body rolls without slipping or skidding?

Consider a wheel of radius R which is rotating without slipping or skidding and with angular velocity ω. The speed of its centre of mass is V. What's the instantaneous speed of the point of contact with the surface B at the bottom of the wheel? What's the instantaneous speed of the point T at the top of the wheel?

◆

8.3.6 Calculating moments of inertia

We'll now show how to calculate the moment of inertia of typical objects. Here we'll only consider situations where the conditions given in the previous remark (remark 8.8) hold.

Therefore, for a planar object in the x–y plane, we have that

$$I_z = \int r_\perp^2 \, \mathrm{d}m = \int r_\perp^2 \, \sigma(x, y) \, \mathrm{d}x \, \mathrm{d}y, \tag{8.16}$$

where $r_\perp$ represents the perpendicular distance between the point in the rigid body and the axis.

We adopt the conventions for densities mentioned in table 8.1. In the following, unless stated otherwise, we consider bodies with constant densities.

Example 8.3. Ring of mass M and radius R

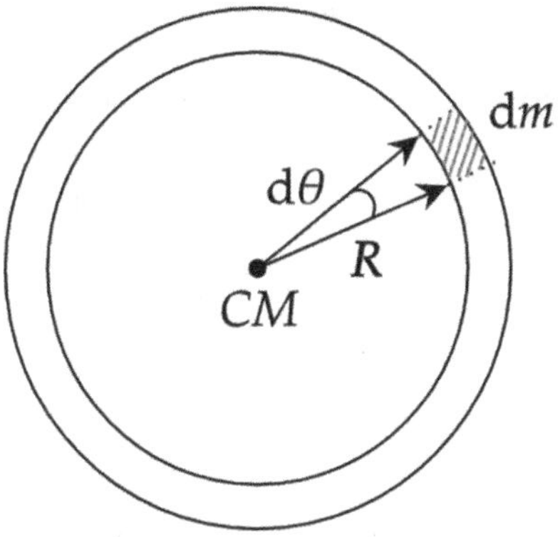

The linear density of a ring of mass M and radius R is given by $\lambda = \frac{M}{2\pi R}$, so an infinitesimal element of mass is given by $\mathrm{d}m = \frac{M}{2\pi R}\,\mathrm{d}r$, where $\mathrm{d}r = R\,\mathrm{d}\theta$ (as shown in the image).

All the points have the same distance R from the centre, therefore $r_\perp = R$. Thus,

$$I = \int r_\perp^2 \, \mathrm{d}m = \int_0^{2\pi} R^2 \frac{M}{2\pi \cancel{R}} \cancel{R} \, \mathrm{d}\theta = \frac{MR^2}{2\pi} \int_0^{2\pi} \mathrm{d}\theta = MR^2.$$

The result we have obtained above is the same as the moment of inertia calculated at a distance R from a point-like object with the same mass M.

Example 8.4. Uniform disk with mass M and radius R

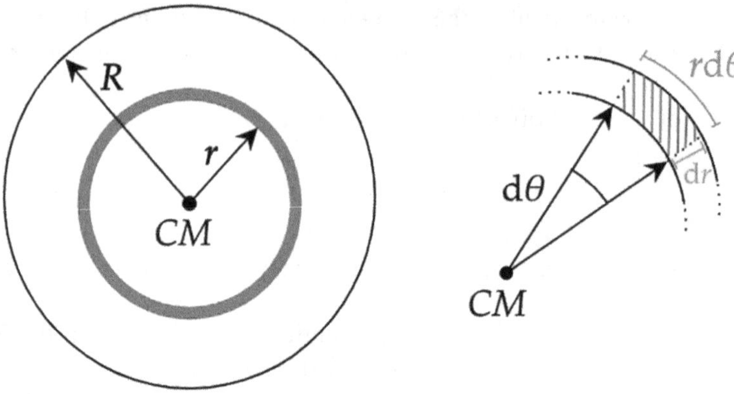

Now we are not only considering a circumference, i.e. a ring, but rather a whole circle, i.e. a 2D disk as seen in the image.

Suppose a uniform surface density σ. In polar coordinates, we have that $dm = \sigma\, r\, dr\, d\theta$. Therefore, the moment of inertia of the body is given by

$$I = \int r^2\, dm = \int_0^{2\pi} \int_0^R r^2\, \sigma\, r\, dr\, d\theta$$
$$= \sigma \int_0^{2\pi} d\theta \int_0^R r^3\, dr = \sigma \times 2\pi \times \frac{R^4}{4} = \frac{1}{2}\pi\sigma R^4.$$

Since the mass of the body is given by

$$M = \int dm = \sigma \int_0^{2\pi} d\theta \int_0^R r\, dr = \sigma \times 2\pi \times \frac{R^2}{2} = \pi\sigma R^2,$$

we can rewrite the moment of inertia as $I = \frac{MR^2}{2}$.

Note that, in the above case, we could have found the mass of the body without integrating, by simply using the definition of uniform surface density. However, if the density is not uniform, we need to integrate to find an expression for the total mass.

Example 8.5. Thin uniform rod of mass M and length L

We will consider two different cases. It is important to remember that we can *choose* where to put the *origin of our coordinate system*. Choosing a convenient point can greatly simplify the treatment of the problem.

Case 1. Rotation around one of the ends

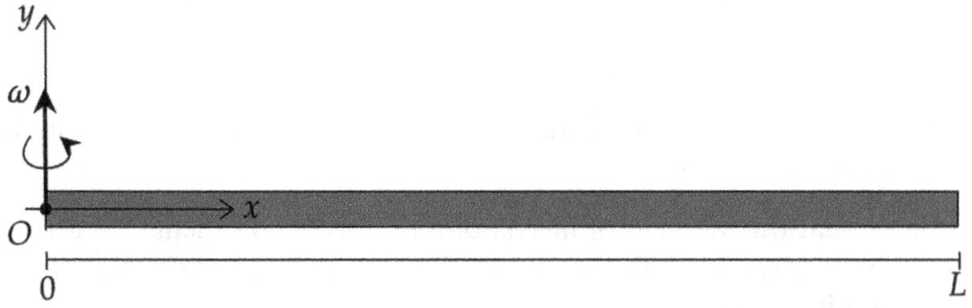

We conveniently choose the origin of our Cartesian coordinate system to coincide with one end of the rod, which extends in the positive x direction (going from $x = 0$ to $x = L$). The rotation is thus around the y-axis. The uniform linear density is $\lambda = M/L$. Therefore,

$$I = \int x^2 \, dm = \frac{M}{L}\int_0^L x^2 \, dx = \frac{M}{L} \times \frac{L^3}{3} = \frac{1}{3}ML^2.$$

Case 2. Rotation around the centre of mass

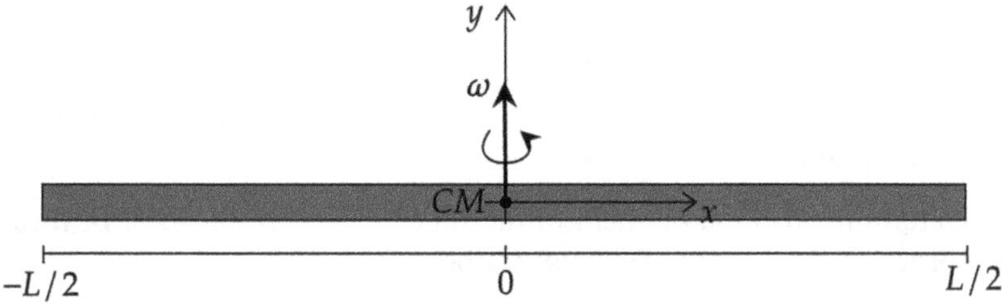

Since the density is uniform, the centre of mass will coincide with the centre of the rod as seen in exercise 8.1. Thus, we put the origin of our coordinate system on the centre of mass, such that the thin rod goes from $x = -L/2$ to $x = L/2$. The uniform linear density is always $\lambda = \frac{M}{L}$. We have that

$$I_{\text{CM}} = \frac{M}{L}\int_{-L/2}^{L/2} x^2 \, dx = \frac{M}{L}\left[\frac{x^3}{3}\right]_{-L/2}^{L/2} = \frac{M}{L} \times \frac{L^3}{12} = \frac{1}{12}ML^2.$$

Notice that the first case could be obtained from the second using the parallel axis theorem. The moment of inertia about the end of the stick is given by the sum of the one about the centre of mass plus the total mass multiplied by the distance squared. So:

$$I = I_{\text{CM}} + M(L/2)^2 = \frac{1}{12}ML^2 + \frac{1}{4}ML^2 = \frac{1}{3}ML^2.$$

Discussion: why does the first case have a higher moment of inertia?

We notice that in the first case we obtain a higher value for the moment of inertia. Why does this happen? The moment of inertia represents the resistance that a body opposes to a change in its movement. Since in the first case there is more mass far away from the centre of rotation, there is more resistance and therefore a higher moment of inertia.

Exercise 8.5.

(i) Write the definition of the moment of inertia, I, of a rigid body, in the form of an integral.

(ii) Show that the moment of inertia of a uniform rod of length L and mass M about an axis through the end of the rod and perpendicular to the plane containing the rod is

$$I = \frac{1}{3}ML^2.$$

The rod has a negligible height and width.

(iii) Show that the moment of inertia of a uniform square plane of mass M and side L about an axis O passing through its centre and at right angles to the plane is

$$I_0 = \frac{ML^2}{6}.$$

(iv) Find also the moment of inertia of the square plane described in the previous part about a parallel axis A through the middle of one side (as in the diagram).

(v) Determine also the moment of inertia about a parallel axis, B, through the corner of the square (see diagram below).

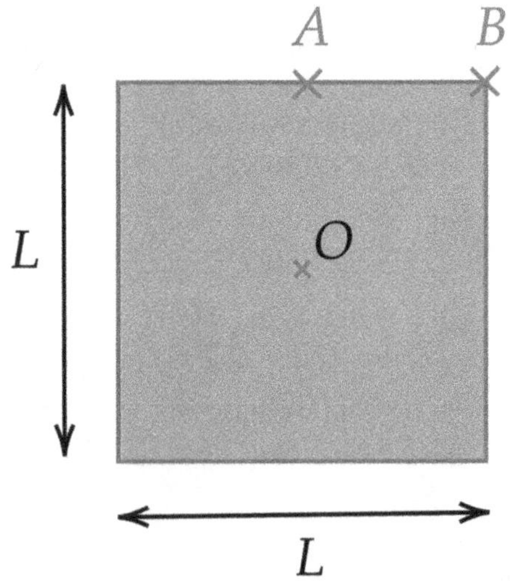

Exercise 8.6. A ring is suspended in the vertical plane from an axis passing through the point A, as seen in the figure below.

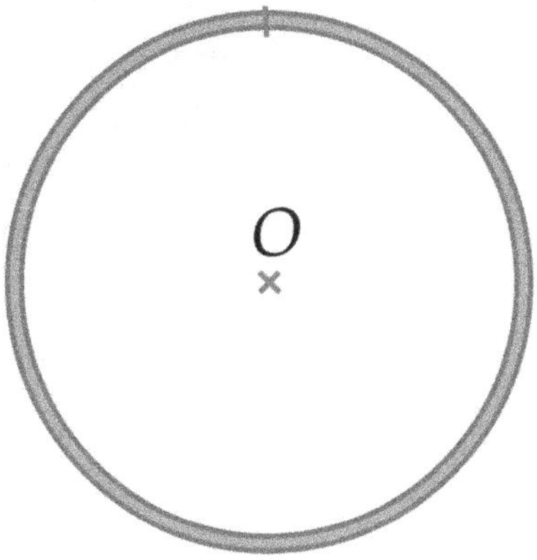

Determine, by integration, the moment of inertia of the ring about its centre O.

Considering that the ring is fixed at point A determine the moment of inertia about this axis.

Exercise 8.7. A stick with negligible thickness and length 2ℓ has a density that can be described as $\lambda = ax^2$, where x is the distance from the centre of mass and a is a constant.

(i) Calculate the mass and the momentum of inertia of the stick around its centre of mass.

(ii) The stick feels a force F for the total duration of 1 second, perpendicular to the length of the stick, and applied at one extreme. Calculate the momentum of the stick, and its angular momentum about its centre of mass, after the application of the force, neglecting the movement of the system during the time the force is applied.

◆

8.4 General motion of a non-planar object in 3D space

Let's now study the most general motion for a non-planar rigid body, i.e. *the axis of rotation is not parallel to the z-axis.*

We already know from theorem 8.1.2 that the velocity of a point rotating with angular speed ω' and at position $\boldsymbol{r}'$ is given by $\boldsymbol{v}' = \boldsymbol{\omega}' \times \boldsymbol{r}'$.

Since the most general motion for a rigid body is the combination of a translation with uniform velocity $\boldsymbol{V}$ and a rigid rotation about some axis, we can generalize the previous equation to

$$\boldsymbol{v} = \boldsymbol{V} + \boldsymbol{\omega}' \times \boldsymbol{r}'. \tag{8.17}$$

Remark 8.10. It is important to remember that $\boldsymbol{r}'$ is the position vector of a point in the rigid body *with respect to the centre of rotation*. See remark 8.5.

8.4.1 Angular momentum and the inertia matrix

As usual, we begin by considering a pure rotation around an axis through the origin, and then move on to extend our treatment to a general motion, a motion composed of a translation of the centre of mass and a rotation around an axis through the centre of mass itself.

Rotation around an axis through the origin

As previously shown, the angular momentum of the entire body is given by $\boldsymbol{L} = \int \boldsymbol{r} \times \boldsymbol{v}\, dm$. By then expressing the velocity as $\boldsymbol{v} = \boldsymbol{\omega} \times \boldsymbol{r}$, we obtain

$$\boldsymbol{L} = \int \boldsymbol{r} \times (\boldsymbol{\omega} \times \boldsymbol{r})\, \mathrm{d}m, \tag{8.18}$$

where the integration runs over the volume of the body.

For a discrete system of point masses, we would need to replace the integral with a sum:

$$\boldsymbol{L} = \sum_i m_i \boldsymbol{r}_i \times (\boldsymbol{\omega}_i \times \boldsymbol{r}_i). \tag{8.19}$$

To express the vector triple product in a more convenient form, we need to use the vector triple product expansion.

Remark 8.11. Given three vectors $\boldsymbol{a}$, $\boldsymbol{b}$, $\boldsymbol{c}$, we have that $\boldsymbol{a} \times (\boldsymbol{b} \times \boldsymbol{c}) = (\boldsymbol{a} \cdot \boldsymbol{c})\boldsymbol{b} - (\boldsymbol{a} \cdot \boldsymbol{b})\boldsymbol{c}$.

Let's apply this general formula to our case. Then,

$$\boldsymbol{r} \times (\boldsymbol{\omega} \times \boldsymbol{r}) = (\boldsymbol{r} \cdot \boldsymbol{r})\boldsymbol{\omega} - (\boldsymbol{r} \cdot \boldsymbol{\omega})\boldsymbol{r} = r^2\boldsymbol{\omega} - (\boldsymbol{r} \cdot \boldsymbol{\omega})\boldsymbol{r}.$$

Substituting this expression in equation (8.18), we obtain

$$\boldsymbol{L} = \int [r^2\boldsymbol{\omega} - (\boldsymbol{r} \cdot \boldsymbol{\omega})\boldsymbol{r}]\, \mathrm{d}m. \tag{8.20}$$

To try to make sense of this equation, we can express $\boldsymbol{r}$ and $\boldsymbol{\omega}$ in terms of their components, i.e.

$$\boldsymbol{r} = x\hat{\boldsymbol{i}} + y\hat{\boldsymbol{j}} + z\hat{\boldsymbol{k}}$$
$$\boldsymbol{\omega} = \omega_x\hat{\boldsymbol{i}} + \omega_y\hat{\boldsymbol{j}} + \omega_z\hat{\boldsymbol{k}},$$

and then let's consider one component of angular momentum, say L_z:

$$\begin{aligned} L_z = \boldsymbol{L} \cdot \hat{z} &= \int [r^2\omega_z - (x\omega_x + y\omega_y + z\omega_z)z]\,\mathrm{d}m \\ &= \int [(r^2 - z^2)\omega_z - xz\omega_x - yz\omega_y]\,\mathrm{d}m. \end{aligned}$$

If we repeat this calculation for every component of the angular momentum, we will obtain the following matrix equation:

$$\begin{pmatrix} L_x \\ L_y \\ L_z \end{pmatrix} = \int \begin{pmatrix} (r^2 - x^2) & -xy & -xz \\ -xy & (r^2 - y^2) & -yz \\ -xz & -yz & (r^2 - z^2) \end{pmatrix} \begin{pmatrix} \omega_x \\ \omega_y \\ \omega_z \end{pmatrix} \mathrm{d}m. \tag{8.21}$$

Definition 8.4.1: Angular momentum and the moment of inertia matrix

The angular momentum of a rigid body rotating around an axis passing through the origin is given by

$$\boldsymbol{L} = \mathbf{I}\,\boldsymbol{\omega}, \tag{8.22}$$

where $\mathbf{I}$ is called the **moment of inertia matrix**[1] and is defined as

$$\mathbf{I} \equiv \begin{pmatrix} \int (r^2 - x^2) & -\int xy & -\int xz \\ -\int xy & \int (r^2 - y^2) & -\int yz \\ -\int xz & -\int yz & \int (r^2 - z^2) \end{pmatrix}. \tag{8.23}$$

Note that $\mathrm{d}m$ *has been omitted from the integrals for the sake of clarity.* Remember that an infinitesimal element of mass can be expressed as $\mathrm{d}m = \rho\,\mathrm{d}V$, where ρ and V are, respectively, the density and the volume of the body.

[1] Technically, it is a tensor, but all tensors of order 2 can be represented by a matrix.

Remark 8.12. There are three important things to point out here:

(i) No assumptions have been made regarding the shape of the rigid body.

(ii) If the rotation is, for instance, only around the z-axis, and we just want the z component of the angular momentum, then we have $L_z = I_{zz}\,\omega_z$.

(iii) The angular momentum vector and the angular velocity vector are not necessarily parallel, but throughout this course we will only consider situations in which they are.

Every matrix with nonzero determinant can be diagonalised, namely it is possible to find a coordinate system where only the diagonal terms are nonzero:

$$\mathbf{I} = \begin{pmatrix} I_{xx} & 0 & 0 \\ 0 & I_{yy} & 0 \\ 0 & 0 & I_{zz} \end{pmatrix}. \tag{8.24}$$

The axes of the reference frame where the matrix of inertia is diagonal are called **principal axes**. If a body rotates about a principal axis, only the relevant component of the moment of inertia matters. So, if the object rotates around the z-axis, we have

$$L_z = I_{zz}\omega_z. \tag{8.25}$$

If a system has a symmetry, the axis of symmetry will be a principal axis. This justifies the fact that in previous chapters we have considered for instance the motion of a cylinder about its axis using a scalar moment of inertia instead of the full matrix: in all cases considered so far, the bodies were rotating about a principal axis.

Let's now find an expression for the kinetic energy of the body.

We have that

$$K = \int \frac{1}{2} v^2 \,\mathrm{d}m = \frac{1}{2}\int (\boldsymbol{\omega} \times \boldsymbol{r}) \cdot (\boldsymbol{\omega} \times \boldsymbol{r}) \,\mathrm{d}m.$$

Recalling that for the properties of vectors $v^2 \equiv \boldsymbol{v} \cdot \boldsymbol{v}$ and using the definition for velocity given in theorem 8.1.2.

Remark 8.13. We can express this in a better way using *Lagrange's identity* in three dimensions, i.e. given two vectors $\boldsymbol{a}$, $\boldsymbol{b}$, we have that $(\boldsymbol{a} \times \boldsymbol{b}) \cdot (\boldsymbol{a} \times \boldsymbol{b}) = |\boldsymbol{a} \times \boldsymbol{b}|^2 = (\boldsymbol{a} \cdot \boldsymbol{a})(\boldsymbol{b} \cdot \boldsymbol{b}) - (\boldsymbol{a} \cdot \boldsymbol{b})^2$.

Applying this identity to our case, we obtain

$$(\boldsymbol{\omega} \times \boldsymbol{r}) \cdot (\boldsymbol{\omega} \times \boldsymbol{r}) = (\boldsymbol{\omega} \cdot \boldsymbol{\omega})(\boldsymbol{r} \cdot \boldsymbol{r}) - (\boldsymbol{\omega} \cdot \boldsymbol{r})^2 = \omega^2 r^2 - (\boldsymbol{\omega} \cdot \boldsymbol{r})^2.$$

Let's rewrite this in a slightly different way:

$$(\omega \times r) \cdot (\omega \times r) = r^2\omega^2 - (r \cdot \omega)(\omega \cdot r).$$

By comparing the equation above with the integrand of equation (8.20), you will realise that the two equations can be related by taking the dot product of $\boldsymbol{\omega}$ with the latter. Therefore, we have

$$(\omega \times r) \cdot (\omega \times r) = \omega \cdot [r^2\omega - (r \cdot \omega)r].$$

Substituting the previous equation into the expression found for the kinetic energy we have

$$K = \frac{1}{2}\int \omega \cdot [r^2\omega - (r \cdot \omega)r]\, \mathrm{d}m = \frac{1}{2}\omega \cdot \int [r^2\omega - (r \cdot \omega)r]\, \mathrm{d}m.$$

Again, comparing with equation (8.20), we finally obtain

$$K = \frac{1}{2}\omega \cdot L = \frac{1}{2}\omega \cdot \mathbf{I}\omega.$$

Definition 8.4.2: Kinetic energy

The kinetic energy of a rigid body rotating around an axis passing through the origin is given by

$$K = \frac{1}{2}\omega \cdot L = \frac{1}{2}\omega \cdot \mathbf{I}\omega. \tag{8.26}$$

Notice that in the general case $\mathbf{I}$ is the matrix of inertia, and ω is a vector, so the above expression is the scalar product between the vector ω and the product between the matrix of inertia $\mathbf{I}$ and again the angular momentum ω. But if the motion happens around a principal axis, the above expression can again be simplified as the product between three scalars.

8.4.2 General motion in 3D

At this point, extending our treatment to a general motion, i.e. a sum of translation and rotation, is fairly simple. The procedure is almost the same as in subsection 8.3.2, the only differences being in the rotational part, which has just been discussed in the previous subsection.

Thus, we will not repeat the whole derivation, but simply state the results for a non-planar rigid body undergoing a general motion and rotating about any axis. Note that, in doing so, we are using theorem 8.1.1. However, while it's true we could choose any point P in the body for the theorem, the only one that gives us some useful results is the centre of mass.

Note also that we keep using the same conventions as in remark 8.5.

Definition 8.4.3: Angular momentum and kinetic energy

The angular momentum and the kinetic energy of a non-planar rigid body undergoing a general motion are, respectively, given by

$$\boldsymbol{L} = \boldsymbol{R} \times \boldsymbol{P} + \boldsymbol{L}_{\text{CM}} = \boldsymbol{R} \times \boldsymbol{P} + \mathbf{I}'\boldsymbol{\omega}', \tag{8.27}$$

$$K = \frac{1}{2}MV^2 + \frac{1}{2}\boldsymbol{\omega}' \cdot \boldsymbol{L}_{\text{CM}} = \frac{1}{2}MV^2 + \frac{1}{2}\boldsymbol{\omega}' \cdot \mathbf{I}'\boldsymbol{\omega}'; \tag{8.28}$$

where $\boldsymbol{L}_{\text{CM}}$ and $\mathbf{I}'$ are, respectively, the angular momentum and the moment of inertia matrix relative to the centre of mass.

Exercise 8.8. A rigid body is composed of N particles with the ith particle having a mass m_i and at a distance from the axis of rotation (A) r_i. Give an expression for the moment of inertia measured about A in terms of m_i and r_i, and for the rotational kinetic energy, given that the angular speed of rotation is ω.

What is the moment of inertia of a uniform ring of mass M and radius R about an axis through the centre of the ring and perpendicular to the plane of the ring?

Determine the moment of inertia about another axis which is also perpendicular to the plane of the ring, but passes through the ring itself, displaced by a distance R from the centre of the ring.

◆

8.4.3 Correspondences between translation and rotation

There are interesting similarities between translational and rotational motions of the rigid body, so different variables have similar meanings for the two cases:

Translational motion		Rotational motion	
Mass	m	I	Moment of inertia
Velocity	v	ω	Angular velocity
Linear momentum	$\boldsymbol{p} = m\boldsymbol{v}$	$\boldsymbol{L} = \boldsymbol{r} \times \boldsymbol{p} = \mathbf{I} \cdot \boldsymbol{\omega}$	Angular momentum
Linear kinetic energy	$\frac{1}{2}mv^2$	$\frac{1}{2}\boldsymbol{\omega} \cdot \mathbf{I} \cdot \boldsymbol{\omega}$	Rotational kinetic energy
Force	$\boldsymbol{F} = \frac{\mathrm{d}\boldsymbol{p}}{\mathrm{d}t}$	$\boldsymbol{\tau} = \frac{\mathrm{d}\boldsymbol{L}}{\mathrm{d}t}$	Torque
Force	$\boldsymbol{F} = m\frac{\mathrm{d}\boldsymbol{v}}{\mathrm{d}t}$	$\boldsymbol{F} = I\frac{\mathrm{d}\omega}{\mathrm{d}t}$	Force

8.4.4 Compound pendulum

Up to this point, we have considered an idealised version of the pendulum. We'll now look at a situation a bit more realistic: a pendulum that is extended and has a non-negligible mass (figure 8.6).

Consider an object of any shape rotating around a pivot point in the body, say O, and oscillating around this point. The body has mass M and its centre of mass is at a

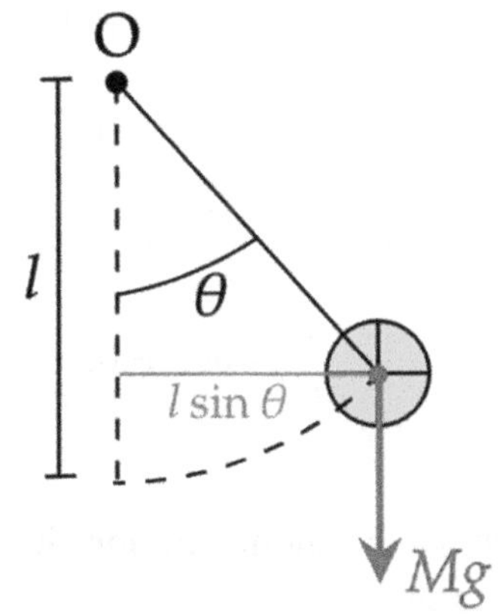

Figure 8.6. Schematic representation of a compound pendulum with mass M subject to gravity.

distance l from the pivot point. The position vector of the centre of mass relative to the pivot point O forms an angle θ with the equilibrium position. Moreover, the object is subject to gravity.

To solve this problem, we'll need to use the concept of torque, which was introduced in chapter 5.

Recall: If we consider a force $\boldsymbol{F}$ acting on the particle, the moment of force or **torque** about the centre is defined as $\boldsymbol{\tau} = \frac{\mathrm{d}\boldsymbol{L}}{\mathrm{d}t} = \boldsymbol{r} \times \boldsymbol{F}$.

The torque produced by gravity on the body is given by

$$\tau = |\boldsymbol{r} \times \boldsymbol{F}| = -Mg\, l \sin\theta.$$

> Remark 8.14. In the previous equation there is a negative sign because the torque is in the direction of decrease of the angle θ.

Assuming we know the moment of inertia of the body around O, we can also express the torque as

$$\tau = \frac{\mathrm{d}L}{\mathrm{d}t} = \frac{\mathrm{d}}{\mathrm{d}t}(I\omega) = I\frac{\mathrm{d}\omega}{\mathrm{d}t} = I\frac{\mathrm{d}^2\theta}{\mathrm{d}t^2}.$$

Therefore, the equation of motion for rotation about the axis through O is

$$I\frac{\mathrm{d}^2\theta}{\mathrm{d}t^2} = -Mg\, l \sin\theta. \tag{8.29}$$

If the angle is always small, then $\sin\theta \simeq \theta$ (in radians). It follows that

$$I\frac{d^2\theta}{dt^2} + Mg\,l\,\theta = 0$$
$$\frac{d^2\theta}{dt^2} + \frac{Mg\,l\,\theta}{I} = 0$$
$$\frac{d^2\theta}{dt^2} + \omega^2\theta = 0,$$

where $\omega = \sqrt{\frac{Mg\,l}{I}}$.

This is the equation for simple harmonic motion (in θ) with period

$$T = \frac{2\pi}{\omega} = 2\pi\sqrt{\frac{I}{Mg\,l}}. \tag{8.30}$$

Let's express the moment of inertia of the body around O as $I = Mk^2$, where $k \equiv \sqrt{\frac{I}{M}}$ is called the **radius of gyration**. Then,

$$T = 2\pi\sqrt{\frac{Mk^2}{Mg\,l}} = 2\pi\sqrt{\frac{k^2}{g\,l}}. \tag{8.31}$$

Let's look at how the previous equation is simplified in the case of a simple pendulum, i.e. a point mass attached at the end of a massless inextensible string. In this case, given a mass M and a string of length l, the moment of inertia is $I = Ml^2$, thus $k = l$. The period becomes

$$T = 2\pi\sqrt{\frac{l^2}{g\,l}} = 2\pi\sqrt{\frac{l}{g}}.$$

as seen in the classic point-like treatment.

8.4.5 Effect of external force

What will happen if we try to apply an external force $\boldsymbol{F}$ to a *free* rigid body, given that the force does not act on its centre of mass?

As shown in figure 8.7, to understand the motion that will follow, we can add a pair of equal and opposite forces, $\boldsymbol{F}$ and $-\boldsymbol{F}$, that are, respectively, parallel and anti-parallel to the original force, and act on the centre of mass. Note that we can do this because the sum of the two forces acting on the centre of mass is zero, since they have the same magnitude but act in opposite directions.

In this new (but equivalent!) situation, we have that the force $\boldsymbol{F}$ through the centre of mass accelerates the body according to

$$F = M\frac{d^2R}{dt^2}. \tag{8.32}$$

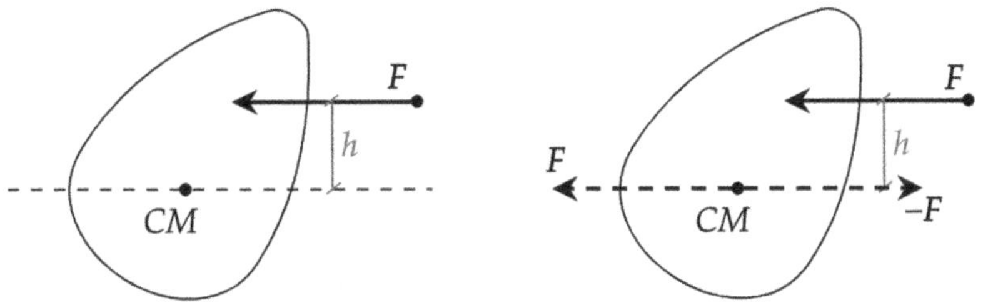

Figure 8.7. External force acting on a rigid body not through its CM.

Instead, the two remaining forces—the original force $\boldsymbol{F}$ and the force $-\boldsymbol{F}$ acting on the centre of mass—form a couple of forces. Supposing there is a distance h between these two forces, as shown in figure 8.7, then the couple produces a torque of $\tau = Fh$ and an angular acceleration of

$$\tau = \frac{\mathrm{d}L}{\mathrm{d}t} = I_0 \frac{\mathrm{d}\omega}{\mathrm{d}t} \tag{8.33}$$

about an *axis through the centre of mass* and *perpendicular* to the plane defined by the direction of the force and the centre of mass. I_0 and ω are, respectively, the moment of inertia of the body around the centre of mass and the angular velocity of the body.

8.4.6 Simple theory of the gyroscope

> **Definition 8.4.4: Gyroscope**
> A gyroscope is a device formed by a wheel or disk that can spin around an axis perpendicular to it, which is also free to rotate.

Gyroscopes are very particular objects that under certain conditions seem to defy gravity. Before talking about them, let's first think about a top, which is a more familiar object, to help us understand their unusual behaviour. If a top is not spinning and you try to put in a vertical position, it will fall due to gravity. However, if we spin it, it won't fall anymore. If we observe a top spinning, we'll notice there are two rotations taking place simultaneously: the top is spinning around its central axis but is also spinning around an axis perpendicular to the plane on which the top is moving.

A gyroscope behaves in a very similar way. If we try to make it remain in a vertical position, it will fall. But, analogously to the top, if we make the gyroscope's disk spin, the whole object will start rotating around an axis perpendicular to the surface on which we put it, and thus defy gravity.

So, how come tops and gyroscopes behave in such a way? Both objects are trying to conserve angular momentum. They are resisting a change in their motion as they

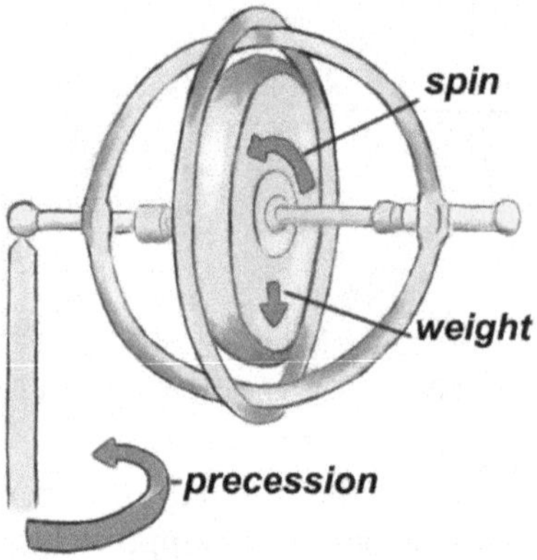

Figure 8.8. A spinning gyroscope set horizontally with one end on a support.

would for linear momentum. Additionally, the gravitational force acting on the bodies forces them to begin another rotation, as mentioned previously.

Let's get more quantitative. Consider a gyroscope whose disk has a moment of inertia of I_0 and which is spinning about the perpendicular axis, as in figure 8.8. *To be clear, the disk does not slide along the axis, it just spins around it.*

If we imagine putting one end of the axis horizontally on a support at a distance x from the centre of the disk, then the torque of the weight about the pivot is $\tau = Mg\ x$. The direction of the torque vector $\boldsymbol{\tau}$ is into the paper. Since $\tau = \frac{\mathrm{d}L}{\mathrm{d}t}$, then we can write that $\mathrm{d}L = \tau \mathrm{d}t$. This quantity $\mathrm{d}L$ represents the change in the angular momentum of the spinning body in time $\mathrm{d}t$ in the direction of the vector $\boldsymbol{\tau}$.

What is happening here is that the spin axis, coinciding with the angular momentum vector, is rotating around the pivot because of the torque acting on the system. *Note that, in general, the angular velocity of the disk rotating around the axis and the angular velocity of the axis, and with it the whole body, rotating around the pivot are different.* It is also important to note that the axis remains horizontal with respect to the support at one end. Moreover, the angular momentum's magnitude remains the same. Therefore, what is changing is only the direction of angular momentum with respect to the pivot.

Take a moment to notice how surprising this behaviour is! The object is defying gravity, it does not fall even though only one of its ends is set on a support. Instead, it remains horizontal and starts rotating. This would not happen if the disk in the gyroscope were not spinning in the first place. This is because there wouldn't be any angular momentum whose direction could change and so allow it to reach stability. The object would simply fall.

Thus, viewed from above we obtain the vector diagram in figure 8.9. $\boldsymbol{L}(t)$ is the angular momentum at time t. As can be seen from the diagram, the angular momentum vector $\boldsymbol{L}$ is rotated through an angle $\mathrm{d}\alpha$ in time $\mathrm{d}t$.

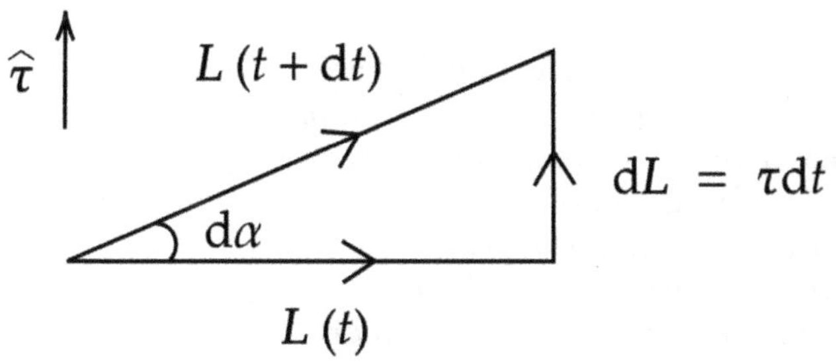

Figure 8.9. Vector diagram of the gyroscope's angular momentum as seen from above at two different times, t and $t + \mathrm{d}t$.

> Remark 8.15. Recall that the value of an angle in radians is given by the ratio of the length of the arc subtended by the angle and the radius of the arc, i.e. $\alpha = \frac{l}{R}$. For infinitesimal quantities, we have that the arc becomes a straight line.

In this case, as shown in the aforementioned figure, the arc subtended by the angle is the vector $\mathrm{d}\boldsymbol{L}$, while the radius is the magnitude of the angular momentum vector, L. Therefore, the infinitesimal angle is given by

$$\mathrm{d}\alpha = \frac{\mathrm{d}L}{L} = \frac{\tau\,\mathrm{d}t}{L} = \frac{Mg\,x\,\mathrm{d}t}{I_0\,\omega}.$$

Thus, the rate of rotation of $\boldsymbol{L}$, or **rate of precession**, is given by

$$\omega_p = \frac{\mathrm{d}\alpha}{\mathrm{d}t} = \frac{Mg\,x}{I_0\,\omega}. \tag{8.34}$$

ω_p is also called the **precessional frequency** and is the frequency with which the axis, and thus the whole body, rotates around the pivot.

> Remark 8.16. This theory holds only if $\omega_p \ll \omega$, i.e. when the disk spins rapidly.

Gyroscopes are widely used in many different fields. Their particular—so unusual—characteristics allow us to check the stability of objects. For instance, they are used for airplane orientation, but also by robots such as the Mars rover.

Exercise 8.9. Consider a solid cylinder and a hollow cylinder (only a cylindrical shell of negligible thickness), both of mass M, radius R and length L. Assume constant densities throughout this exercise. Calculate the cylinders' moments of inertia about their central axis. Why does the hollow cylinder have a higher moment of inertia than the solid cylinder?

Now consider two solid cylinders rolling down a plane inclined with an angle θ with respect to the horizontal. The cylinders are undergoing a pure roll motion, i.e. they are not skidding or slipping. Additionally, one of the cylinders has mass M and the other one has mass $30M$. For all other aspects, the cylinders are identical. Which one will arrive first at the end of the plane, if their starting positions are the same? You can ignore air resistance. [*Hint*: calculate the torque with respect to the point of contact between the plane and the cylinders, and equate it to the product between the moment of inertia and angular acceleration.]

Now, instead, consider two cylinders, one solid and one hollow, exactly as described in part (i). Suppose again that the two cylinders are rolling down a plane inclined with an angle α to the horizontal. As before, the cylinders are undergoing a pure roll motion, have the same starting point, and you can ignore air resistance. Which one will arrive first and why? [*Hint*: use the same approach as in part (ii) and think about the different moments of inertia of the two bodies.]

Exercise 8.10. A wheel of mass M, radius R and moment of inertia about its centre of I_0 is rolling, *without skidding*, across a rough surface.

(i) State the relationship between the linear velocity v and the angular velocity ω for the condition described above.

(ii) Determine the total kinetic energy of the wheel, **only** in terms of I_0, R, v and M.

(iii) The wheel is then placed on an inclined plane at an angle θ with the horizontal, and is released from rest. Draw a diagram of the forces acting on the wheel in the case of it rolling without sliding.

(iv) Use the conservation of mechanical energy to determine the wheel's linear velocity v, once the centre of mass will have travelled a distance l down the surface of the plane. [Note: the ball rolls because there is a frictional force creating a torque. Thus, mechanical energy is not truly conserved. However, here we assume that the amount of energy dissipated into heat is small enough that mechanical energy can be considered to be conserved.]

Exercise 8.11. A cat can be approximated as a cylinder of radius $r = 0.1$ m and mass $m = 3$ kg, plus a tail that points perpendicularly outwards from the axis of the cylinder, i.e. it is perpendicular to the curved surface of the cylinder. The tail starts at the edge of the cylinder (0.1 m from the axis) and can be approximated as a stick of length $\ell = 0.2$ m and mass $m_t = 0.2$ kg.

(i) State the parallel-axes theorem for moments of inertia.

(ii) Hence, calculate the total moment of inertia of the cat about the axis of symmetry of the cylinder, indicating separately the contributions from the main body and the tail.

(iii) We assume the tail and the main body of the cat can independently rotate around the axis of the cylinder. If the whole cat is at rest (with angular momentum zero) and the tail starts spinning in the anti-clockwise direction with respect to this axis with angular velocity ω_t, calculate the angular velocity ω_c that the main body of the cat (the cylinder) must acquire to conserve total angular momentum (for instance, to be able to fall back on its paws).

Exercise 8.12. A compound pendulum is made of a metal stick of length ℓ and mass m. This stick can rotate vertically around a pivot point P placed at distance d from the centre of the stick.

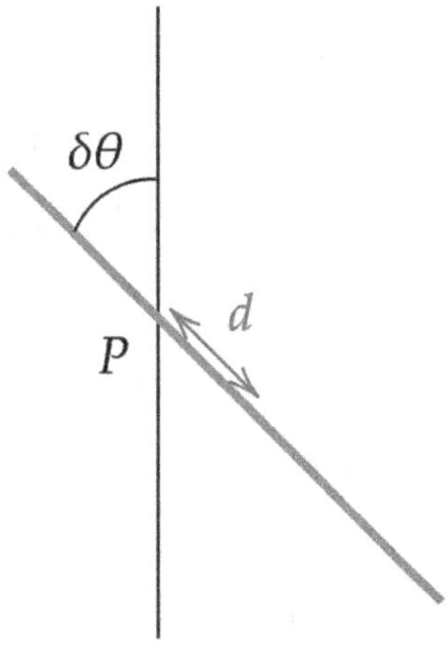

(i) Show that the moment of inertia of the stick about a perpendicular axis through the pivot point is

$$I = \frac{1}{12}m\ell^2 + md^2.$$

(ii) Derive the torque acting about the pivot point when the stick is displaced from its vertical equilibrium position through a small angle $\delta\theta$.

(iii) Hence, derive the frequency of small oscillations of the stick about the vertical position.

Exercise 8.13. A hand spinner is a toy with an equilateral triangle structure (whose mass can be neglected), able to rotate without friction around its centre (see diagram).

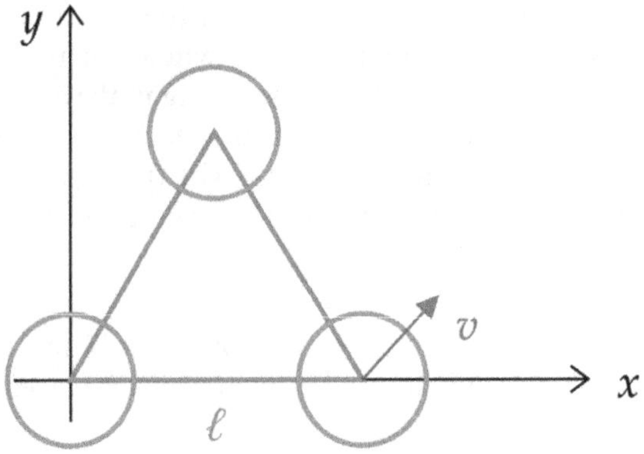

At the corners of the triangle (with side ℓ), three small discs with equal mass m are attached.

(i) Determine the expression for the moment of inertia of the system for two cases:

a) when the discs are considered as point-like objects;

b) when they are assumed to have radius $\ell/4$ and a uniform mass distribution.

It is convenient to use a Cartesian coordinate system with one disc at the origin and a second one on the x-axis, as shown in the figure. [*Hint:* you may use the formula for the moment of inertia about an axis passing through the centre for a disc with mass m, radius r, and a uniform mass density:

$$I = \frac{mr^2}{2},$$

and then apply the *parallel axis theorem*.]

(ii) Now, consider the point-like disc approximation. Holding the spinner firmly in the centre, one of the discs is given a speed v, in the direction perpendicular to the line connecting it to the centre. Determine the expression for the angular velocity and angular momentum of the system. Remember that the system is rigid, so all masses move at the same speed.

(iii) One of the discs is removed from the spinner (to simplify calculations, it is suggested to remove the disc at the top of the figure). Calculate, always in the point-like approximation, the new position of the centre of mass of the system when the disc is removed, and the moment of inertia about an axis parallel to z and passing through the geometrical centre, as well as another axis, parallel to z and passing through the new centre of mass.

(iv) The same linear velocity v perpendicular to the line connecting it to the centre is given to one disc. Calculate the angular momentum in the configuration with the disc removed, as well as the centrifugal force acting on the centre of the spinner due to the rotation. [*Hint:* for the last point, instead of using the equations of a solid body, consider the system made of two point-like objects, each of them rotating with uniform angular speed. Then you can either calculate the vector sum of the centrifugal forces acting on each of the two discs, or you can consider the force acting on the centre of mass of the two discs, as if the masses of both discs were concentrated there.]

Exercise 8.14. A windmill can be described as six identical blades of length ℓ, mass m and negligible width, equally separated by 60°, rotating around a pivot point, located at height $\ell + h$ above the ground. In the beginning, the windmill is not moving and there are no external forces. The lowest blade is vertical, so its lowest point will be at height h (initially unknown) above ground. All frictions should be neglected.

(i) Derive the moment of inertia of the windmill about the pivot.

(ii) A boy is sitting on the ground, at the side of the windmill, at a distance d from its base, along the direction defined by the plane containing the blades. He throws a stone with mass m_s with initial velocity v at an angle α with respect to the horizontal plane. Derive an expression for the distance d at which the boy should be sitting, such that the stone hits the bottom of the lowest blade when the stone is at its maximal height (so the vertical component of the stone's velocity is zero). Also derive an expression for h as a function of α and v.

(iii) The stone will hit the blade, and stick to it, putting the windmill in motion. Derive an expression for the initial angular velocity of the windmill after the collision (also considering the moment of inertia of the stone).

(iv) Derive an expression for the kinetic energy of the system before and after the collision. Neglecting the air resistance, find the maximal height that the stone will reach. What is the condition for the stone to reach the top of the windmill and do a full tour?

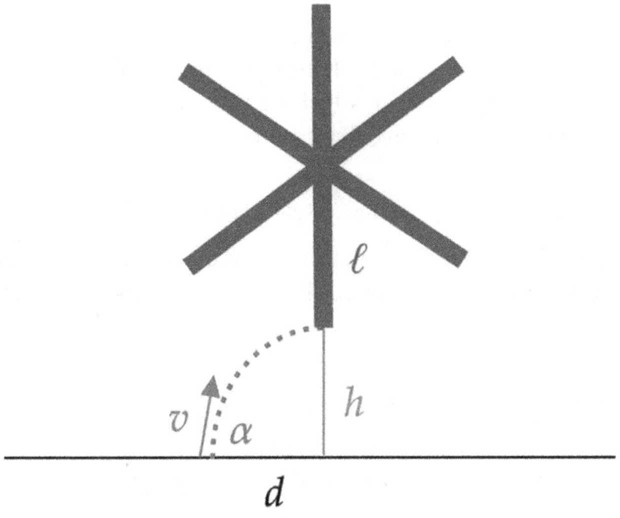

Chapter 9

Accelerating frames of reference

9.0 Accelerating reference frames

9.0.1 Inertial and non-inertial frames

Newton's laws, which have been discussed in chapter 2, only work for inertial frames. To treat accelerating and rotating frames of reference, which are **non-inertial** we will have to make some alterations to these principles.

> Definition 9.0.1: Non–inertial frame of reference
> This is a reference frame which has a non-uniform or accelerated motion with respect to a fixed one (or one moving at constant speed), so Newton's laws (in particular the principle of inertia) appear to be wrong.

In order to keep using Newton laws in non-inertial frames we need to introduce some new forces which are called: 'fictitious forces'. These are forces that an observer in an accelerating frame has to take into account in order to use $\boldsymbol{F} = m\boldsymbol{a}$ and get a correct answer for the acceleration $\boldsymbol{a}$.

9.0.2 Accelerating and rotating frames of reference

We can mathematically derive these fictitious forces, in their most general forms, by relating the coordinates in the accelerating frame to those in an inertial frame of reference.

9.0.3 Derivation of fictitious forces

We start by considering two coordinate systems:

- the first, inertial, with origin O_1 and axes $\hat{x}_1$, $\hat{y}_1$ and $\hat{z}_1$,
- the second, non-inertial frame, with origin O and axes $\hat{x}$, $\hat{y}$ and $\hat{z}$.

doi:10.1088/978-0-7503-2690-2ch9

These are the coordinate systems of our two frames of reference, the first with origin O_1 for the inertial frame of reference while the second possibly accelerating, will be the coordinate system of the non-inertial frame or reference.

> **Remark 9.1.** Notice that the origin and axes describing the non-inertial frame are allowed to change their position with respect to the inertial frame if it undergoes acceleration or rotation, i.e. they are not fixed such that acceleration and/or rotation are allowed.

Our goal is to find a mathematical model to describe the axes of the non-inertial frame as functions of the inertial axes.

To start relating coordinates we need to define a vector $\boldsymbol{R}$ from origin O_1 to origin O. Now let the position of a body be defined by a vector $\boldsymbol{r}_1$ with respect to origin O_1 and a vector $\boldsymbol{r}$ with respect to origin O, as shown in figure 9.1. (Notice how can this be easily applied to the 3D case.)

Now let's write these vectors in terms of the two reference frames:

$$\boldsymbol{R} = X\hat{x}_1 + Y\hat{y}_1 + Z\hat{z}_1 \tag{9.1}$$

$$\boldsymbol{r}_1 = x_1\hat{x}_1 + y_1\hat{y}_1 + z_1\hat{z}_1 \tag{9.2}$$

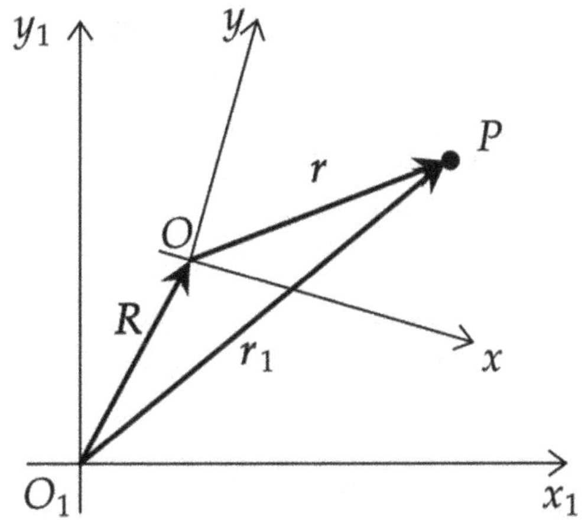

Figure 9.1. Schematic representation of the relation between two frames of reference, one with origin O_1 at rest with point P, and one translated and possibly accelerating or rotating with origin O.

$$\boldsymbol{r} = x\hat{\boldsymbol{x}} + y\hat{\boldsymbol{y}} + z\hat{\boldsymbol{z}}. \tag{9.3}$$

From figure 9.1 notice how we can write $\boldsymbol{r}_1$ as the sum of two vectors

$$\boldsymbol{r}_1 = \boldsymbol{R} + \boldsymbol{r}. \tag{9.4}$$

This relation can be written in terms of coordinates as

$$x_1\hat{\boldsymbol{x}}_1 + y_1\hat{\boldsymbol{y}}_1 + z_1\hat{\boldsymbol{z}}_1 = (X\hat{\boldsymbol{x}}_1 + Y\hat{\boldsymbol{y}}_1 + Z\hat{\boldsymbol{z}}_1) + (x\hat{\boldsymbol{x}} + y\hat{\boldsymbol{y}} + z\hat{\boldsymbol{z}}). \tag{9.5}$$

Now that we know the relations between position vectors in the two different frames of reference, we can start focusing on how to find the fictitious forces. To do this we need to find the accelerations in the non-inertial reference frame, by taking the second derivative of the position, and then interpreting the result in the form $\boldsymbol{F} = m\boldsymbol{a}$.

By taking the second derivative of $\boldsymbol{r}_1$ we find the acceleration of the body with respect to the inertial frame of reference. In this frame of reference Newton laws hold, hence we can write $\boldsymbol{F} = m\ddot{\boldsymbol{r}}_1$. Similarly, we can take the second derivative of $\boldsymbol{R}$ to get an expression for the acceleration of the origin of the non-inertial frame of reference. However, what we are interested in is finding an expression for the second derivative of $\boldsymbol{r}$ in terms of the coordinates of the *accelerating frame* (x,y,z). But here is the tricky part, we can't just take the second derivative of the points (x, y, z) since the whole expression for the vector reads $\boldsymbol{r} = x\hat{\boldsymbol{x}} + y\hat{\boldsymbol{y}} + z\hat{\boldsymbol{z}}$ and, as we mentioned above, these axes, $\hat{\boldsymbol{x}}$, $\hat{\boldsymbol{y}}$, $\hat{\boldsymbol{z}}$, and the origin O are not fixed nor inertial, so may undergo acceleration or rotation.

For instance, even in case (x,y,z) are fixed, which means that $\boldsymbol{r}$ doesn't move with respect to the moving system, vector $\boldsymbol{r}$ could still be changing with respect to the inertial system if the origin O or the axes $\hat{\boldsymbol{x}}$, $\hat{\boldsymbol{y}}$, $\hat{\boldsymbol{z}}$ are moving.

Calculation of $\frac{\mathrm{d}^2\boldsymbol{r}}{\mathrm{d}t^2}$

Our goal is to obtain $\frac{\mathrm{d}^2\boldsymbol{r}}{\mathrm{d}t^2}$ in terms of the coordinates in the non-inertial frame of reference, since we want to be able to describe the force $\boldsymbol{F} = m\boldsymbol{a}$ from the perspective of an observer in the non-inertial frame, without having to consider the inertial frame at all.

We start by taking the first derivative of the whole expression of $\boldsymbol{r}$, i.e. equation (9.3), using the product rule $\left(\frac{\mathrm{d}}{\mathrm{d}x}(\boldsymbol{ab}) = \frac{\mathrm{d}\boldsymbol{a}}{\mathrm{d}x}\boldsymbol{b} + \boldsymbol{a}\frac{\mathrm{d}\boldsymbol{b}}{\mathrm{d}x}\right)$

$$\frac{\mathrm{d}\boldsymbol{r}}{\mathrm{d}t} = \left(\frac{\mathrm{d}x}{\mathrm{d}t}\hat{\boldsymbol{x}} + \frac{\mathrm{d}y}{\mathrm{d}t}\hat{\boldsymbol{y}} + \frac{\mathrm{d}z}{\mathrm{d}t}\hat{\boldsymbol{z}}\right) + \left(x\frac{\mathrm{d}\hat{\boldsymbol{x}}}{\mathrm{d}t} + y\frac{\mathrm{d}\hat{\boldsymbol{y}}}{\mathrm{d}t} + z\frac{\mathrm{d}\hat{\boldsymbol{z}}}{\mathrm{d}t}\right). \tag{9.6}$$

If we take a closer look at equation (9.6) we can see the first group of terms in the brackets describes the rate of change of $\boldsymbol{r}$ with respect to the accelerating frame. This is the quantity mentioned before that doesn't take into account the motion of the axes. From now till the rest of the derivation this quantity will be called $\frac{\delta \boldsymbol{r}}{\delta t}$.

The second group of terms instead arises from the motion of the new axes (translation and/or rotation) with respect to the inertial ones. The distance between the two frames, $\boldsymbol{R}$, describes the translation; the rotation of the axes will be about some axis ω through origin O. Bear in mind that the axis ω can change with time, however, at a given instant there will be a unique axis of rotation describing this movement. This information will be useful when we calculate the second derivative of $\boldsymbol{r}$.

In chapter 5 we already discussed that an arbitrary vector $\boldsymbol{A}$ with a fixed length that rotates with angular velocity $\boldsymbol{\omega} = \omega\hat{\boldsymbol{\omega}}$ changes at a rate

$$\frac{\mathrm{d}\boldsymbol{A}}{\mathrm{d}t} = \boldsymbol{v} = \boldsymbol{\omega} \times \boldsymbol{A}.$$

Similarly, the rotation of the axes of the accelerating frame will change at a rate:

$$\mathrm{d}\hat{\boldsymbol{x}}/\mathrm{d}t = \boldsymbol{\omega} \times \hat{\boldsymbol{x}} \tag{9.7}$$

$$\mathrm{d}\hat{\boldsymbol{y}}/\mathrm{d}t = \boldsymbol{\omega} \times \hat{\boldsymbol{y}} \tag{9.8}$$

$$\mathrm{d}\hat{\boldsymbol{z}}/\mathrm{d}t = \boldsymbol{\omega} \times \hat{\boldsymbol{z}}. \tag{9.9}$$

How can this information be used? We can start from a single coordinate from the second group of terms in equation (9.6), for example $x\left(\frac{\mathrm{d}\hat{\boldsymbol{x}}}{\mathrm{d}t}\right)$; this term equals $x(\boldsymbol{\omega} \times \hat{\boldsymbol{x}}) = \boldsymbol{\omega} \times (x\hat{\boldsymbol{x}})$. We can apply this to all three coordinates:

$$\left(x\frac{\mathrm{d}\hat{\boldsymbol{x}}}{\mathrm{d}t} + y\frac{\mathrm{d}\hat{\boldsymbol{y}}}{\mathrm{d}t} + z\frac{\mathrm{d}\hat{\boldsymbol{z}}}{\mathrm{d}t}\right) = \tag{9.10}$$

$$(x(\omega \times \hat{\boldsymbol{x}}) + y(\omega \times \hat{\boldsymbol{y}}) + z(\omega \times \hat{\boldsymbol{z}})) = \tag{9.11}$$

$$(\boldsymbol{\omega} \times (x\hat{\boldsymbol{x}}) + \boldsymbol{\omega} \times (y\hat{\boldsymbol{y}}) + \boldsymbol{\omega} \times (z\hat{\boldsymbol{z}})) = \tag{9.12}$$

$$\Rightarrow \boldsymbol{\omega} \times \boldsymbol{r}. \tag{9.13}$$

Where in the last passage we have combined $x\hat{\boldsymbol{x}} + y\hat{\boldsymbol{y}} + z\hat{\boldsymbol{z}} = \boldsymbol{r}$.

We can now use $\frac{\delta r}{\delta t}$ and equation (9.13) to rewrite equation (9.6)

$$\frac{\mathrm{d}\boldsymbol{r}}{\mathrm{d}t} = \frac{\delta \boldsymbol{r}}{\delta t} + \boldsymbol{\omega} \times \boldsymbol{r}. \tag{9.14}$$

This is just the first derivative; to find $\mathrm{d}^2\boldsymbol{r}/\mathrm{d}t^2$ we need to derive one more time. Using again the product rule we obtain

$$\frac{\mathrm{d}^2\boldsymbol{r}}{\mathrm{d}t^2} = \frac{\mathrm{d}}{\mathrm{d}t}\left(\frac{\delta \boldsymbol{r}}{\delta t}\right) + \frac{\mathrm{d}\boldsymbol{\omega}}{\mathrm{d}t} \times \boldsymbol{r} + \boldsymbol{\omega} \times \frac{\mathrm{d}\boldsymbol{r}}{\mathrm{d}t}. \tag{9.15}$$

The first term on the right, $\frac{\mathrm{d}}{\mathrm{d}t}\left(\frac{\delta r}{\delta t}\right)$, is the derivative of the velocity vector with respect to the acceleration frame (remember $\frac{\delta r}{\delta t} = \left(\frac{\mathrm{d}x}{\mathrm{d}t}\hat{\boldsymbol{x}} + \frac{\mathrm{d}y}{\mathrm{d}t}\hat{\boldsymbol{y}} + \frac{\mathrm{d}z}{\mathrm{d}t}\hat{\boldsymbol{z}}\right)$). Hence, we can use the same approach we used to calculate $\mathrm{d}\boldsymbol{r}/\mathrm{d}t$, the only difference this time is that instead of calculating the derivative of the *position* with respect to the moving frame we are calculating the derivative of the *velocity* with respect to the non-inertial frame. We can apply equation (9.14) finding

$$\frac{\mathrm{d}}{\mathrm{d}t}\left(\frac{\delta \boldsymbol{r}}{\delta t}\right) = \frac{\delta^2 \boldsymbol{r}}{\delta t^2} + \boldsymbol{\omega} \times \frac{\delta \boldsymbol{r}}{\delta t}. \tag{9.16}$$

To deal with the third term on the right of equation (9.15) we just plug in expression (9.14):

$$\boldsymbol{\omega} \times \frac{\mathrm{d}\boldsymbol{r}}{\mathrm{d}t} = \boldsymbol{\omega} \times \left(\frac{\delta \boldsymbol{r}}{\delta t} + \boldsymbol{\omega} \times \boldsymbol{r}\right). \tag{9.17}$$

At this point equation (9.14) can be rewritten and then rearranged as follows

$$\frac{\mathrm{d}^2\boldsymbol{r}}{\mathrm{d}t^2} = \left(\frac{\delta^2 \boldsymbol{r}}{\delta t^2} + \boldsymbol{\omega} \times \frac{\delta \boldsymbol{r}}{\delta t}\right) + \left(\frac{\mathrm{d}\boldsymbol{\omega}}{\mathrm{d}t} \times \boldsymbol{r}\right) + \left(\boldsymbol{\omega} \times \left(\frac{\delta \boldsymbol{r}}{\delta t} + \boldsymbol{\omega} \times \boldsymbol{r}\right)\right) \tag{9.18}$$

$$= \left(\frac{\delta^2 \boldsymbol{r}}{\delta t^2} + \boldsymbol{\omega} \times \frac{\delta \boldsymbol{r}}{\delta t}\right) + \left(\frac{\mathrm{d}\boldsymbol{\omega}}{\mathrm{d}t} \times \boldsymbol{r}\right) + \boldsymbol{\omega} \times \frac{\delta \boldsymbol{r}}{\delta t} + \boldsymbol{\omega} \times (\boldsymbol{\omega} \times \boldsymbol{r}) \tag{9.19}$$

$$= \frac{\delta^2 \boldsymbol{r}}{\delta t^2} + \boldsymbol{\omega} \times (\boldsymbol{\omega} \times \boldsymbol{r}) + 2\left(\boldsymbol{\omega} \times \frac{\delta \boldsymbol{r}}{\delta t}\right) + \frac{\mathrm{d}\boldsymbol{\omega}}{\mathrm{d}t} \times \boldsymbol{r}. \tag{9.20}$$

All that is left now is to use the expressions of velocity and acceleration with respect to the non-inertial frame:

- $\boldsymbol{v} \equiv \frac{\delta \boldsymbol{r}}{\delta t} = \left(\frac{\mathrm{d}x}{\mathrm{d}t}\hat{\boldsymbol{x}} + \frac{\mathrm{d}y}{\mathrm{d}t}\hat{\boldsymbol{y}} + \frac{\mathrm{d}z}{\mathrm{d}t}\hat{\boldsymbol{z}}\right)$
- $\boldsymbol{a} \equiv \frac{\delta^2 \boldsymbol{r}}{\delta t^2} = \left(\frac{\mathrm{d}^2x}{\mathrm{d}t^2}\hat{\boldsymbol{x}} + \frac{\mathrm{d}^2y}{\mathrm{d}t^2}\hat{\boldsymbol{y}} + \frac{\mathrm{d}^2z}{\mathrm{d}t^2}\hat{\boldsymbol{z}}\right)$.

And we obtain our final expression for the second derivative of $\boldsymbol{r}$ as measured by someone in the accelerating and/or rotating frame

$$\frac{\mathrm{d}^2\boldsymbol{r}}{\mathrm{d}t^2} = \boldsymbol{a} + \boldsymbol{\omega} \times (\boldsymbol{\omega} \times \boldsymbol{r}) + 2\boldsymbol{\omega} \times \boldsymbol{v} + \frac{\mathrm{d}\boldsymbol{\omega}}{\mathrm{d}t} \times \boldsymbol{r}. \tag{9.21}$$

This might look strange at first since we have found that $\mathrm{d}^2\boldsymbol{r}/\mathrm{d}t^2 \neq \boldsymbol{a}$. However, it should be clear from our derivation how $\mathrm{d}^2\boldsymbol{r}/\mathrm{d}t^2$ is the acceleration of the position vector $\boldsymbol{r}$ measured by someone in the *inertial* frame of reference, which takes into account the rotation and/or acceleration of the axes of the non-inertial frame, while $\boldsymbol{a}$ is the second derivative of only the points (x, y, z) as measured by someone in the moving frame.

Equation (9.21) is a very powerful statement about vectors which is purely the result of the mathematical description of the vectors defined previously in figure 9.1. We will see in the next section how this equation leads to the fictitious forces we mentioned at the beginning of this section, which are essential to solve any problem involving a non-inertial system.

9.1 Fictitious forces

In the section above we obtained $\mathrm{d}^2\boldsymbol{r}/\mathrm{d}t^2$, the acceleration of a body as measured by an observer in the **non-inertial** frame of reference. We are now going to manipulate equation (9.21), in order to express the accelerations as consequence of a series of forces.

We will start by using **Galilean transformations**, when observing our two reference frames. Looking at figure 9.1 we can use the transformation to conclude that

$$\frac{\mathrm{d}^2\boldsymbol{r}}{\mathrm{d}t^2} = \frac{\mathrm{d}^2\boldsymbol{r}_1}{\mathrm{d}t^2} - \frac{\mathrm{d}^2\boldsymbol{R}}{\mathrm{d}t^2}. \tag{9.22}$$

We also know that when viewing the body from the *inertial frame*, the **resulting external force** acting on the object must be equal to

$$F_{ext} = m\frac{\mathrm{d}^2\boldsymbol{r}_1}{\mathrm{d}t^2}. \tag{9.23}$$

First we can multiply equation (9.22) by the mass m of the object

$$m\frac{\mathrm{d}^2\boldsymbol{r}}{\mathrm{d}t^2} = m\frac{\mathrm{d}^2\boldsymbol{r}_1}{\mathrm{d}t^2} - m\frac{\mathrm{d}^2\boldsymbol{R}}{\mathrm{d}t^2}.$$

We calculated before what $\mathrm{d}^2\boldsymbol{r}/\mathrm{d}t^2$ is, so we can substitute this into the equation above and obtain

$$m\boldsymbol{a} + m\boldsymbol{\omega} \times (\boldsymbol{\omega} \times \boldsymbol{r}) + 2m\boldsymbol{\omega} \times \boldsymbol{v} + m\frac{\mathrm{d}\boldsymbol{\omega}}{\mathrm{d}t} \times \boldsymbol{r} = m\frac{\mathrm{d}^2\boldsymbol{r}_1}{\mathrm{d}t^2} - m\frac{\mathrm{d}^2\boldsymbol{R}}{\mathrm{d}t^2}.$$

By substituting equation (9.23) and rearranging we obtain

$$\begin{aligned} m\boldsymbol{a} = F_{ext} &- m\frac{\mathrm{d}^2\boldsymbol{R}}{\mathrm{d}t^2} - m\boldsymbol{\omega} \times (\boldsymbol{\omega} \times \boldsymbol{r}) \\ &- 2m\boldsymbol{\omega} \times \boldsymbol{v} - m\frac{\mathrm{d}\boldsymbol{\omega}}{\mathrm{d}t} \times \boldsymbol{r}. \end{aligned} \tag{9.24}$$

This is the new form of **Newton's second law for non-inertial frames**. The four terms seen after the F_{ext} aren't real acting forces. These are the **fictitious forces** due to the fact that the observer is placed in an accelerating and rotating frame of reference.

> Remark 9.2. Notice that in the equation above the quantities $\boldsymbol{r}$, $\boldsymbol{v}$ and $\boldsymbol{a}$, are measured by the observer in the accelerating frame of reference, whilst the quantities $\boldsymbol{\omega}$ and $\frac{d^2R}{dt^2}$ must be given to the observer as they are *properties of the accelerating frame of reference*, measured in the **inertial frame**.

In conclusion, we can say that if a person in an accelerating frame wanted to find their **proper acceleration** $\boldsymbol{a}$ in an external inertial frame they would have to apply a new form of Newton's second law, seen in equation (9.24). This can be more easily rewritten as

Definition 9.1.1: Newton's second law for non–inertial frames

If an observer in a **non-inertial** frame calculates its *proper acceleration* $\boldsymbol{a}$ in an inertial frame, the following equation should be used:

$$m\boldsymbol{a} = \sum F_{ext} + \sum F_{\text{fictitious}} \tag{9.25}$$

Where the fictitious forces are;

$$\sum F_{\text{fictitious}} = F_{\text{tran}} + F_{\text{centrifugal}} + F_{\text{coriolis}} + F_{\text{azm}} \tag{9.26}$$

Now that we have fully derived the four fictitious forces, we can look into what each one actually represents. In the following sections we will introduce all the forces accompanied by examples of their real life applications.

9.1.1 Translation force

Definition 9.1.2: Translational force

Is the force felt when there is a linear acceleration of the frame of reference.

$$F_{\text{tran}} = -m\frac{\mathrm{d}^2\boldsymbol{R}}{\mathrm{d}t^2} \tag{9.27}$$

For example when you feel pushed backwards as you accelerate in a car.

Let's consider a real life example of the translational force. The translational force is the most common of the fictitious forces, and is experienced on a daily basis. An example is that of being in an elevator. If you are in an elevator which is accelerating upwards you will feel heavier, whilst if you are accelerating downwards you will feel lighter on your feet.

Example 9.1. Weight in an elevator

Let's assume a man is standing on top of a scale in an elevator at rest. He has a mass of $m = 70$ kg. The normal force with which the scale pushes up on the person is read on the scale, and gives the *apparent weight* of the man. When the elevator is at rest, the scale will read a weight of

$$\text{weight} = mg = 686 \text{ N}$$

where we have assumed the gravitational acceleration to be $g = 9.81 \text{ m s}^{-2}$.

When the elevator starts accelerating upwards there will be an additional acceleration a of the person and scale in the same direction. From an external observer's point of view, we see that to accelerate the man's body the scale will have to exert an additional upwards force on him $\boldsymbol{F}_s = ma\hat{\boldsymbol{k}}$.

This will result in an addition of force acting down on the scale and the man will read a greater weight on the screen. This can be seen schematically from the image below

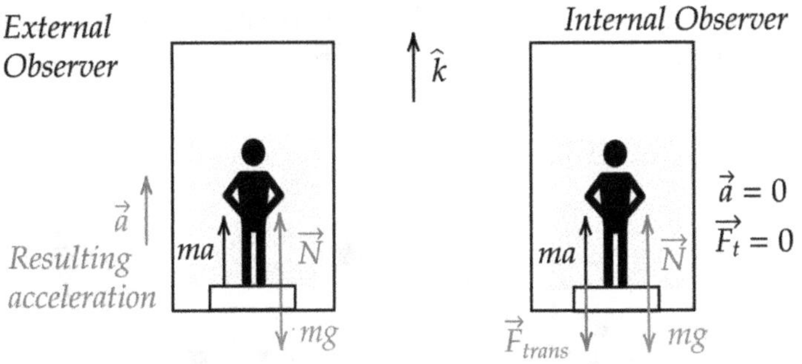

The person however, is in the non-inertial reference frame of the elevator. He will only see the weight on the scale increase and feel his body pushed downwards as a consequence of his reference frame accelerating upwards.

Inside the elevator the man appears to be stationary, for this reason he thinks there must be an additional force balancing the larger force read on the scale. According to Newton's second law, for the person to be stationary the sum of the forces must be zero. Therefore, to balance the greater apparent weight, the man must introduce a force pushing him downwards, which is the translational force.

◆

The above demonstrates how fictitious forces are introduced, to be able to apply Newton's second law to non-inertial reference frames, like an accelerating elevator.

9.1.2 Centrifugal force

The second of the fictitious forces we have introduced is the Coriolis force.

Definition 9.1.3: Centrifugal force

The apparent force pushing outwards when the frame has an angular acceleration.

$$F_{\text{centrifugal}} = -m\boldsymbol{\omega} \times (\boldsymbol{\omega} \times \boldsymbol{r}) \equiv -mr_{\perp}\omega^2. \tag{9.28}$$

This can be felt on a merry-go-round, as the force pushing one outwards.

A real life example of the centrifugal force is the geostationary satellite, i.e. a satellite that has the same rotational period of the Earth, so it is always above the same position along the equator. The following example is introduced to show that the result calculated in an external inertial frame, is equivalent to that calculated in a non-inertial frame, provided the fictitious forces are considered.

Example 9.2. Geostationary satellite
An observer on Earth is looking at a geostationary satellite and calculates its orbital radius. The observer is standing on the Earth, which is rotating, hence he is in the same non-inertial frame as the satellite. He sees the satellite as stationary, so the sum of the forces acting on the satellite must be zero.

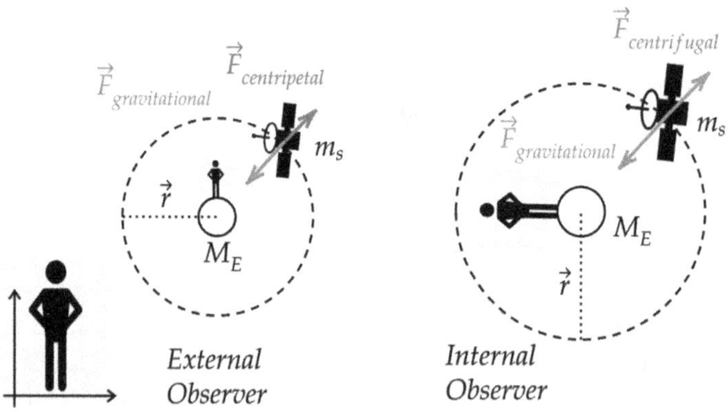

The observer knows that the force of gravity, due to the Earth's attraction, is acting downwards on the satellite with an equation

$$\boldsymbol{F}_t = -\,G\frac{M_e m_s}{r^2}\hat{\boldsymbol{r}}$$

$$\boldsymbol{F}_t \equiv 0.$$

The above won't make any sense if the observer on the Earth doesn't introduce the centrifugal force, to balance the gravitational one: without the fictitious force, the satellite should quickly fall to Earth. Introducing the centrifugal force, the equation above becomes

$$0 = -G\frac{M_e m_s}{r^2}\hat{\boldsymbol{r}} - m_s\boldsymbol{\omega} \times (\boldsymbol{\omega} \times \boldsymbol{r})$$

$$0 = -G\frac{M_e m_s}{r^2}\hat{\boldsymbol{r}} - m_s\omega^2\boldsymbol{r}_\perp$$

where $\boldsymbol{r}_\perp \equiv \boldsymbol{r}$

$$r = \left(\frac{GM_e}{\omega^2}\right)^{\frac{1}{3}}$$

Now the same system is seen by an external observer in a stationary spaceship. He too wishes to find the radius of the geostationary orbit. From his point of view the satellite is moving in uniform circular motion with the same angular velocity ω as the Earth's rotation. In this case the gravitational force will act as the *centripetal force* of the circular motion, giving exactly the same result:

$$-mr\omega^2\hat{r} = -G\frac{M_e m_s}{r^2}\hat{r}$$

$$r = \left(\frac{GM_e}{\omega^2}\right)^{\frac{1}{3}}.$$

So, fictitious forces can be used to find results directly in non-inertial reference frames, consistent with those obtained (often with much more complicated calculations) in an inertial frame.

Exercise 9.1. A particle of mass m is moving in a circular orbit of radius r. Firstly, explain what is meant by *centripetal force* and draw a diagram indicating the direction in which the centripetal force will act. Assuming that the time taken for one revolution T is constant, find an expression for the centripetal force in terms of T, m and r.

Exercise 9.2. A geostationary satellite is orbiting the Earth with an angular velocity of $\omega = \frac{2\pi}{\text{day}} = 7.27 \times 10^{-5}\,\text{s}^{-1}$. Given this information find the radius of the orbit.

Exercise 9.3. Knowing that the Sun has a mass of $M_s = 1.99 \times 10^{30}$ kg, determine the radius, in *km* of the Earth's orbit around the Sun. You assume that the only gravitational force acting on the Earth is produced by the Sun, and the orbit is circular.

◆

9.1.3 Coriolis force

A very important fictitious force is the Coriolis force, defined below.

Definition 9.1.4: Coriolis force

The force acting in a direction perpendicular to the velocity of a body that has a radial component with respect to the rotating centre O of the accelerating frame of reference.

$$F_{\text{coriolis}} = -2m\hat{\omega} \times v. \tag{9.29}$$

This can be felt for instance when walking radially towards the centre of a rotating carousel, as the force pushing sideways.

The Coriolis force is important for movements at planetary scales (like trade winds or ocean currents), since our planet is rotating; its effect can however be observed at a much smaller scale using a rotating pendulum, as shown by Leon Foucault in 1851. This is because, due to the rotation of the Earth the *plane of the swinging pendulum* rotates slowly, at a frequency which can be calculated.

Remark 9.3. Polar and azimuthal angles, latitude and longitude

In many problems involving the Coriolis force you will need to use either the polar angle or the latitude. This angle is essential to find the horizontal and vertical components of the angular velocity of the rotating Earth or any other planet. These components are then used to find the Coriolis force.

Conventionally the geographical latitude angle is the angle **with respect to the equator**, it has values between −90° to 90°. Negative values describe positions below the equator, i.e. in the south hemisphere. Other conventions can be used when referring to latitudes in the south hemisphere, but they will not be used here.

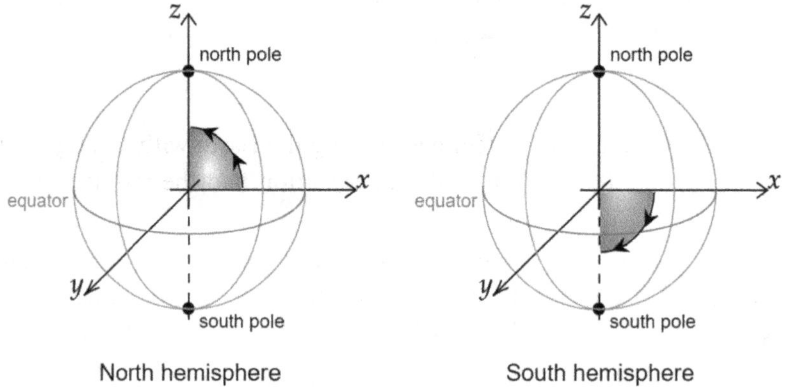

The polar angle is simply the angle θ that we've been using in spherical polar coordinates. This has values from 0° to 180°, where 0° points to the North pole and 180° to the South pole.

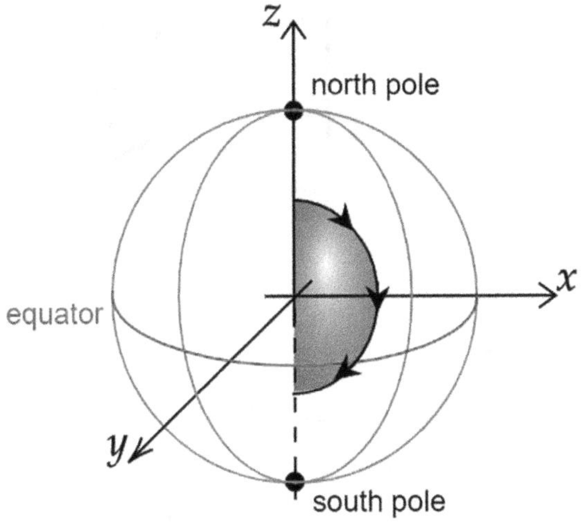

The longitude and the azimuthal angle are instead equivalent.

Example 9.3. In this example we are going to study the effect of the Coriolis force on an aircraft and calculate the inclination of the wings needed to cancel this effect.

A commercial aircraft has an average speed of 800 km h^{-1}. Suppose it is travelling from London to Moscow, we can approximate the velocity direction to be towards East and at a latitude of 55° N. We can assume the only forces in the system are the force due to the weight of the aircraft, the lift from the wings and the Coriolis force. To make calculations easier we can define a new orthogonal basis ($\hat{e}_1$, $\hat{e}_2$, $\hat{e}_3$) where $\hat{e}_1$ points inside the page, $\hat{e}_2$ is tangent to the Earth and $\hat{e}_3$ is pointing radially outwards the Earth.

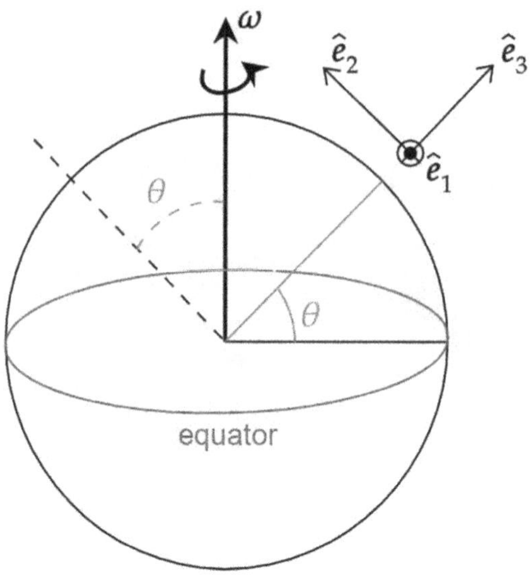

The equation for the Coriolis force is

$$\boldsymbol{F}_{COR} = -2m(\boldsymbol{\omega} \times \boldsymbol{v})$$

where $\boldsymbol{\omega}$ is the angular velocity of the Earth's rotation (period T is 24 h), and $\boldsymbol{v}$ is the velocity of the aircraft. Let's write $\boldsymbol{v}$ and $\boldsymbol{\omega}$, as well as the force due to the weight of the aircraft, $\boldsymbol{F}_w$, in the new coordinate system.

$$\begin{aligned} \boldsymbol{v} &= v\hat{e}_1 \\ \boldsymbol{\omega} &= \omega(\cos\theta\hat{e}_2 + \sin\theta\hat{e}_3) \\ \boldsymbol{F}_w &= mg(-\hat{e}_3) \end{aligned}$$

where θ is the latitude in the North hemisphere.

In this form the Coriolis force becomes

$$\begin{aligned} \boldsymbol{F}_{COR} &= -\,2m(\boldsymbol{\omega} \times \boldsymbol{v}) \\ &= -\,2m\omega v\,[\cos\theta(\hat{e}_2 \times \hat{e}_1) + \sin\theta(\hat{e}_3 \times \hat{e}_1)] \\ &= 2m\omega v[\,\cos\theta\hat{e}_3 - \sin\theta\hat{e}_2] \end{aligned}$$

where we used the fact that, using the right-hand rule, in the coordinate system defined $\hat{e}_2 \times \hat{e}_1 = -\hat{e}_3$ and $\hat{e}_3 \times \hat{e}_1 = \hat{e}_2$.

In order to cancel the effect of the Coriolis force, the wings should tilt such that the resultant force, $\boldsymbol{F}_R = \boldsymbol{F}_{COR} + \boldsymbol{F}_w$, is perpendicular to the surface of the wings. The resultant force can be written as

$$\begin{aligned} \boldsymbol{F}_R &= \boldsymbol{F}_{COR} + \boldsymbol{F}_w \\ &= 2m\omega v[\,\cos\theta\hat{e}_3 - \sin\theta\hat{e}_2] + mg(-\hat{e}_3) \\ &= (-2m\omega v\sin\theta)\hat{e}_2 + (-mg + 2m\omega v\cos\theta)\hat{e}_3 \end{aligned}$$

Here's a schematic representation of the situation:

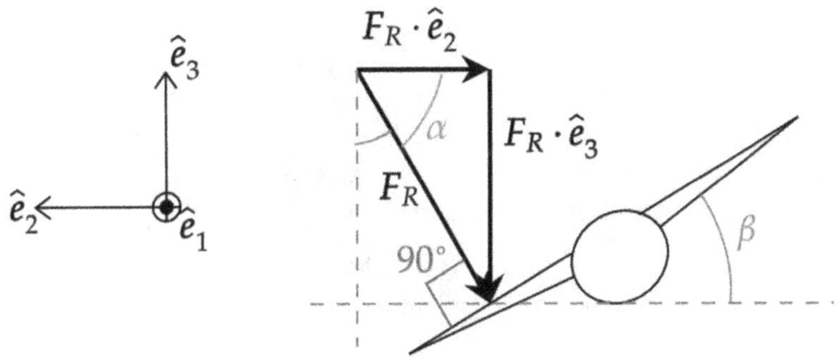

Where $F_R \cdot \hat{e}_2$ is the component of F_R along unit vector $\hat{e}_2$ and $F_R \cdot \hat{e}_3$ along $\hat{e}_3$.

Our goal is to find the angle β of inclination of the wings. Notice how $\alpha + \beta = 90°$. α can be found from the trigonometric relation

$$\tan \alpha = \frac{F_R \cdot \hat{e}_3}{F_R \cdot \hat{e}_2} = \frac{-mg + 2m\omega v \cos \theta}{-2m\omega v \sin \theta}.$$

Therefore, to find β

$$\beta = 90 - \tan^{-1}\left(\frac{\cancel{m}g - 2\cancel{m}\omega v \cos \theta}{2\cancel{m}\omega v \sin \theta}\right).$$

Before calculating β we just need to find the magnitude of $\boldsymbol{\omega}$ using the period of rotation of the Earth

$$\omega = \frac{2\pi}{T} \tag{9.30}$$

$$= \frac{2\pi}{24 \times 3600} \tag{9.31}$$

$$\Rightarrow \omega = 7.27 \times 10^{-5}\ \text{s}^{-1} \tag{9.32}$$

and convert the speed of the aircraft in m s^{-1}

$$v = 800\ \text{km h}^{-1} = \frac{800 \times 10^3}{3600} = \frac{800}{3.6} = 222.22\ \text{m s}^{-1}.$$

Now sub in the equation to find β

$$\beta = 90 - \tan^{-1}\left(\frac{9.81 - 2 \times 7.27 \times 10^{-5} \times 222.22 \times \cos(55)}{2 \times 7.27 \times 10^{-5} \times 222.22 \times \sin(55)}\right)$$

$$\Rightarrow \beta = 0.15°$$

Exercise 9.4. A spaceship of mass m is moving along the surface of Mars at a latitude of 45° at constant velocity v. The trajectory of the spaceship is traced using a stationary satellite in space, and is measured as a curve. Supposing Mars is rotating with a uniform angular velocity of $\boldsymbol{\omega}$, write down the two fictitious forces which cause the curve of the spaceship's trajectory.

Exercise 9.5. An electron is subject to a magnetic field $\boldsymbol{B}$ acting in the vertical $+\hat{\boldsymbol{k}}$ direction. It moves in a circular orbit of radius $R_0 = 50$ cm in the horizontal $(\hat{\boldsymbol{r}}, \hat{\boldsymbol{\theta}})$ plane with an angular velocity of ω.

The centre of the electron's orbit is at latitude of 41.84° N and is subject to a Coriolis force due to the Earth's rotation. State the direction in which the Coriolis force acts, and determine the magnitude of such force.

You can neglect the change in latitude as the electron orbits where the electron charge $q = -1.6 \times 10^{-19}$ C and $m = 9.1 \times 10^{-31}$ kg.

◆

9.1.4 Azimuthal force

The final fictitious force, and the least common is the azimuthal force, also called *Euler's force*.

Definition 9.1.5: Azimuthal force

This is a force induced by a change in the angular velocity of the rotating and accelerating frame of reference.

$$F_{\text{azm}} = -m\frac{\mathrm{d}\boldsymbol{\omega}}{\mathrm{d}t} \times \boldsymbol{r} \tag{9.33}$$

This can be experienced when one is being pushed faster on a merry-go-round, if its angular velocity increases.

When a spinning ice skater brings their arms close to their body, their angular speed increases. This can be easily explained using conservation of angular momentum from an external viewers perspective. Let's now look at how we can explain this case from the ice skater's point of view, using fictitious forces.

As the ice skater brings in her arms at a speed v, there will be a Coriolis force acting on the arms, which will be pushing in the same direction as the spinning. Since in her reference frame the ice skater remains stationary, there must be an additional force, in the opposite direction of the spinning, which will balance the Coriolis force. This additional force is our final fictitious one, the azimuthal force.

Remark 9.4. Working with fictitious forces

When starting on an exercise with fictitious forces, select a frame in which to work. Once this frame has been selected, ensure that throughout the problem you work consistently in the same frame. *Switching frames can often cause confusion and lead to the addition of forces which are counted twice, both from an inertial and non-inertial point of view.*

Exercise 9.6. Anthony the engineer has built a new jet plane in Norway, at 70° N latitude, and is flight-testing it. His plane has a mass of $m = 1000$ kg. Find the horizontal component, *parallel to the ground*, of the centrifugal force acting on the plane, as a result of the Earth's rotation. *You can assume the radius of the Earth to be approximately* 6400 *km*.

While Anthony is running his flight test, he puts the plane into a vertical dive at a speed of $v = 20$ m s^{-1}. Determine the magnitude and direction of the Coriolis force on the jet plane during the dive.

Exercise 9.7. In his experiments to measure the effects of gravity, Galileo started by dropping coins from the top of a ladder. From his house in Pisa at 44° N, he extended his arm horizontally at a height of 1.7 m above the ground. Find the magnitude of the Coriolis force acting on the coin of mass $m = 0.01$ kg, in the instant before it hit the ground. The coin will have travelled 1.7 m, under the acceleration of gravity of $g = 9.81$ m s^{-2}.

Neglect all forms of resistances and the change in motion of the coin due to the fictitious forces. Remember the coin will be subject to a centrifugal force given by the Earth's rotation.

Exercise 9.8. A cylindrical bucket of radius 10 cm is rotated about its vertical symmetry axis. It contains water, and is rotated until the level of the water at the centre is 3 cm lower than that at the edge. Using this information find the frequency of rotation.

Exercise 9.9. An explorer is standing on the Pole. He/she see a polar bear at a distance of 100 m, and, scared, throws an arrow, with horizontal velocity of 10 m s^{-1}

(the vertical velocity of the arrow is not relevant for this problem, so it should be ignored), aiming directly at the centre of the head of the bear.

(i) If the bear's head has a width of 14 cm, and neglecting the effect of the air on the arrow, will the bear's head be hit? Solve the problem in the rotating reference frame, assuming the velocity of the arrow used for the calculation of the Coriolis force is always the same.

(ii) Answer the same question considering instead the same problem as seen from an observer external to the Earth, that sees the bear moving because of the Earth's rotation.

Exercise 9.10. A satellite is in a geostationary orbit, namely its rotational period is one day.

(i) Show that the radial distance r_0 from the centre of the Earth as a function the Earth mass M_E and of Newton's constant G can be written as

$$r_0^3 = \frac{GM_E}{\omega^2}.$$

(ii) An explosion occurs, and the satellite is divided in two parts, the first with mass m, and the second with mass $2m$. The explosion happens in the direction orthogonal to the direction of rotation, such that in the rotating frame the initial velocity of the larger mass is directed towards the Earth, and the initial velocity of the smaller mass is directed away from the Earth. If the kinetic energy of the system increases by an amount E, calculate the radial velocities of both masses immediately after the explosion.

(iii) In the approximation of quasi-circular motion, show that the motion in the rotating frame can be approximated as that of a harmonic oscillator with elastic constant

$$k = \frac{GM_E m}{r_0^3} = \omega^2 m.$$

[*Hint:* given the initial angular momentum $L = m\omega r_0^2$, take a first-order Taylor expansion of the centrifugal force $F_C = \frac{L^2}{mr^3}$ and of the gravitational force $F_G = \frac{GM_E m}{r^2}$ around r_0.]

Exercise 9.11. Imagine being on your favourite carousel, riding some nice carriage at the very edge of the carousel. The carousel has radius R and it is rotating with angular speed ω in the anti-clockwise direction, as shown below.

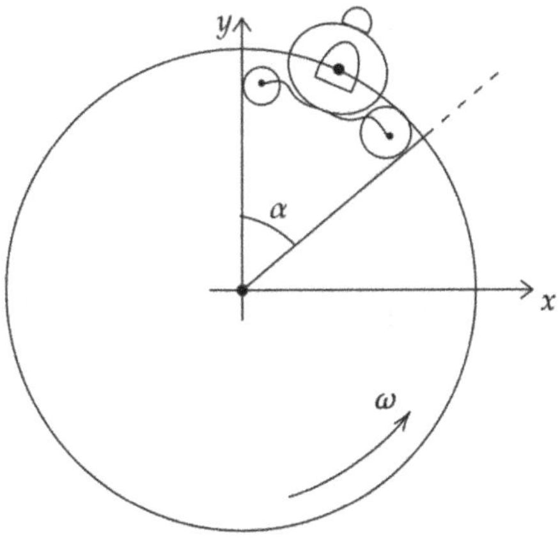

(i) When the front wheels cross the y-axis, your best friend Sarah, who happens the be in the centre of the carousel, throws a ball with speed v along the y-axis. Define angle α as the angle between the front wheels and the back wheels of the carriage. What is the minimal angle α for the ball to hit the carriage when it reaches the end of the carousel? Is it easier to solve the problem using the fixed Cartesian system shown in the picture (external observer point of view) or using a rotating frame of reference (observer rotating with the carousel)?

(ii) In the rotating frame of reference, calculate the Coriolis force experienced by the ball at the beginning of the motion. What is the angle between the varying velocity and the Coriolis force? Will the Coriolis force be able to do work on the ball?

(iii) While sitting on the carriage at the edge of the carousel, you lose your shoe and the carousel stops because of safety reasons. What fictitious force do you experience when the carousel goes from angular speed ω to angular speed zero? Write an expression for this force and state its direction.

IOP Publishing

Classical Mechanics
A professor-student collaboration
Mario Campanelli

Chapter 10

Fluid mechanics

10.0 Introduction

In this chapter, we will not delve into the intricacies and gory details of fluid mechanics, but rather give an overview of the most important results from this branch of mechanics, as well as, some interesting applications. We'll look at fluids at rest (hydrostatics) and fluids in motion (hydrodynamics).

Firstly, we'll see how pressure varies with depth and introduce buoyancy and Archimedes' principle. In the second part, well define what an ideal fluid is, we'll analyse the equation of continuity, Bernoulli's equation and the Venturi meter.

Definition 10.0.1: Fluid

A fluid is a substance that can flow from one point of space to another, without keeping a fixed shape. In other words, has the property of *fluidity*.

A fluid conforms to the boundaries of any container in which it is put. It does so because a fluid cannot sustain a force that is tangential to its surface, or, in other words, cannot withstand a shearing stress. It can, however, exert a force in the direction perpendicular to its surface. Both liquids and gases are fluids.

In our treatment of the mechanics of fluids, we will be applying principles and analysis models that we have already discussed. First, we consider the mechanics of a fluid at rest, i.e. **hydrostatics**, and then study fluids in motion, i.e. **hydrodynamics**.

doi:10.1088/978-0-7503-2690-2ch10

10.1 Hydrostatics

Definition 10.1.1: Pressure
Pressure is the normal force per unit area exerted on a solid surface immersed in the fluid (microscopic origin is the collision of the molecules in the fluid with the surface). In formulae,

$$P \equiv \frac{F}{A}.$$

Pressure is defined as a scalar quantity.

If the pressure varies over an area, the infinitesimal force $\mathrm{d}F$ on an infinitesimal surface element of area $\mathrm{d}A$ is

$$\mathrm{d}F = P\,\mathrm{d}A \tag{10.1}$$

where P is the pressure at the location of the area $\mathrm{d}A$. To calculate the total force exerted on a surface of a container, we must integrate the equation above over the surface.

10.1.1 Variation of pressure with depth

Consider the equilibrium of a small cube of liquid in a tank under gravity as in figure 10.1. The density of the liquid is ρ.

There is no horizontal net force so forces on each pair of opposite parallel vertical faces must cancel. Hence if P_1 is average pressure on face 1, and P_2 is average pressure on face 2, then force on face 1 is

$$\mathrm{d}\boldsymbol{F}_1 = P_1\mathrm{d}y\,\mathrm{d}z\hat{\boldsymbol{\imath}} \tag{10.2}$$

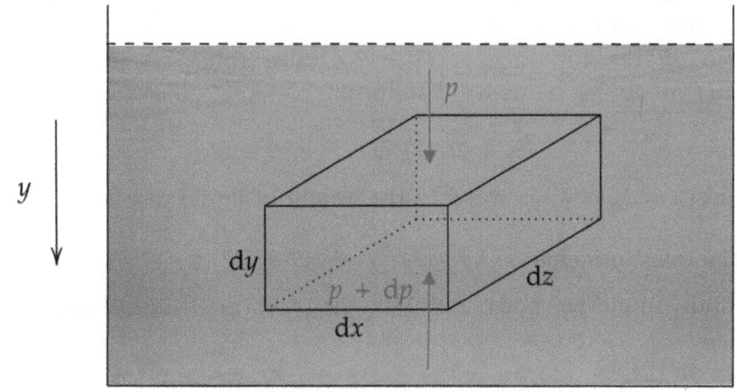

Figure 10.1. Infinitesimal fluid element.

and force of face 2 *is*

$$\mathrm{d}\boldsymbol{F}_2 = -P_2 \mathrm{d}y\, \mathrm{d}z\hat{\boldsymbol{i}} \tag{10.3}$$

but $\mathrm{d}F_1 + \mathrm{d}F_2 = 0$, so $P_1 = P_2$. Therefore **pressure is the same at all points of same depth**.

Consider forces on upper and lower horizontal faces. The situation is now different since the weight of the small cube will only act on the bottom face. The downward force on top face is

$$\mathrm{d}\boldsymbol{F}_5 = -P\, \mathrm{d}x\, \mathrm{d}z\hat{\boldsymbol{k}}. \tag{10.4}$$

The upward force on bottom face is

$$\mathrm{d}\boldsymbol{F}_6 = (p + \mathrm{d}p)\, \mathrm{d}x\, \mathrm{d}z\hat{\boldsymbol{k}}. \tag{10.5}$$

Therefore, for equilibrium of mass of liquid $\rho(\mathrm{d}x\, \mathrm{d}y\, \mathrm{d}z)$,

$$\mathrm{d}\boldsymbol{F}_6 + \mathrm{d}\boldsymbol{F}_5 = -\rho(\mathrm{d}x\, \mathrm{d}y\, \mathrm{d}z)g\hat{\boldsymbol{k}}, \tag{10.6}$$

$$\mathrm{d}P\, \mathrm{d}x\, \mathrm{d}z = \rho(\mathrm{d}x\, \mathrm{d}y\, \mathrm{d}z)g \tag{10.7}$$

$$\frac{\mathrm{d}p}{\mathrm{d}z} = -\rho g. \tag{10.8}$$

On integrating (provided ρ and g are constant) we obtain

$$p = -\rho g z + p_0, \tag{10.9}$$

where p_0 is pressure at $y = 0$. Hence pressure in a static liquid under gravity varies only with the vertical height. Note it increases with increasing depth (i.e. as y decreases).

Pressure is measured in pascals (Pa), 1 Pa = 1 N m^{-2}.

10.1.2 Buoyancy

Consider a cylinder of mass m with end-face area A floating partially submerged in a liquid of density ρ as in figure 10.2.

The force on the top face is $F_T = P_0 A$. The force on the bottom face is $F_B = P_1 A = (P_0 + \rho g y)A$. Thus for equilibrium

$$mg = F_B - F_T = \rho g y A = [(yA)\rho]g, \tag{10.10}$$

i.e. the upthrust $= F_B - F_T$ is equal to the weight of liquid displaced.

10.1.3 Archimedes' principle

Consider a fully immersed body. Let N be the reaction of the cylinder on legs as in figure 10.3.

The force on the top face is

$$F_T = P_1 A. \tag{10.11}$$

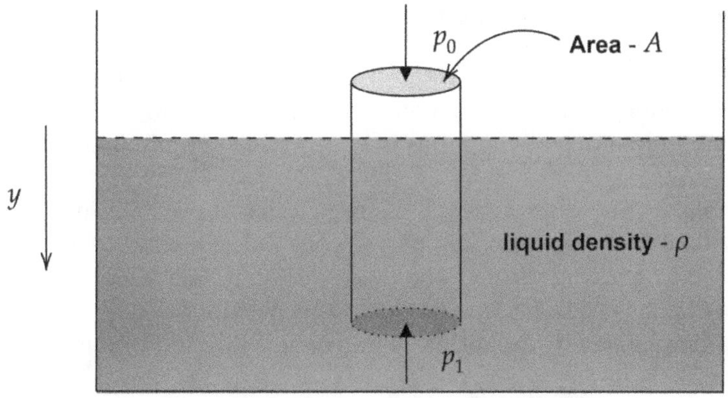

Figure 10.2. Floating cylinder.

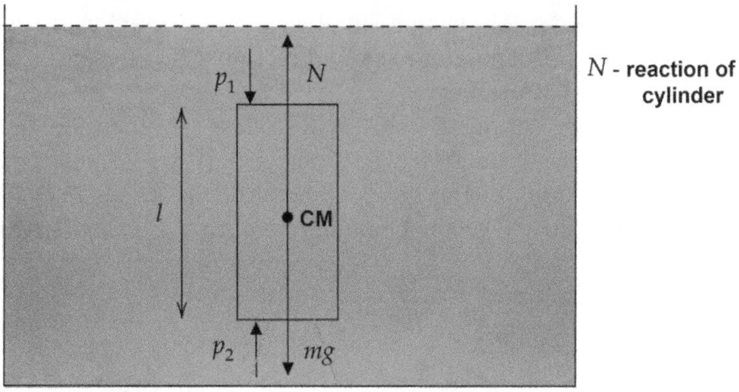

Figure 10.3. Submerged cylinder.

The force on the bottom face is

$$F_B = P_2A + \rho g(lA). \tag{10.12}$$

Thus upthrust $F_B - F_T = \rho(lA)g$. Again, the upthrust equals the weight of liquid displaced. For equilibrium

$$mg = \text{upthrust} + N. \tag{10.13}$$

Whether the body is partially or fully immersed there is an upthrust equal to weight of fluid displaced. This is **Archimedes' principle**. The result applies to any shape of body.

10.2 Hydrodynamics—fluids in motion

10.2.1 The ideal fluid flow

Because the motion of real fluids is very complex and not fully understood, we make the following simplifying assumptions in our approach to obtain the model of **ideal fluid flow**:

1. **The fluid is incompressible**, i.e. its density has a constant, uniform value.
2. **The fluid is non-viscous.** In a non-viscous fluid, internal friction is neglected. An object moving through the fluid experiences no viscous force.
3. **The fluid has constant temperature throughout.**
4. **The flow is steady**, i.e. all particles passing through a point have the same velocity.
5. **The flow is irrotational.** In irrotational flow, the fluid has no angular momentum about any point.

10.2.2 Streamlines

At each point with coordinates (x, y, z) the liquid has a velocity vector v and pressure p (scalar). Consider following a small element of liquid as it travels along.

The velocity may change in magnitude and direction along the path. This path of a fluid element is called a **streamline**.

Consider another fluid element. This will follow another streamline. Two streamlines can never cross because it would mean that the velocity vector would have two different directions at the same point which is impossible. Flow along stream lines, with the velocity vector changing smoothly is called **streamline flow** (figure 10.4).

10.2.3 The equation of continuity

Consider a thin bundle of adjacent streamlines as in figure 10.5 forming a **streamtube**.

Because streamlines cannot cross, any fluid that enters a streamtube at end A, must exit through end B. There is no loss of fluid through the side of the streamtube.

If density of fluid at A is ρ_1 and density at B is ρ_2, mass of fluid entering the tube at A in time t is $m_1 = \rho_1 v_1 t dA_1$ and the mass of fluid leaving through end B is $m_2 = \rho_2 v_2 t dA_2$. Since mass is conserved, $m_1 = m_2$ and

$$\rho_1 v_1 t dA_1 = \rho_2 v_2 t dA_2, \tag{10.14}$$

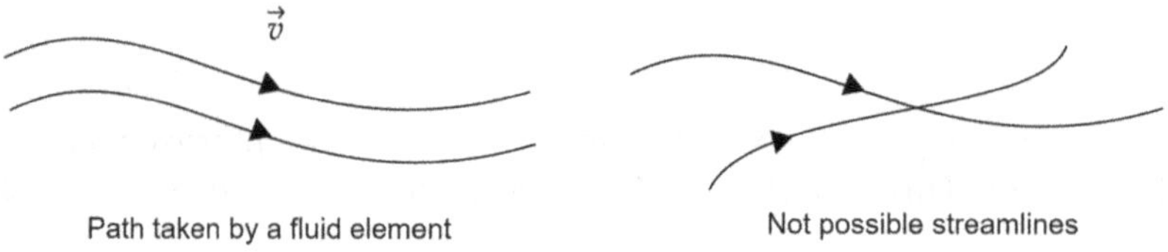

Figure 10.4.

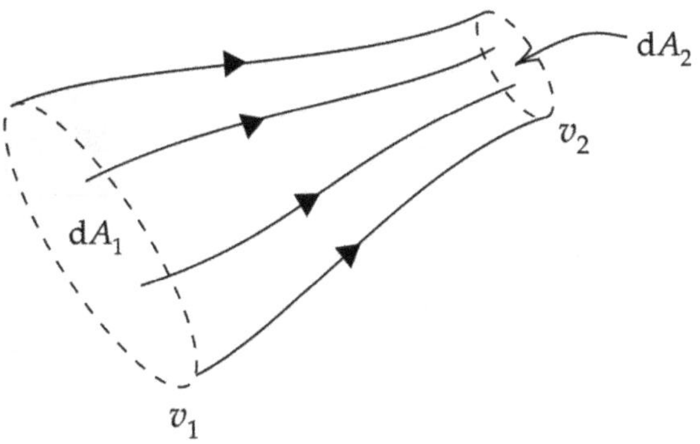

Figure 10.5. A streamtube.

$$\rho_1 v_1 \mathrm{d}A_1 = \rho_2 v_2 \mathrm{d}A_2. \tag{10.15}$$

This is the **equation of continuity**. This is true for liquids and gases. For an ideal liquid which is incompressible, so $\rho_1 = \rho_2$, then

$$v_1 \mathrm{d}A_1 = v_2 \mathrm{d}A_2. \tag{10.16}$$

Example 10.1. Imagine any past experience you have with a hose or a pipe using either your thumb or a nozzle, which can be attached to the end of the hose. If you decrease the nozzle area by putting your thumb on it, the velocity will increase. This can be proved by using the equation of continuity.

Hosepipe nozzle of area A_2; volume rate of flow of water along the pipe of cross-sectional area A_1 is

$$v_1 A_1 = v_2 A_2 \implies v_2 = v_1 \frac{A_1}{A_2}$$

at the nozzle, and hence $v_2 > v_1$ since $A_2 < A_1$.

◆

10.2.4 Bernoulli's equation

Consider an ideal liquid flowing along under gravity as in figure 10.6. Assume there is no viscosity (i.e. no dissipative forces) so that the total energy of a small volume of liquid as it flows along a streamtube must be conserved.

Consider the work done by the pressure in time $\mathrm{d}t$. Work done **by** pressure at end A_1 is

$$\mathrm{d}W_1 = (p_1 \mathrm{d}A_1) v_1 \mathrm{d}t, \tag{10.17}$$

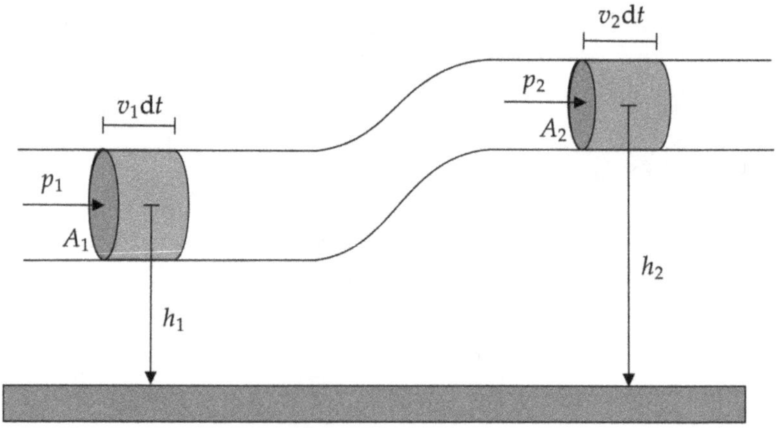

Figure 10.6. Diagram for Bernoulli's equation.

and work done **against** pressure at end A_2 is

$$\mathrm{d}W_2 = (p_2\mathrm{d}A_2)v_2\mathrm{d}t. \tag{10.18}$$

By the continuity equation, the mass of a liquid element dm is (note $\rho_1 = \rho_2 = \rho$ as liquid is incompressible)

$$\mathrm{d}m = \rho\mathrm{d}A_1v_1\mathrm{d}t = \rho\mathrm{d}A_2v_2\mathrm{d}t. \tag{10.19}$$

Change of kinetic energy of this mass from end A_1 to end A_2 is

$$\mathrm{d}W_3 = \frac{1}{2}\mathrm{d}m(v_2^2 - v_1^2) \tag{10.20}$$

$$= \frac{1}{2}\rho\mathrm{d}A_1v_1\mathrm{d}t(v_2^2 - v_1^2). \tag{10.21}$$

Work done against gravity in raising this mass of liquid from h_1 to h_2 is

$$\mathrm{d}W_4 = \mathrm{d}m\, g(h_2 - h_1) \tag{10.22}$$

$$= \rho\mathrm{d}A_1v_1\mathrm{d}t\, g(h_2 - h_1). \tag{10.23}$$

Therefore, from conservation of energy

$$\mathrm{d}W_1 - \mathrm{d}W_2 = \mathrm{d}W_3 + \mathrm{d}W_4 \tag{10.24}$$

$$(p_1\mathrm{d}A_1)v_1\mathrm{d}t - (p_2\mathrm{d}A_2)v_2\mathrm{d}t = \frac{1}{2}\rho\mathrm{d}A_1v_1\mathrm{d}t\left(v_2^2 - v_1^2\right) + \rho\mathrm{d}A_1v_1\mathrm{d}t\, g(h_2 - h_1). \tag{10.25}$$

The continuity equation gives $dA_1v_1 = dA_2v_2$, so

$$p_1 - p_2 = \frac{1}{2}\rho\left(v_2^2 - v_1^2\right) + \rho g(h_2 - h_1) \tag{10.26}$$

$$p_1 + \frac{1}{2}\rho v_1^2 + \rho g h_1 = p_2 + \frac{1}{2}\rho v_2^2 + \rho g h_2. \tag{10.27}$$

Thus in general along a streamtube

$$p + \frac{1}{2}\rho v^2 + \rho g h = \text{constant}. \tag{10.28}$$

This is Bernoulli's equation. If fluid is not moving, $v = 0$, then Bernoulli's equation becomes

$$p + \rho g h = \text{constant}, \tag{10.29}$$

i.e. pressure increases with depth as derived earlier (remember h is measured in an upwards sense).

Example 10.2. A hole in a tank
Consider the flow of water through a small hole in a tank, as in the figure below.

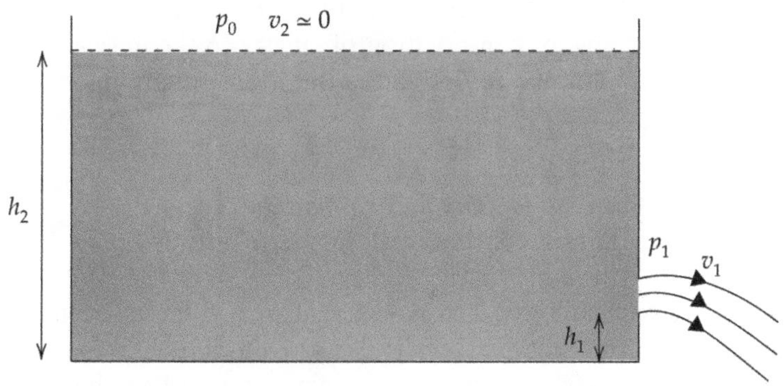

At the surfaces $p_0 = p_1$ is the atmospheric pressure and we can assume $v \simeq 0$ as the water level falls slowly; So

$$p_0 + \frac{1}{2}\rho v_1^2 + \rho g h_1 = p_0 + 0 + \rho g h_2 \tag{10.30}$$

$$v_1^2 = 2g(h_2 - h_1). \tag{10.31}$$

This is the same speed as free fall under gravity through a height $h = h_2 - h_1$.
Rate of flow (volume per second) through area A is Av.

Exercise 10.1. Derive Bernoulli's equation

The diagram below shows a streamtube in the flow of an ideal incompressible fluid of density ρ.

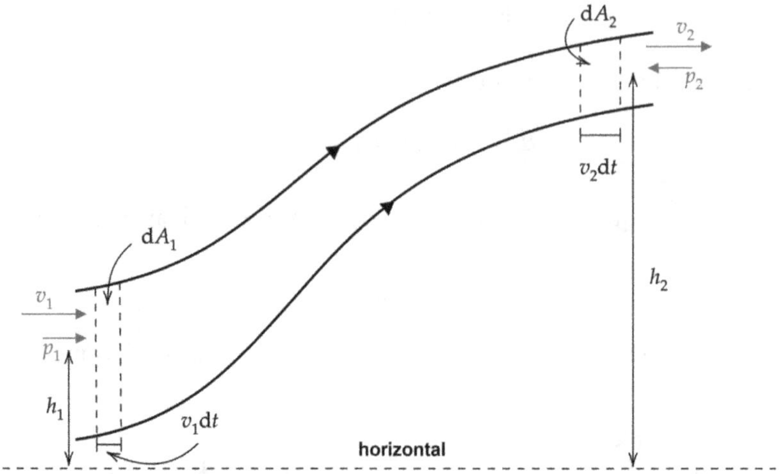

(i) State the mass dm of fluid entering end 1 and time dt.
(ii) Apply the principle of conservation of energy to the mass of liquid flowing through the streamtube to derive the Bernoulli equation.

Exercise 10.2.

(i) Explain the terms *streamline* and *streamtube.*
(ii) State the equation of continuity along a streamtube for an incompressible fluid.
(iii) A hosepipe with an internal cross-sectional area A_1 has at its end a nozzle with a hole whose cross-sectional area is A_2 ($A_2 < A_1$). If the speed of the water as it emerges from the nozzle is v_0, determine its speed in the hosepipe. Also, determine the excess pressure inside the hosepipe, which is horizontal. [*You may recall that Bernoulli's equation, in the usual notation, is* $p + \frac{1}{2}\rho v^2 + \rho gh =$ constant *along a streamtube.*]
(iv) By considering the conservation of energy applied to a small volume of an ideal incompressible liquid of density ρ as it flows along a streamtube, derive the Bernoulli equation

$$p + \frac{1}{2}\rho v^2 + \rho gh = \text{constant}$$

along a streamline, where p is the pressure, v is the speed of flow and h is the vertical height above some convenient level.

Exercise 10.3. Water is flowing smoothly along a straight horizontal pipe of circular cross section with radius R. However, in a short section of its length the radius of the pipe smoothly reduces to r and then smoothly increases to R again. The water pressure is p_1 in the wide section of the pipe and p_2 in the narrowest section. A pressure-measuring device is connected between the wide and the narrow sections of the pipe and a pressure difference $\Delta p = (p_1 - p_2)$ is measured.

(i) Use Bernoulli's equation, together with the equation of continuity, to determine the water's speed of flow v_0 in the wide section of the pipe, expressing your results in terms of R, r and Δp.

(ii) Hence, determine the volume rate of flow of water through the pipe.

◆

Venturi meter

If the flow is horizontal so with no change in height then

$$p + \frac{1}{2}\rho v^2 = \text{constant} \tag{10.32}$$

and pressure is greater where the speed of flow is smaller. This is the basis of the Venturi meter as illustrated in figure 10.7. This instrument is widely used to measure the flow of a fluid passing through a pipe, by measuring the pressure drop between two sections of the tube with different thickness.

From the diagram

$$p_1 + \frac{1}{2}\rho v_1^2 = p_2 + \frac{1}{2}\rho v_2^2 \tag{10.33}$$

and

$$p_1 - p_2 = \rho' g h \tag{10.34}$$

so

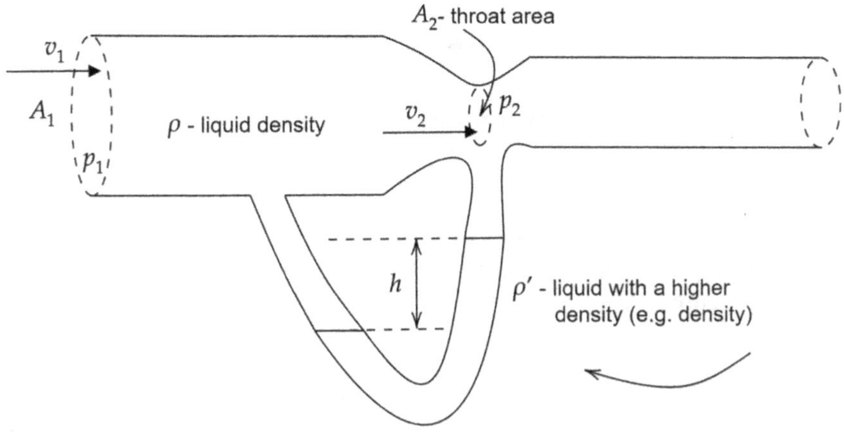

Figure 10.7. Diagram showing a Venturi meter.

$$\rho' gh = \frac{1}{2}\rho v_2^2 - \frac{1}{2}\rho v_1^2. \quad (10.35)$$

But the continuity equation is $v_1 A_1 = v_2 A_2$ so $v_2 = v_1 A_1 / A_2$ so substituting for v_2

$$\rho' gh = \frac{1}{2}\rho v_1^2\left(\frac{A_1^2}{A_2^2} - 1\right) \quad (10.36)$$

$$v_1 = \sqrt{\frac{2\rho' gh A_2^2}{\rho(A_1^2 - A_2^2)}} \quad (10.37)$$

and volume rate of flow along the pipe is

$$A_1 v_1 = \sqrt{\frac{2\rho' gh A_1^2 A_2^2}{\rho(A_1^2 - A_2^2)}}. \quad (10.38)$$

Example 10.3. Flow of air over an aircraft wing as in figure 10.8.

Air flow over the top of the wing, v_1 is faster than under the bottom, v_2 so pressure above the wing, p_1 is less than that under the wing p_2 which gives rise to the upward force (lift) on the wing.

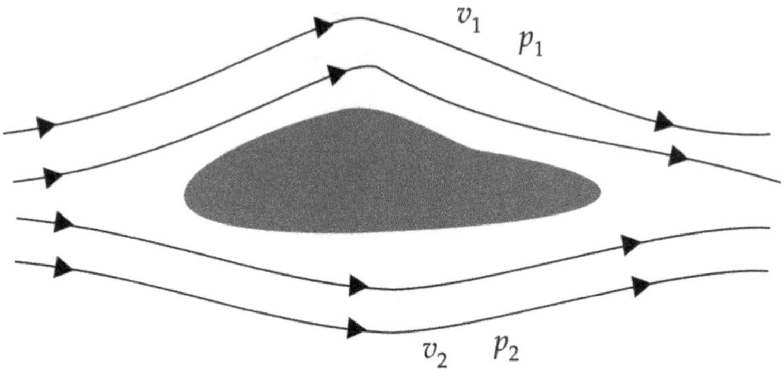

Figure 10.8. Flow over an aerofoil.

Example 10.4. Stability of football suspended on an air jet. The air speed $v_1 > v_2$ so $p_1 < p_2$ producing a resultant force pushing the ball towards the centre of the air jet, as seen in the figure below.

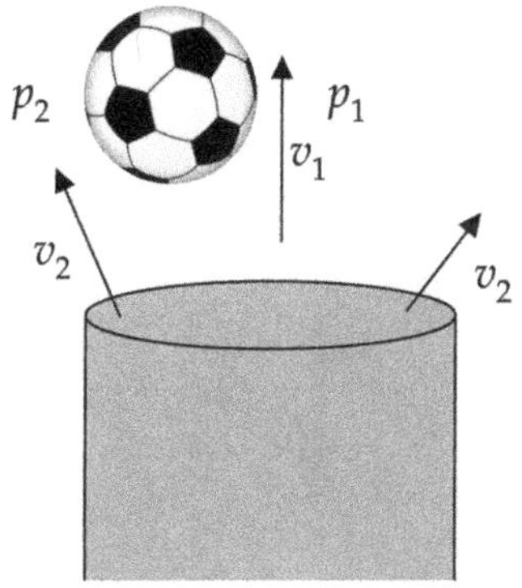

Example 10.5. Deflection of a spinning rough ball. Ball moving at speed v through the air and spinning about a vertical axis as seen below.

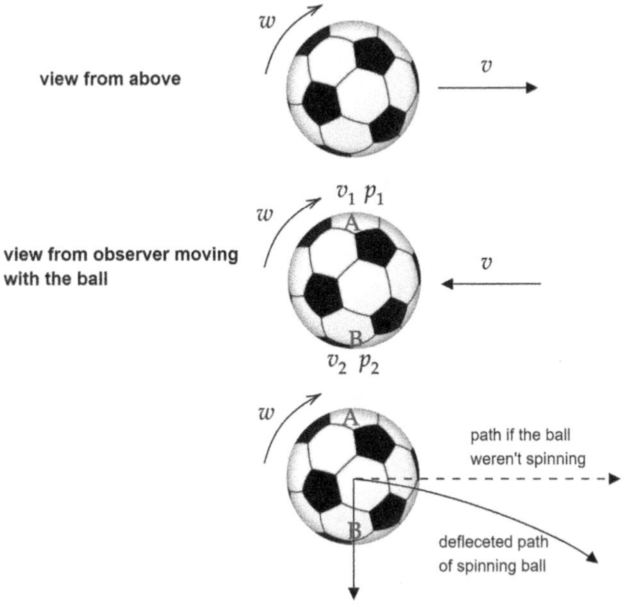

As the ball is rough it slows down air at A and tends to increase speed of the air at B, such that $v_1 < v_2$. This means that we have $p_1 > p_2$, thus producing a resultant force in the direction from A to B. The ball is deflected horizontally.

Note, the ball must be rough, a smooth ball would not drag the air round with it and the effect described above would not be observed for a smooth surface.

Exercise 10.4. Ball on a vertical jet
In the figure below we see a football balancing on a vertical jet of upward moving air.

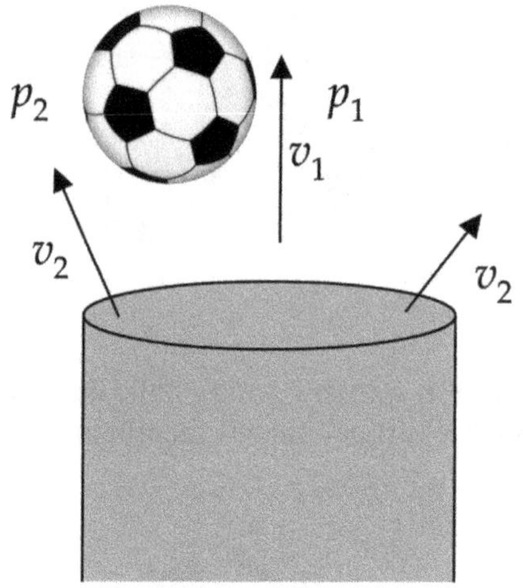

Explain:

(i) Why is this possible?
(ii) Why the trajectory of a ball with rough surface that is spinning is deflected relative to what it would be if the ball were not spinning.

Chapter 11

Solutions to Chapter 1: Mathematical preliminaries

Solution: Exercise 1.1. Given $\mathbf{A} = \begin{pmatrix} 1 & 2 \\ 3 & 1 \end{pmatrix}$ and $\mathbf{I}_2 = \begin{pmatrix} 1 & 0 \\ 0 & 1 \end{pmatrix}$, we want to prove $\mathbf{A}^2 = 2\mathbf{A} + 5\mathbf{I}_2$.
Let's compute the lhs:

$$\mathbf{A}^2 = \mathbf{A} \times \mathbf{A} = \begin{pmatrix} 1 & 2 \\ 3 & 1 \end{pmatrix} \times \begin{pmatrix} 1 & 2 \\ 3 & 1 \end{pmatrix} = \begin{pmatrix} a & b \\ c & d \end{pmatrix}.$$

- a is the total of the first row multiplied by the first column.
$$a = 1 \times 1 + 2 \times 3 = 7.$$
- b is the total of the first row multiplied by the second column.
$$b = 1 \times 2 + 2 \times 1 = 4.$$
- c is the total of the second row multiplied by the first column.
$$c = 3 \times 1 + 1 \times 3 = 6.$$
- d is the total of the second row multiplied by the second column.
$$c = 3 \times 2 + 1 \times 1 = 7.$$

Hence,

$$\mathbf{A}^2 = \begin{pmatrix} 7 & 4 \\ 6 & 7 \end{pmatrix}.$$

doi:10.1088/978-0-7503-2690-2ch11

Now, let's compute the rhs.

$$2\mathbf{A} = 2 \times \begin{pmatrix} 1 & 2 \\ 3 & 1 \end{pmatrix} = \begin{pmatrix} 2 \times 1 & 2 \times 2 \\ 2 \times 3 & 2 \times 1 \end{pmatrix} = \begin{pmatrix} 2 & 4 \\ 6 & 2 \end{pmatrix},$$

and

$$5\mathbf{I}_2 = 5 \times \begin{pmatrix} 1 & 0 \\ 0 & 1 \end{pmatrix} = \begin{pmatrix} 5 \times 1 & 5 \times 0 \\ 5 \times 0 & 5 \times 1 \end{pmatrix} = \begin{pmatrix} 5 & 0 \\ 0 & 5 \end{pmatrix}.$$

Thus,

$$2\mathbf{A} + 5\mathbf{I}_2 = \begin{pmatrix} 2 & 4 \\ 6 & 2 \end{pmatrix} + \begin{pmatrix} 5 & 0 \\ 0 & 5 \end{pmatrix} = \begin{pmatrix} 7 & 4 \\ 6 & 7 \end{pmatrix} = \mathbf{A}^2.$$

Thus, lhs = rhs as required.

Solution: Exercise 1.2.

(i)

$$\begin{aligned}\det\{\mathbf{A}\} &= \begin{vmatrix} 2 & 1 & -1 \\ 1 & 0 & 4 \\ -4 & 2 & 1 \end{vmatrix} = 2 \times \begin{vmatrix} 0 & 4 \\ 2 & 1 \end{vmatrix} - 1 \times \begin{vmatrix} 1 & 4 \\ -4 & 1 \end{vmatrix} - 1 \times \begin{vmatrix} 1 & 0 \\ -4 & 2 \end{vmatrix} \\ &= 2 \times (0 \times 1 - 4 \times 2) - 1 \times (1 \times 1 - 4 \times (-4)) - 1 \times (1 \times 2 - 0 \times (-4)) \\ &= 2 \times (-8) - 1 \times (1 + 16) - 1 \times 2 \\ &= -16 - 17 - 2 = -35.\end{aligned}$$

(ii) Given $\det\{\mathbf{B}\} = 2$, we want to find the value of α.

$$\begin{aligned}\det\{\mathbf{B}\} &= \begin{vmatrix} 3 & 1 & 2 \\ \alpha & 4 & 5 \\ 0 & 2 & 3 \end{vmatrix} \\ &= 3 \times (4 \times 3 - 5 \times 2) - 1 \times (\alpha \times 3 - 5 \times 0) + 2 \times (\alpha \times 2 - 4 \times 0) \\ &= 3(12 - 10) - 3\alpha + 4\alpha = 6 + \alpha.\end{aligned}$$

Thus, $6 + \alpha = 2 \Longrightarrow \alpha = -4$.

(iii)

$$\mathbf{A} \times \mathbf{B} = \begin{pmatrix} 2 & 1 & -1 \\ 1 & 0 & 4 \\ -4 & 2 & 1 \end{pmatrix} \times \begin{pmatrix} 3 & 1 & 2 \\ -4 & 4 & 5 \\ 0 & 2 & 3 \end{pmatrix} = \begin{pmatrix} 2 & 4 & 6 \\ 3 & 9 & 14 \\ -20 & 6 & 5 \end{pmatrix}.$$

(iv) We want to verify that $\det\{\mathbf{AB}\} = \det\{\mathbf{A}\}\det\{\mathbf{B}\}$.

$$\det\{\mathbf{AB}\} = \begin{vmatrix} 2 & 4 & 6 \\ 3 & 9 & 14 \\ -20 & 6 & 5 \end{vmatrix} = -70.$$

Since $\det\{\mathbf{A}\} = -35$ and $\det\{\mathbf{B}\} = 2 \Longrightarrow \det\{\mathbf{A}\}\det\{\mathbf{B}\} = -35 \times 2 = -70 = \det\{\mathbf{AB}\}$, as required.

Solution: Exercise 1.3.

(i)

$$\hat{\boldsymbol{a}} = \frac{\boldsymbol{a}}{|\boldsymbol{a}|} = \frac{\hat{\boldsymbol{\imath}} + 2\hat{\boldsymbol{\jmath}} + \hat{\boldsymbol{k}}}{\sqrt{1^2 + 2^2 + 1^2}} = \frac{1}{\sqrt{6}}(\hat{\boldsymbol{\imath}} + 2\hat{\boldsymbol{\jmath}} + \hat{\boldsymbol{k}}).$$

(ii)

$$\boldsymbol{a} \cdot \boldsymbol{b} = (\hat{\boldsymbol{\imath}} + 2\hat{\boldsymbol{\jmath}} + \hat{\boldsymbol{k}}) \cdot (\hat{\boldsymbol{\imath}} - 3\hat{\boldsymbol{\jmath}} + 2\hat{\boldsymbol{k}}) = 1 \times 1 + 2 \times (-3) + 1 \times 2 = 1 - 6 + 2$$
$$= -3.$$

(iii)

$$\boldsymbol{a} \times \boldsymbol{b} = \begin{vmatrix} \hat{\boldsymbol{\imath}} & \hat{\boldsymbol{\jmath}} & \hat{\boldsymbol{k}} \\ 1 & 2 & 1 \\ 1 & -3 & 2 \end{vmatrix}$$
$$= \hat{\boldsymbol{\imath}}(2 \times 2 - 1 \times (-3)) - \hat{\boldsymbol{\jmath}}(1 \times 2 - 1 \times 1) + \hat{\boldsymbol{k}}(1 \times (-3) - 2 \times 1)$$
$$= 7\hat{\boldsymbol{\imath}} - \hat{\boldsymbol{\jmath}} - 5\hat{\boldsymbol{k}}.$$

(iv) The area of the parallelogram formed by vectors $\boldsymbol{a}$ and $\boldsymbol{b}$ is $|\boldsymbol{a} \times \boldsymbol{b}|$. Thus,

$$|\boldsymbol{a} \times \boldsymbol{b}| = \sqrt{7^2 + (-1)^2 + (-5)^2} = \sqrt{49 + 1 + 25} = \sqrt{75}.$$

Solution: Exercise 1.4.

(i)

$$|\boldsymbol{a}| = \sqrt{1^2 + 1^2 + 2^2} = \sqrt{6}$$
$$|\boldsymbol{b}| = \sqrt{1^2 + 0^2 + 7^2} = \sqrt{50}$$
$$|\boldsymbol{a} + 2\boldsymbol{b}| = |(\hat{\boldsymbol{\imath}} + \hat{\boldsymbol{\jmath}} + 2\hat{\boldsymbol{k}}) + 2(\hat{\boldsymbol{\imath}} + 7\hat{\boldsymbol{k}})| = |3\hat{\boldsymbol{\imath}} + \hat{\boldsymbol{\jmath}} + 16\hat{\boldsymbol{k}}| = \sqrt{3^2 + 1^2 + 16^2} = \sqrt{266}.$$

(ii)

$$\boldsymbol{a} \cdot \boldsymbol{b} = 1 \times 1 + 1 \times 0 + 2 \times 7 = 15.$$

(iii) From the geometrical definition of scalar product, we know $\boldsymbol{a} \cdot \boldsymbol{b} = |\boldsymbol{a}||\boldsymbol{b}|\cos\theta$, where θ is the angle between $\boldsymbol{a}$ and $\boldsymbol{b}$. Re-arranging,

$$\cos\theta = \frac{\boldsymbol{a} \cdot \boldsymbol{b}}{|\boldsymbol{a}||\boldsymbol{b}|} = \frac{15}{\sqrt{6}\sqrt{50}} = \frac{\sqrt{3}}{2}.$$

Therefore, the angle between $\boldsymbol{a}$ and $\boldsymbol{b}$ is $\pi/6$, i.e. 30°.

Solution: Exercise 1.5.

(i)

$$|\boldsymbol{v}| = \sqrt{1^2 + (-2)^2 + (-2)^2} = 3.$$

The unit vector $\hat{\boldsymbol{v}} = \boldsymbol{v}/|\boldsymbol{v}|$ points in the same direction as $\boldsymbol{v}$, so we want

$$-\hat{\boldsymbol{v}} = \frac{1}{3}(-\hat{\boldsymbol{i}} + 2\hat{\boldsymbol{j}} + 2\hat{\boldsymbol{k}}).$$

(ii) We want to find a non-zero vector orthogonal to $\boldsymbol{v}$, i.e. a vector $\boldsymbol{w}$ satisfying $\boldsymbol{v} \cdot \boldsymbol{w} = 0$.

Let $\boldsymbol{w} = w_1\hat{\boldsymbol{i}} + w_2\hat{\boldsymbol{j}} + w_3\hat{\boldsymbol{k}}$. Thus,

$$\boldsymbol{v} \cdot \boldsymbol{w} = (\hat{\boldsymbol{i}} - 2\hat{\boldsymbol{j}} - 2\hat{\boldsymbol{k}}) \cdot (w_1\hat{\boldsymbol{i}} + w_2\hat{\boldsymbol{j}} + w_3\hat{\boldsymbol{k}}) = w_1 - 2w_2 - 2w_3 = 0.$$

Without loss of generality, set $w_3 = 0$. Thus, we have

$$w_1 - 2w_2 = 0 \Longrightarrow w_1 = 2w_2 \quad (*).$$

Moreover, we know the length of $\boldsymbol{w}$ must be 2. Hence,

$$\sqrt{w_1^2 + w_2^2} = 2 \Longrightarrow w_1^2 + w_2^2 = 4.$$

From equation (*), we obtain

$$4w_2^2 + w_2^2 = 4 \Longrightarrow w_2^2 = \frac{4}{5}.$$

Thus,

$$w_2 = \frac{2}{\sqrt{5}} \quad \text{and} \quad w_1 = 2w_2 = \frac{4}{\sqrt{5}}.$$

Therefore,

$$w = \frac{2}{\sqrt{5}}(2\hat{i} + \hat{j}).$$

Many other answers are possible (just check that the length is 2 and that the scalar product with v is zero).

Solution: Exercise 1.6. We have $|\boldsymbol{b}| = 2|\boldsymbol{a}|$. Let θ be the angle between $\boldsymbol{a}$ and $\boldsymbol{b}$. Then,

$$\begin{aligned}
0 &= (2\boldsymbol{a} - 5\boldsymbol{b}) \cdot (6\boldsymbol{a} - \boldsymbol{b}) \\
&= 12\boldsymbol{a} \cdot \boldsymbol{a} + 5\boldsymbol{b} \cdot \boldsymbol{b} - 32\boldsymbol{a} \cdot \boldsymbol{b} \\
&= 12|\boldsymbol{a}|^2 + 5|\boldsymbol{b}|^2 - 32|\boldsymbol{a}||\boldsymbol{b}|\cos\theta \\
&= 12|\boldsymbol{a}|^2 + 5 \times 4|\boldsymbol{a}|^2 - 32 \times 2|\boldsymbol{a}|^2 \cos\theta \\
&= (12 + 5 \times 4)|\boldsymbol{a}|^2 - 32 \times 2|\boldsymbol{a}|^2 \cos\theta \\
&= 32(1 - 2\cos\theta).
\end{aligned}$$

Hence, $\cos\theta = \frac{1}{2}$, which shows that the angle between $\boldsymbol{a}$ and $\boldsymbol{b}$ is $\theta = \pi/3$.

Solution: Exercise 1.7. No, it doesn't follow that $\boldsymbol{a} = \boldsymbol{b}$. Consider the case $\boldsymbol{a} = \hat{\boldsymbol{i}}$, $\boldsymbol{b} = \hat{\boldsymbol{j}}$, and $\boldsymbol{c} = \hat{\boldsymbol{k}}$.
Multiplying equation (*) by the non-zero vector $\boldsymbol{c}/|\boldsymbol{c}|^2$, we see that (*) is equivalent to

$$\frac{\boldsymbol{a} \cdot \boldsymbol{c}}{\boldsymbol{c} \cdot \boldsymbol{c}}\boldsymbol{c} = \frac{\boldsymbol{b} \cdot \boldsymbol{c}}{\boldsymbol{c} \cdot \boldsymbol{c}}\boldsymbol{c}$$

which is saying that vector $\boldsymbol{a}$ and $\boldsymbol{b}$ have the same projections on $\boldsymbol{c}$, i.e. $\mathrm{Proj}_c\boldsymbol{a} = \mathrm{Proj}_c\boldsymbol{b}$.

Solution: Exercise 1.8. Since the vectors are non-zero, the scalar products between any two of them is zero if and only if they are perpendicular.

Suppose that $(\boldsymbol{a} \cdot \boldsymbol{c})\boldsymbol{b} = (\boldsymbol{b} \cdot \boldsymbol{c})\boldsymbol{a}$. Thus, if $\boldsymbol{a}$ and $\boldsymbol{c}$ are NOT perpendicular, then $\boldsymbol{a} \cdot \boldsymbol{c} \neq 0$ and so $\boldsymbol{b} = \frac{\boldsymbol{b}\cdot\boldsymbol{c}}{\boldsymbol{a}\cdot\boldsymbol{c}}\boldsymbol{a}$. Thus, $\boldsymbol{b}$ is a non-zero multiple of $\boldsymbol{a}$ and, hence, they are parallel.

On the other hand, if $\boldsymbol{a}$ and $\boldsymbol{c}$ are perpendicular, then, $\boldsymbol{a} \cdot \boldsymbol{c} = 0$, but then $(\boldsymbol{b} \cdot \boldsymbol{c})\boldsymbol{a} = 0$.

Since $\boldsymbol{a} \neq \boldsymbol{0}$, it follows that $\boldsymbol{b} \cdot \boldsymbol{c} = 0$ and, therefore, $\boldsymbol{b}$ and $\boldsymbol{c}$ are also perpendicular.

Solution: Exercise 1.9. Since the area of a parallelogram is the magnitude of the cross product of two vectors, then the area of the triangle defined by two vectors is half the magnitude of their cross product.
In this problem, we can take, e.g.

$$\overrightarrow{AB} = \boldsymbol{b} - \boldsymbol{a} \quad \text{and} \quad \overrightarrow{AC} = \boldsymbol{c} - \boldsymbol{a}.$$

Then, the area of the triangle is:

$$\begin{aligned}\text{Area} &= \frac{1}{2}\left|\vec{AB}\times\vec{AC}\right| \\ &= \frac{1}{2}|(\boldsymbol{b} - \boldsymbol{a}) \times (\boldsymbol{c} - \boldsymbol{a})| \\ &= \frac{1}{2}|\boldsymbol{b} \times \boldsymbol{c} - \boldsymbol{a} \times \boldsymbol{c} - \boldsymbol{b} \times \boldsymbol{a}|.\end{aligned}$$

By anti-commutativity:

$$\text{Area} = \frac{1}{2}|\boldsymbol{b} \times \boldsymbol{c} + \boldsymbol{c} \times \boldsymbol{a} + \boldsymbol{a} \times \boldsymbol{b}|,$$

as required.

Solution: Exercise 1.10.

(i)

(a) $\boldsymbol{v} = \frac{d\boldsymbol{r}}{dt} = \boldsymbol{r}' = 2t\hat{\boldsymbol{i}} + 2\hat{\boldsymbol{j}} + 6t\hat{\boldsymbol{k}}$.

(b) $\boldsymbol{a} = \frac{d\boldsymbol{v}}{dt} = \boldsymbol{r}'' = 2\hat{\boldsymbol{i}} + 6\hat{\boldsymbol{k}}$.

(ii)

(a) $\boldsymbol{v} = \frac{d\boldsymbol{r}}{dt} = \boldsymbol{r}' = -\sin t\hat{\boldsymbol{i}} + \sqrt{3}\cos t\hat{\boldsymbol{j}} + (e^{10t} + 10te^{10t})\hat{\boldsymbol{k}}$.

(b) $\boldsymbol{a} = \frac{d\boldsymbol{v}}{dt} = \boldsymbol{r}'' = -\cos t\hat{\boldsymbol{i}} - 3\sin t + (20e^{10t} + 100te^{10t})\hat{\boldsymbol{k}}$.

(iii)

(a) $\boldsymbol{v} = \frac{d\boldsymbol{r}}{dt} = \boldsymbol{r}' = 2\cos 2t\hat{\boldsymbol{i}} - \sqrt{3}\sin t\sqrt{3}\hat{\boldsymbol{j}} + e^{2t}(1 + 2t)\hat{\boldsymbol{k}}$.

(b) $\boldsymbol{a} = \frac{d\boldsymbol{v}}{dt} = \boldsymbol{r}'' = -4\sin 2t\hat{\boldsymbol{i}} - 3\cos t\sqrt{3}\hat{\boldsymbol{j}} + 4e^{2t}(t + 1)\hat{\boldsymbol{k}}$.

(iv)

(a) $\boldsymbol{v} = \frac{d\boldsymbol{r}}{dt} = \boldsymbol{r}' = -2\sin 2t\hat{\boldsymbol{i}} + 4t\hat{\boldsymbol{j}} + \sqrt{2}e^{t\sqrt{2}}\hat{\boldsymbol{k}}$.

(b) $\boldsymbol{a} = \frac{d\boldsymbol{v}}{dt} = \boldsymbol{r}'' = -4\cos 2t\hat{\boldsymbol{i}} + 4\hat{\boldsymbol{j}} + 2e^{t\sqrt{2}}\hat{\boldsymbol{k}}$.

Solution: Exercise 1.11.

(i) From the expression of the position vector, we have:

$$x = a\cos(\omega t), \quad \text{and} \quad y = b\sin(\omega t),$$

which are parametric equations for an ellipse with semi-major axis of length a and semi-minor axis of length b. Since

$$\frac{x^2}{a^2} + \frac{x^2}{b^2} = \cos^2(\omega t) + \sin^2(\omega t) = 1.$$

(ii) The velocity is

$$v = \frac{\mathrm{d}\boldsymbol{r}}{\mathrm{d}t} = -a\omega \sin(\omega t)\hat{\boldsymbol{i}} + b\omega \cos(\omega t)\hat{\boldsymbol{j}}\,.$$

(iii) The acceleration is

$$\boldsymbol{a} = \frac{\mathrm{d}\boldsymbol{v}}{\mathrm{d}t} = \frac{\mathrm{d}^2\boldsymbol{r}}{\mathrm{d}t^2} = -a\omega^2 \cos(\omega t)\hat{\boldsymbol{i}} - b\omega^2 \sin(\omega t)\hat{\boldsymbol{j}}$$
$$= -\,\omega^2[a \cos(\omega t)\hat{\boldsymbol{i}} + b \sin(\omega t)\hat{\boldsymbol{j}}\,] = -\omega^2\boldsymbol{r},$$

from which it immediately follows that the acceleration is radial, hence directed towards the centre.

Solution: Exercise 1.12.

(i)

$$\frac{\partial f}{\partial x} = y^3 e^{2z} \cos(xy^3)$$
$$\frac{\partial f}{\partial y} = 3xy^2 e^{2z} \cos(xy^3)$$
$$\frac{\partial f}{\partial z} = 2e^{2z} \sin(xy^3).$$

(ii)

$$\frac{\partial^2 f}{\partial x \partial y} = \frac{\partial}{\partial x}\left(\frac{\partial f}{\partial y}\right) = \frac{\partial}{\partial x}\left(3xy^2 e^{2x} \cos(xy^3)\right)$$
$$= 3e^{2z}[y^2 \cos(xy^3) - xy^5 \sin(xy^3)]$$
$$\frac{\partial^2 f}{\partial y \partial x} = \frac{\partial}{\partial y}\left(\frac{\partial f}{\partial x}\right) = \frac{\partial}{\partial y}\left(y^3 e^{2z} \cos(xy^3)\right)$$
$$= e^{2z}[3y^2 \cos(xy^3) - 3xy^5 \sin(xy^3)].$$

Clearly, $\frac{\partial^2 f}{\partial x \partial y} = \frac{\partial^2 f}{\partial y \partial x}$, as required.

(iii)

$$\nabla f\,|_{(0,\,2,\,0)} = \left.\frac{\partial f}{\partial x}\right|_{(0,\,2,\,0)} \hat{\boldsymbol{i}} + \left.\frac{\partial f}{\partial y}\right|_{(0,\,2,\,0)} \hat{\boldsymbol{j}} + \left.\frac{\partial f}{\partial z}\right|_{(0,\,2,\,0)} \hat{\boldsymbol{k}} = 8\hat{\boldsymbol{i}}\,.$$

Solution: Exercise 1.13.

$$\begin{aligned}\nabla(fg) &= \frac{\partial(fg)}{\partial x}\hat{\boldsymbol{i}} + \frac{\partial(fg)}{\partial y}\hat{\boldsymbol{j}} + \frac{\partial(fg)}{\partial z}\hat{\boldsymbol{k}} \\ &= \left(\frac{\partial f}{\partial x}g + f\frac{\partial g}{\partial x}\right)\hat{\boldsymbol{i}} + \left(\frac{\partial f}{\partial y}g + f\frac{\partial g}{\partial y}\right)\hat{\boldsymbol{j}} + \left(\frac{\partial f}{\partial z}g + f\frac{\partial g}{\partial z}\right)\hat{\boldsymbol{k}} \\ &= \left(\frac{\partial f}{\partial x}\hat{\boldsymbol{i}} + \frac{\partial f}{\partial y}\hat{\boldsymbol{j}} + \frac{\partial f}{\partial z}\hat{\boldsymbol{k}}\right)g + f\left(\frac{\partial g}{\partial x}\hat{\boldsymbol{i}} + \frac{\partial g}{\partial y}\hat{\boldsymbol{j}} + \frac{\partial g}{\partial z}\hat{\boldsymbol{k}}\right) \\ &= (\nabla f)g + f(\nabla g).\end{aligned}$$

Solution: Exercise 1.14. Let $f(x) = \sin x$. In order to find the Maclaurin series coefficients, we must evaluate

$$\left(\frac{\mathrm{d}^k}{\mathrm{d}x^k}\sin x\right)\Bigg|_{x=0} \qquad \text{for} \quad k = 0, 1, 2, 3, 4, \ldots$$

First, we calculate the first few coefficients and we look for a pattern:

$$f(0) = \sin 0 = 0 \tag{11.1}$$

$$f'(0) = \cos 0 = 1 \tag{11.2}$$

$$f''(0) = -\sin 0 = 0 \tag{11.3}$$

$$f'''(0) = -\cos 0 = -1 \tag{11.4}$$

$$f^{(4)}(0) = \sin 0 = 0 \tag{11.5}$$

$$f^{(5)}(0) = \cos 0 = 1. \tag{11.6}$$

Hence, we notice that the coefficients alternate between 0, 1, and −1.

Now, we substitute the coefficients into the expansion. We thus obtain the Maclaurin series for $\sin(x)$:

$$\begin{aligned}\sum_{k=0}^{\infty}\frac{f^{(k)}(0)}{k!}(x-0)^k &= f(0) + \frac{f'(0)}{1!}x + \frac{f''(0)}{2!}x^2 + \frac{f'''(0)}{3!}x^3 + \cdots \\ &= 0 + \left(\frac{1}{1!}x\right) + 0 + \left(\frac{-1}{3!}x^3\right) + 0 + \left(\frac{1}{5!}x^5\right) + \cdots \\ &= x - \frac{x^3}{3!} + \frac{x^5}{5!} - \cdots\end{aligned}$$

We can spot a pattern that allows us to derive an expansion for the nth term in the series:

$$\frac{(-1)^n}{(2n+1)!}x^{2n+1}, \quad n = 0, 1, 2, 3, \ldots$$

Substituting this into the formula for Taylor series expansion, we obtain:

$$\sin x = \sum_{k=0}^{\infty} \frac{(-1)^k}{(2k+1)!} x^{2k+1} = x - \frac{x^3}{3!} + \frac{x^5}{5!} - \cdots$$

Next, we want to find the Maclaurin series for $\cos(x)$ at $x = 0$.

$$\begin{aligned}
\cos x &= \frac{\mathrm{d}}{\mathrm{d}x} \sin x \\
&= \frac{\mathrm{d}}{\mathrm{d}x} \sum_{k=0}^{\infty} \frac{(-1)^k}{(2k+1)!} x^{2k+1} \\
&= \frac{\mathrm{d}}{\mathrm{d}x} \left(x - \frac{x^3}{3!} + \frac{x^5}{5!} - \cdots \right) \\
&= 1 - \frac{x^2}{2!} + \frac{x^4}{4!} - \cdots \\
&= \sum_{k=0}^{\infty} \frac{(-1)^k x^{2k}}{(2k)!}.
\end{aligned}$$

Solution: Exercise 1.15. Let $f(x) = \ln(1 + x)$. Then,

$$\begin{aligned}
f'(x) &= (1 + x)^{-1} \\
f''(x) &= -(1 + x)^{-2} \\
f'''(x) &= (-2)(-1)(1 + x)^{-3}
\end{aligned}$$

and, for $n \geqslant 2$,

$$f^{(n)}(x) = (-1)^{n-1}(n-1)(1 + x)^{-n}.$$

Thus, $f(0) = 0$, and, for $n \geqslant 1$,

$$f^{(n)}(0) = (-1)^{n-1}(n-1)!$$

Therefore, the Maclaurin series for $\ln(1 + x)$ is:

$$\begin{aligned}
\ln(1 + x) &= \sum_{n=0}^{\infty} \frac{f^{(n)}(0)}{n!} x^n \\
&= \sum_{n=1}^{\infty} (-1)^{n-1} \frac{(n-1)!}{n!} x^n \\
&= \sum_{n=1}^{\infty} (-1)^{n-1} \frac{x^n}{n}.
\end{aligned}$$

(i) Recall that

$$\cos x = \sum_{k=0}^{\infty} \frac{(-1)^k x^{2k}}{(2k)!} = 1 - \frac{x^2}{2} + \frac{x^4}{24} + \cdots$$

Thus, $\cos(2x) = 1 - \frac{(2x)^2}{2} + \frac{(2x)^4}{24} + \cdots = 1 - 2x^2 + \frac{2x^4}{3} + \cdots$
Therefore,

$$\begin{aligned}
\cos(2x) \cdot \ln(1 + x) &= \left(1 - 2x^2 + \frac{2x^4}{3} + \cdots\right) \cdot \left(x - \frac{x^2}{2} + \frac{x^3}{3} - \frac{x^4}{4} + \cdots\right) \\
&= x - \frac{x^2}{2} + \left(\frac{1}{3} - 2\right)x^3 + \left(-\frac{1}{4} + 1\right)x^4 + \cdots \\
&= x - \frac{x^2}{2} - \frac{5}{3}x^3 + \frac{3}{4}x^4 + \cdots
\end{aligned}$$

(ii)

$$\begin{aligned}
\ln(\cos x) &= \ln\left(1 - \frac{x^2}{2} + \frac{x^4}{24} + \cdots\right) \\
&= \left(-\frac{x^2}{2} + \frac{x^4}{24} + \cdots\right) - \frac{1}{2}\left(-\frac{x^2}{2} + \frac{x^4}{24} + \cdots\right)^2 + \cdots \\
&= -\frac{x^2}{2} + \left[\frac{1}{24} - \frac{1}{2}\left(-\frac{1}{2}\right)^2\right]^2 x^4 + \cdots \\
&= -\frac{x^2}{2} - \frac{x^4}{6} + \cdots
\end{aligned}$$

Solution: Exercise 1.16. Let us denote the observed frequency by f, the source frequency by f_0, the source velocity by v and the velocity of light by our conventional c. Let us now consider the relativistic Doppler shift

$$f = f_0 \sqrt{\frac{1 + \frac{v}{c}}{1 + \frac{v}{c}}} = f_0 \frac{\sqrt{1 - \frac{v^2}{c^2}}}{1 - \frac{v}{c}}.$$

Since we are considering the case $v \ll c$, we can use Taylor expansion up to the first two terms, and neglect higher order terms, i.e.

$$\left[1 - \left(\frac{v}{c}\right)^2\right] = 1 - \frac{1}{2}\frac{v^2}{c^2} + \cdots$$

which leads to

$$f \approx f_0 \frac{1 - \dfrac{v^2}{2c^2}}{1 - \dfrac{v}{c}}.$$

When expressed in terms of the frequency shift, this becomes

$$\frac{f - f_0}{f_0} = \frac{\Delta f}{f_0} \approx \frac{v}{c}.$$

Solution: Exercise 1.17.

(i) Let $I \equiv \int x \sin x \, \mathrm{d}x$. We identify $u = x$ and $v' = \sin x$. Thus $u' = 1$ and $v = -\cos x$. Thus,

$$\begin{aligned} I &= -x \cos x + \int \cos x \, \mathrm{d}x \\ &= -x \cos x + \sin x + C, \end{aligned}$$

where C is a constant of integration.

(ii) Let $I \equiv \int \sin x \sin(\cos x) \, \mathrm{d}x$. Then, let $y = \cos x$. Thus,

$$\frac{\mathrm{d}y}{\mathrm{d}x} = -\sin x \quad \text{and} \quad \mathrm{d}x = -\frac{1}{\sin x}\mathrm{d}y.$$

Thus, $I = -\int \sin y \, \mathrm{d}y$. Now, solving the standard integral

$$\int \sin y \, \mathrm{d}y = -\cos y.$$

We thus have, $I = \cos y$.

'Undoing' the substitution $y = \cos x$, $I = \cos(\cos x)$. Therefore,

$$\int \sin x \sin(\cos x) \, \mathrm{d}x = \cos(\cos x) + C,$$

where C is a constant of integration.

(iii) Let $I = \int x^2 e^{3x} \, \mathrm{d}x$. We identify $u = x^2$ and $v' = e^{3x}$. Thus, $u' = 2x$ and $v = \frac{1}{3}e^{3x}$. Hence,

$$I = \frac{x^2 e^{3x}}{3} - \frac{2}{3}\int x e^{3x} \, \mathrm{d}x.$$

Now, let $J = \int x e^{3x} \, \mathrm{d}x$. Integrating by parts again, we identify $u = x$ and $v' = e^{3x}$. Thus, $u' = 1$ and $v = \frac{1}{3}e^{3x}$. Hence,

$$J = \frac{x e^{3x}}{3} - \frac{1}{3}\int e^{3x} \, \mathrm{d}x.$$

Now, solving

$$\frac{1}{3}\int e^{3x}\,\mathrm{d}x = \frac{1}{9}e^{3x}.$$

Thus,

$$J = \frac{xe^{3x}}{3} - \frac{e^{3x}}{9}.$$

Further,

$$\frac{2}{3}\int xe^{3x}\,\mathrm{d}x = \frac{2}{3}J = \frac{2xe^{3x}}{9} - \frac{2e^{3x}}{27}$$

and

$$I = \frac{x^2e^{3x}}{3} - \frac{2xe^{3x}}{9} + \frac{2e^{3x}}{27}.$$

Therefore,

$$\int x^2e^{3x}\,\mathrm{d}x = \frac{x^2e^{3x}}{3} - \frac{2xe^{3x}}{9} + \frac{2e^{3x}}{27} + C = \frac{(9x^2 - 6x + 2)e^{3x}}{27} + C,$$

where C is a constant of integration.

Solution: Exercise 1.18.

(i) Using integration by parts $u = t^x$ and $v = e^{-t}$, we have

$$\begin{aligned}
\Gamma(x+1) &= \int_0^\infty t^x e^{-t}\,\mathrm{d}t \\
&= -\int_0^\infty u\frac{\mathrm{d}v}{\mathrm{d}t}\,\mathrm{d}t \\
&= -\lim_{b\to\infty} b^x e^{-b} + \lim a \to 0^+ a^x e^{-a} + \int_0^\infty x t^{x-1} e^{-t}\,\mathrm{d}t \\
&= 0 + 0 + x\int_0^\infty t^{x-1}e^{-t}\,\mathrm{d}t \\
&= 0 + x\Gamma(x),
\end{aligned}$$

as required.

(ii) Now,

$$\begin{aligned}
\Gamma(1) &= \int_0^\infty t^0 e^{-t}\,\mathrm{d}t \\
&= \lim_{b\to\infty}\int_0^b e^{-t}\,\mathrm{d}t \\
&= -\lim_{b\to\infty} e^{-b} + 1 = 1.
\end{aligned}$$

Thus,

$$\begin{aligned}
\Gamma(1) &= 1 \\
\Gamma(2) &= 1 \times \Gamma(1) = 1 \\
\Gamma(3) &= 2 \times \Gamma(2) = 2 \\
\Gamma(4) &= 3 \times \Gamma(3) = 3 \times 2 = 3! \\
\Gamma(5) &= 4 \times \Gamma(4) = 4 \times 3! = 4!
\end{aligned}$$

By induction, since $\Gamma(1) = 1$ and assuming $\Gamma(k) = (k - 1)!$, we have that

$$\Gamma(k + 1) = k \times \Gamma(k) = k \times (k - 1)! = k!.$$

[*There's no need for the full formal induction proof, but you do need to show that* $\Gamma(1) = 1$.]

Solution: Exercise 1.19.

(i) It is a first order separable ordinary differential equation. Re-write in a form which is easier to solve:

$$\frac{1}{x^2 + 1}\frac{\mathrm{d}x}{\mathrm{d}t} = t^2.$$

Integrating both sides

$$\int \frac{1}{x^2 + 1}\frac{\mathrm{d}x}{\mathrm{d}t}\,\mathrm{d}t = \int t^2\,\mathrm{d}t$$

leads to

$$\tan^{-1}(x) = \frac{t^3}{3} + C,$$

where C is a constant of integration.
Isolating x,

$$x = \tan\left(\frac{t^3}{3} + C\right).$$

(ii) It is a first order separable ordinary differential equation. Re-write in a form which is easier to solve:

$$\frac{1}{x + 1}\frac{\mathrm{d}x}{\mathrm{d}t} = t^2.$$

Integrating both sides

$$\int \frac{1}{x + 1}\frac{\mathrm{d}x}{\mathrm{d}t}\,\mathrm{d}t = \int t^2\,\mathrm{d}t$$

leads to

$$\ln(x + 1) = \frac{t^3}{3} + C,$$

where C is a constant of integration.

Isolating x,

$$x = \exp\left(\frac{t^3}{3} + C\right) - 1.$$

Solution: Exercise 1.20.

(i) First, find the complementary solution, that is, the solution of the homogeneous equation

$$x''_h + 2x''_h + 2x_h = 0.$$

The characteristic equation

$$\lambda^2 + 2\lambda + 2 = 0$$

has complex roots

$$-1 \pm i.$$

Thus, two solutions of the homogeneous equation are the real and imaginary parts of

$$\exp[(-1 + i)t] = e^{-t}e^{it} = e^{-t}(\cos t + i \sin t).$$

Hence, the general solution of the homogeneous equation is

$$x_h = e^{-t}(c_1 \cos t + c_2 \sin t),$$

where c_1 and c_2 are constants. Now, we need to find a particular solution to the original equation. First, we re-write the equation as

$$\frac{\mathrm{d}^2x}{\mathrm{d}t^2} + 2\frac{\mathrm{d}x}{\mathrm{d}t} + 2x = \frac{1}{2}[1 - \cos(2t)],$$

and we look for a particular solution of the form

$$x_p = \alpha \cos(2t) + \beta \sin(2t) + \gamma.$$

This gives

$$2\gamma + (-4\alpha + 4\beta + 2\alpha)\cos(2t) + (-4\beta - 4\alpha + 2\beta)\sin(2t) = \frac{1}{2} - \frac{1}{2}\cos(2t).$$

Thus, the constant terms give

$$\gamma = \frac{1}{4},$$

the coefficients of cos(2t) give

$$\alpha - 2\beta = \frac{1}{4},$$

and the coefficients of sin(2t) give

$$2\alpha + \beta = 0.$$

Therefore, we obtain

$$\alpha = \frac{1}{20} \quad \text{and} \quad \beta = -\frac{1}{10}$$

and the general solution of the original equation is

$$x(t) = x_p + x_h = \frac{1}{20}\cos(2t) - \frac{1}{10}\sin(2t) + \frac{1}{4} + e^{-t}(c_1 \cos t + c_2 \sin t).$$

Now,

$$2 = x(0) = \frac{1}{20} + \frac{1}{4} + c_1 \Longrightarrow c_1 = \frac{17}{10}.$$

Also,

$$-1 = x'(0) = -\frac{1}{5} - \frac{17}{10} + c_2 \Longrightarrow c_2 = \frac{9}{10}.$$

Therefore, the solution is

$$x(t) = \frac{1}{20}[\cos(2t) - 2\sin(2t) + 5 + e^{-t}(34\cos t + 18\sin t)].$$

(ii) The characteristic equation corresponding to the homogeneous equation is

$$\lambda^2 - 2\lambda + 1 = 0,$$

which has the repeated root

$$\lambda = 1.$$

Thus, the general solution of the homogeneous equation is

$$x_h = (c_1 + c_2 t)e^t.$$

Re-write the inhomogeneous equation as

$$\begin{aligned}\frac{d^2x}{dt^2} - 2\frac{dx}{dt} + x = \cosh t \cosh(2t) &= \frac{1}{4}(e^t + e^{-t})(e^{2t} + e^{-2t}) \\ &= \frac{1}{4}(e^{3t} + e^t + e^{-t} + e^{-3t}).\end{aligned}$$

Note that both e^t and te^t solve the homogeneous equation, so we should look for a particular solution of the form

$$x_p(t) = \alpha e^{3t} + \beta t^2 e^t + \gamma e^{-t} + \delta e^{-3t}.$$

Substituting into the equation gives

$$(9 - 6 + 1)\alpha e^{3t} + 2\beta e^t + (1 + 2 + 1)\gamma e^{-t} + (9 + 6 + 1)\delta e^{-3t}$$
$$= \frac{1}{4}(e^{3t} + e^t + e^{-t} + e^{-3t}).$$

Thus,

$$\alpha = \frac{1}{16}, \quad \beta = \frac{1}{8}, \quad \gamma = \frac{1}{16}, \quad \text{and} \quad \delta = \frac{1}{64}.$$

Therefore,

$$x_p(t) = \frac{1}{64}(4e^{3t} + 8t^2 e^t + 4e^{-t} + e^{-3t}).$$

The general solution of the original equation is

$$x(t) = \frac{1}{64}(4e^{3t} + 8t^2 e^t + 4e^{-t} + e^{-3t}) + (c_1 + c_2 t)e^t.$$

The initial conditions give

$$2 = x(0) = \frac{9}{64} + c_1 \Longrightarrow c_1 = \frac{119}{64}.$$

Further,

$$-1 = x'(0) \Longrightarrow c_2 = -\frac{47}{64}.$$

Therefore,

$$x(t) = \frac{1}{64}[4e^{3t} + (119 - 188t + 8t^2)e^t + 4e^{-t} + e^{-3t}]$$
$$= \frac{1}{16}e^{3t} + \left(\frac{119}{64} - \frac{47}{16}t + \frac{1}{8}t^2\right)e^t + \frac{1}{16}e^{-t} + \frac{1}{64}e^{-3t}.$$

(iii) Solution of the homogeneous equation:

$$x_h = c_1 e^t + c_2 e^{-t}.$$

Look for a particular solution of the form

$$x_p = \alpha t^2 + \beta t + \gamma,$$

which gives

$$(-\alpha)t^2 + (-\beta)t + (2\alpha - \gamma) = 7t^2 + 3t.$$

Equating coefficients gives

$$\alpha = -7, \quad \beta = -3, \quad \text{and} \quad \gamma = 2\alpha = -14.$$

Hence, the general solution of the equation is

$$x(t) = -7t^2 - 3t - 14 + c_1 e^t + c_2 e^{-t},$$

giving

$$x'(t) = -14t - 3 + c_1 e^t - c_2 e^{-t}.$$

Therefore, the initial conditions become

$$2 = x(0) = -14 + c_1 + c_2$$

and

$$-1 = x'(0) = -3 + c_1 - c_2,$$

giving

$$c_1 = 9 \quad \text{and} \quad c_2 = 7,$$

therefore the solution of the initial value problem is

$$x(t) = -7t^2 - 3t - 14 + 9e^t + 7e^{-t}.$$

Solution: Exercise 1.21.

(i)

$$x(t) = e^{\lambda t}$$
$$x'(t) = \lambda e^{\lambda t}$$
$$x''(t) = \lambda^2 e^{\lambda t}.$$

Substituting this into the differential equation gives

$$(\lambda^2 - 2\alpha\lambda + \alpha^2)e^{\lambda t} = 0.$$

Since $e^{\lambda t} \neq 0$, we have the *characteristic equation*

$$\lambda^2 - 2\alpha\lambda + \alpha^2 = (\lambda - \alpha)^2 = 0.$$

Hence, $\lambda = \alpha$ is a repeated root.

(ii) Now, set $x(t) = te^{\alpha t}$. Thus,

$$x'(t) = (1 + \alpha t)e^{\alpha t}$$
$$x''(t) = (2\alpha + \alpha^2 t)e^{\alpha t}.$$

Substituting this into the lhs of the differential equation gives

$$(2\alpha + \alpha^2 t)e^{\alpha t} - 2\alpha(1 + \alpha t)e^{\alpha t} + \alpha^2 t e^{\alpha t} = 0.$$

Hence, $x(t) = te^{\alpha t}$ is a solution of the differential equation. Furthermore,

$$x(0) = 0e^{0\alpha} = 0$$
$$x'(0) = (1 + 0\alpha)e^{0\alpha} = 1.$$

Therefore, x satisfies the initial value problem.

(iii)

(a) The characteristic equation is

$$0 = \lambda^2 - (2\alpha + \varepsilon)\lambda + \alpha(\alpha + \varepsilon) = (\lambda - \alpha)(\lambda - \alpha - \varepsilon).$$

Thus, the characteristic equation has the (distinct) roots

$$\lambda_1 = \alpha \quad \text{and} \quad \lambda_2 = \alpha + \varepsilon.$$

The general solution of the differential equation is

$$X(t) = c_1 e^{\alpha t} + c_2 e^{(\alpha+\varepsilon)t}.$$

Therefore,

$$0 = X(0) = c_1 + c_2$$
$$1 = X'(0) = c_1\alpha + c_2(\alpha + \varepsilon).$$

Solving for c_1 and c_2 gives

$$c_1 = -\varepsilon^{-1} \quad \text{and} \quad c_2 = \varepsilon^{-1}.$$

Hence,

$$X(t) = \varepsilon^{-1}e^{\alpha t}(-1 + e^{\varepsilon t}) = e^{\alpha t}\left(\frac{e^{\varepsilon t} - 1}{\varepsilon}\right).$$

(b) Now,

$$e^{\varepsilon t} = 1 + \varepsilon t + \frac{(\varepsilon t)^2}{2} + \cdots \Longrightarrow \lim_{\varepsilon \to 0} \frac{e^{\varepsilon t} - 1}{\varepsilon} = t.$$

Therefore,

$$\lim_{\varepsilon \to 0} X(t) = e^{\alpha t} \lim_{\varepsilon \to 0}\left(\frac{e^{\varepsilon t} - 1}{\varepsilon}\right) = e^{\alpha t} t,$$

as required.

IOP Publishing

Classical Mechanics
A professor-student collaboration
Mario Campanelli

Chapter 12

Solutions to Chapter 2: Newton's laws

Solution: Exercise 2.1
BOOKWORK.

Solution: Exercise 2.2: An old friend

(i) From Newton's second law, $\boldsymbol{F} = m\boldsymbol{a}$, and, since $m = 1$kg, we have $\hat{\boldsymbol{F}} = \hat{\boldsymbol{a}}$. At $t = 0$,

$$\boldsymbol{a}(t = 0) = 2\hat{\boldsymbol{i}} + 6\hat{\boldsymbol{k}}.$$

Thus, the unit vector defining the direction of the force $\boldsymbol{F}$ is

$$\hat{\boldsymbol{F}}(t = 0) = \hat{\boldsymbol{a}}(t = 0) = \frac{\boldsymbol{a}}{|\boldsymbol{a}|} = \frac{2\hat{\boldsymbol{i}} + 6\hat{\boldsymbol{k}}}{\sqrt{2^2 + 6^2}} = \frac{1}{\sqrt{10}}(\hat{\boldsymbol{i}} + 3\hat{\boldsymbol{k}}).$$

(ii) From Newton's second law, $\boldsymbol{F} = m\boldsymbol{a}$, and, since $m = 1$ kg, we have $\hat{\boldsymbol{F}} = \hat{\boldsymbol{a}}$. At $t = 0$,

$$\boldsymbol{a}(t = 0) = -\hat{\boldsymbol{i}} + 120\hat{\boldsymbol{k}}.$$

Thus, the unit vector defining the direction of the force $\boldsymbol{F}$ is

$$\hat{\boldsymbol{F}}(t = 0) = \hat{\boldsymbol{a}}(t = 0) = \frac{\boldsymbol{a}}{|\boldsymbol{a}|} = \frac{-\hat{\boldsymbol{i}} + 120\hat{\boldsymbol{k}}}{\sqrt{(-1)^2 + 120^2}} = \frac{1}{\sqrt{14\,401}}(-\hat{\boldsymbol{i}} + 120\hat{\boldsymbol{k}}).$$

doi:10.1088/978-0-7503-2690-2ch12

(iii) From Newton's second law, $\boldsymbol{F} = m\boldsymbol{a}$, and, since $m = 1$ kg, we have $\hat{\boldsymbol{F}} = \hat{\boldsymbol{a}}$. At $t = 0$,

$$\boldsymbol{a}(t = 0) = -4\hat{\boldsymbol{\imath}} + 4\hat{\boldsymbol{\jmath}} + 2\hat{\boldsymbol{k}}.$$

Thus, the unit vector defining the direction of the force $\boldsymbol{F}$ is

$$\hat{\boldsymbol{F}}(t = 0) = \hat{\boldsymbol{a}}(t = 0) = \frac{\boldsymbol{a}}{|\boldsymbol{a}|} = \frac{-4\hat{\boldsymbol{\imath}} + 4\hat{\boldsymbol{\jmath}} + 2\hat{\boldsymbol{k}}}{\sqrt{(-4)^2 + 4^2 + 2^2}} = \frac{1}{3}(-2\hat{\boldsymbol{\imath}} + 2\hat{\boldsymbol{\jmath}} + \hat{\boldsymbol{k}}).$$

(iv) From Newton's second law, $\boldsymbol{F} = m\boldsymbol{a}$, and, since $m = 1$ kg, we have $\hat{\boldsymbol{F}} = \hat{\boldsymbol{a}}$. At $t = 0$,

$$\boldsymbol{a}(t = 0) = -3\hat{\boldsymbol{\jmath}} + 4\hat{\boldsymbol{k}}.$$

Thus, the unit vector defining the direction of the force $\boldsymbol{F}$ is

$$\hat{\boldsymbol{F}}(t = 0) = \hat{\boldsymbol{a}}(t = 0) = \frac{\boldsymbol{a}}{|\boldsymbol{a}|} = \frac{-3\hat{\boldsymbol{\jmath}} + 4\hat{\boldsymbol{k}}}{\sqrt{3^2 + 4^2}} = \frac{1}{5}(-3\hat{\boldsymbol{\jmath}} + 4\hat{\boldsymbol{k}}).$$

Solution: Exercise 2.3

(i) Since $m = 1$ kg, the vertical force of gravity is

$$F = mg = g.$$

The component perpendicular to the plane is

$$F_{\perp} = g\cos\alpha.$$

The component parallel to the plane is

$$F_{\parallel} = g\sin\alpha.$$

(ii) For the body to start moving, the parallel component must be greater than the friction force.

Friction force is

$$F_f = \mu F_{\perp} = \mu g\cos\alpha.$$

Thus,

$$F_{\parallel} > F_f$$
$$g\sin(\alpha_m) > \mu g\cos(\alpha_m)$$

where α_m is the minimal value of α for which the parallel component of the gravitational force is stronger than friction, and the body starts moving.

Hence,

$$\tan(\alpha_m) > \mu = 0.2 \Longrightarrow \alpha_m = 0.197 \text{ rad}.$$

(iii) When the mass starts moving, the friction coefficient will be reduced to $\mu' = 0.1$. Therefore, along the parallel component, the force will be

$$F = F_{\parallel} - F'_f = g\sin(\alpha_m) - \mu' g\cos(\alpha_m) = 0.975 \text{ N}$$

and, since $m = 1$ kg, the acceleration is

$$a = 0.957 \text{ m s}^{-2}.$$

Solution: Exercise 2.4

(i) The component of the gravitational force along the plane is

$$F_{\parallel} = mg\sin\alpha.$$

Clearly, without any friction,

$$a = g\sin\alpha.$$

Integrating once and twice w.r.t. t,[1] we obtain

$$v = g\sin(\alpha)t$$
$$s = \frac{1}{2}g\sin(\alpha)t^2.$$

The normal force is

$$F_{\perp} = mg\cos\alpha.$$

(ii) The total force with the plane's friction but without air friction is

$$F_s = F_{\parallel} - F_f = F_{\parallel} - \mu F_{\perp} = mg\sin\alpha - \mu mg\cos\alpha.$$

In the presence of air friction,

$$F = m\frac{dv}{dt} = mg(\sin\alpha - \mu\cos\alpha) - \beta v.$$

As the mass's speed $v(t)$ is a function of time, we cannot equate integration dt to multiplication by t, i.e. $\int v\, dt \neq vt$. We must collect terms in v by dividing the lhs by the rhs:

[1] Recall that the angle α is constant.

$$\frac{m}{mgc - \beta v}\frac{dv}{dt} = 1$$

where we have written $c = (\sin\alpha - \mu\cos\alpha)$, for convenience. Hence,

$$\int \frac{m}{mgc - \beta v}\, dv = \int dt.$$

This integral becomes simple upon making the substitution $y = (mgc - \beta v)$, and using the rule $\int \frac{1}{y}\, dy = \ln y + \text{constant}$.[2] Since $dy = -\beta dv$, we have

$$\int \frac{dy}{y} = -\frac{\beta}{m}\int dt.$$

Thus,

$$\ln y = -\frac{\beta}{m}t + \tilde{A} \Longrightarrow y = A\exp\left(-\frac{\beta}{m}t\right).$$

Undoing substitution,

$$mg(\sin\alpha - \mu\cos\alpha) - \beta v = A\exp\left(-\frac{\beta}{m}t\right).$$

Therefore, isolating v,

$$v = \frac{1}{\beta}\left[mg(\sin\alpha - \mu\cos\alpha) - A\exp\left(-\frac{\beta}{m}t\right)\right].$$

Since at the instant $t = 0$ the mass is not moving, we deduce our initial condition $v(t = 0) = 0$. Thus,

$$A = mg(\sin\alpha - \mu\cos\alpha).$$

Hence,

$$v = \frac{1}{\beta}mg(\sin\alpha - \mu\cos\alpha)\left[1 - \exp\left(-\frac{\beta}{m}t\right)\right].$$

Solution: Exercise 2.5

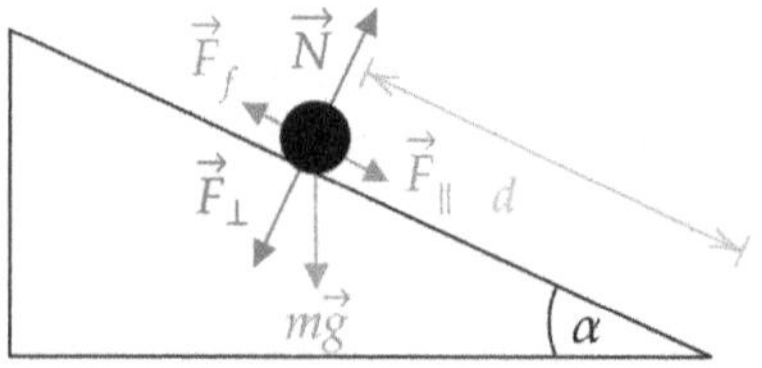

[2] It is worth noting that it is also possible to use an 'integrating factor' approach, or substitute $v = dx/dt$ and solve for x, though these methods are arguably more complicated.

(i) The forces acting on the system are:
- gravity, mg, which has two components (one parallel and one perpendicular to the plane);
- reaction (or normal) force of the plane, which exactly compensates the component of the gravitational force perpendicular to the plane;
- friction, which is proportional to the reaction force, but perpendicular to it, and pointing in the direction opposite to that of motion.

The combination of the two components parallel to the plane gives

$$F_{\parallel} = mg(\sin\alpha - \mu\cos\alpha).$$

(ii) A body under a constant acceleration, clearly $a = g(\sin\alpha - \mu\cos\alpha)$, and zero initial speed will have a displacement given by

$$s = \frac{1}{2}at^2.$$

The distance d will be travelled in a time

$$t = \sqrt{\frac{2s}{a}} = \sqrt{\frac{2d}{g(\sin\alpha - \mu\cos\alpha)}},$$

as required.

Solution: Exercise 2.6

(i) The first mass is acted upon by two forces: one acting along the direction of the plane given by

$$F_{\parallel} = m_1 g\sin\alpha - T$$

(where T is the rope tension), and another one, which is perpendicular to it, given by

$$F_{\perp} = m_1 g\cos\alpha.$$

The perpendicular component is compensated by the normal force of the plane itself (but will be important to determine the friction of the plane).

Let r be the ratio of the two masses such that $r \equiv m_2/m_1$, and $m \equiv m_1$. The force acting on the second mass will be the force of gravity minus the rope tension, i.e.

$$F_2 = rmg - T$$

along the perpendicular direction.

Since the length of the rope does not change, the parallel movement of mass m_1 and the vertical movement of mass m_2 are connected. Thus, we can write the total force (considering as positive the direction where the second mass falls) as

$$F_t = rmg - mg \sin\alpha = mg(r - \sin\alpha).$$

At equilibrium, this total force is zero, so the tension of the cable exactly compensates the gravitational force of the second mass, i.e.

$$T = rmg$$

and $F_t = 0$ for the ratio $r = \sin\alpha$.

(ii) In the case of friction, its absolute value will be the product between the normal component of gravity and the friction coefficient, but the sign will be opposite to the direction of the prevailing force:

$$F_f = \mu|F_\perp| = \mp\mu mg \cos\alpha.$$

The negative signs applies when the system would tend to move towards the positive direction, the positive otherwise.

(iii) Hence, the system will move when the total force (without friction) is equal to the absolute value of friction, i.e.

$$|F_t| = |mg(r - \sin\alpha)| = |\mu mg \cos\alpha|.$$

The two solutions correspond to taking opposite signs, i.e.

$$mg(r - \sin\alpha) = \mu mg \cos\alpha$$
$$mg(r - \sin\alpha) = -\mu mg \cos\alpha.$$

Dividing both sides of both equations by mg, we obtain:

$$r - \sin\alpha = \mu \cos\alpha$$
$$r - \sin\alpha = -\mu \cos\alpha.$$

Therefore, combining both equations we can write

$$r = \sin\alpha \pm \mu \cos\alpha.$$

(iv) If the system moves towards m_1 (negative direction), the force in the direction parallel to the plane is

$$F_t = mg(r - \sin\alpha).$$

The acceleration, which is parallel to the inclined plane, will be

$$a = g(r - \sin\alpha)(r + 1).$$

The acceleration with respect to the x-axis will be the horizontal component

$$a_x = \frac{g\cos\alpha(r - \sin\alpha)}{r+1} = \frac{g\cos\alpha(m_2 - m_1\sin\alpha)}{m_1 + m_2},$$

which is indeed the acceleration of the plane needed to keep the masses at rest in the accelerated system.

Solution: Exercise 2.7
BOOKWORK.

Solution: Exercise 2.8

(i) N2Law for the system is

$$m\boldsymbol{a} = m\boldsymbol{g} - \beta\boldsymbol{v}.$$

All the forces are in the $\hat{\boldsymbol{j}}$-direction, where positive values are taken to be pointing downwards (towards the Earth).
We can therefore write N2Law as:

$$ma\hat{\boldsymbol{j}} = mg\hat{\boldsymbol{j}} - \beta v\hat{\boldsymbol{j}}.$$

Another way to express this solution may be

$$\sum F_y = ma_y = mg - \beta v.$$

(ii) Now, we re-write our N2Law ignoring the vector notation since it's clear from the set-up that we are dealing with a one-dimensional problem,

$$ma = mg - \beta v$$
$$\Longrightarrow m\frac{\mathrm{d}v}{\mathrm{d}t} = mg - \beta v.$$

We are dealing with a separable variables first order ODE, which we are not asked to solve but rather verify that solution given by equation (*) satisfies our equation. Thus, we only need to differentiate (*), i.e.

$$\frac{\mathrm{d}v}{\mathrm{d}t} = -\frac{A}{\beta}\left(-\frac{\beta}{m}\right)e^{-\beta t/m} = \frac{A}{m}e^{-\beta t/m} \Longrightarrow m\frac{\mathrm{d}v}{\mathrm{d}t} = Ae^{-\beta t/m}.$$

Let's now check that rhs equals lhs. Our lhs can be re-written as

$$mg - \beta v = mg - \beta\left(\frac{mg}{\beta} - \frac{A}{\beta}e^{-\beta t/m}\right) = Ae^{-\beta t/m} \Longrightarrow \text{lhs} = \text{rhs}.$$

(iii) This is just a substitution game. If we set $t = 0$ and we are assuming that the body starts from rest, i.e. $v = 0$ at $t = 0$, we have that equation (*) becomes

$$v(t = 0) = \frac{mg}{\beta} - \frac{A}{\beta}e^{-\beta(0)/m} = 0 \Longrightarrow \frac{mg}{\beta} - \frac{A}{\beta} = 0 \Longrightarrow A = mg.$$

Solution: Exercise 2.9

(i) The total force acting on the body in air, taking the downwards direction of the vertical axis to be positive, is

$$F = mg - \beta v.$$

Thus, the terminal velocity is

$$v = \frac{mg}{\beta}.$$

The ball falling in vacuum will have reached this velocity at time

$$t_f = \frac{m}{\beta}.$$

(ii) The equation of motion will be

$$m\frac{\mathrm{d}v}{\mathrm{d}t} = mg - \beta v.$$

Dividing both sides by rhs, we obtain

$$\frac{m}{mg - \beta v}\frac{\mathrm{d}v}{\mathrm{d}t} = 1,$$

which can be re-written as

$$\frac{\mathrm{d}v}{mg - \beta v} = \frac{\mathrm{d}t}{m}.$$

Multiplying both sides by $-\beta$, we have

$$\frac{\mathrm{d}v}{v - \frac{mg}{\beta}} = -\frac{\beta}{m}\mathrm{d}t.$$

Integrating both sides,

$$\ln\left(v - \frac{mg}{\beta}\right) = -\frac{\beta}{m}t + \tilde{A} \Longrightarrow v - \frac{mg}{\beta} = A\exp\left(-\frac{\beta}{m}t\right).$$

Isolating v and with the initial condition of speed and position zero at $t = 0$, $A = -mg/\beta$,

$$v = \frac{mg}{\beta}\left[1 - \exp\left(-\frac{\beta}{m}t\right)\right].$$

(iii) From the result above, at time t_f, its speed will be

$$v = \frac{mg}{\beta}(1 - e^{-1}) = \left(1 - \frac{1}{e}\right)\frac{mg}{\beta},$$

reduced by 30% w.r.t. the frictionless case.

Solution: Exercise 2.10: Baking cup

(i) During the fall, the total force acting on the baking cup has got an intensity of $F = W - F_f$, where W is the weight of the baking cup and F_f is air friction.

If air friction scales up with velocity, during the fall the intensity of the force will decrease until it will be neglectable so that one could assume $F_f = W$. In such a set-up, we can consider the motion to be uniform linear motion with speed $v_L = 1.4 \text{ m s}^{-1}$.

During the fall, air friction is negligible and the motion occurs as a result of weight producing an acceleration equal to g.

We now apply to the function $v(t)$ the two conditions described in the problem. The first condition is satisfied if

$$\lim_{t\to+\infty} \alpha\frac{e^{\beta t} - 1}{e^{\beta t} + 1} = v_L,$$

which can be found by applying De L'Hopital, i.e.

$$\lim_{t\to+\infty} \alpha\frac{e^{\beta t} - 1}{e^{\beta t} + 1} = \lim_{t\to+\infty} \alpha\frac{\beta e^{\beta t}}{\beta e^{\beta t}} = \alpha.$$

Thus,

$$\alpha = v_L = 1.4 \text{ m s}^{-1}.$$

The acceleration is described by the function

$$a(t) = \frac{\mathrm{d}v}{\mathrm{d}t} = \alpha\frac{\beta e^{\beta t}(e^{\beta t} + 1) - \beta e^{\beta t}(e^{\beta t} - 1)}{(e^{\beta t} + 1)^2} = 2\alpha\beta\frac{e^{\beta t}}{(e^{\beta t} + 1)^2}.$$

For the acceleration of the baking cup to be equal to the gravitational acceleration g, one must have

$$a(t=0) = \frac{2\alpha\beta}{4} = g$$

whence

$$\alpha\beta = 2g \Longrightarrow \beta = \frac{2g}{\alpha} = \frac{2g}{v_L} = \frac{2 \times 9.8 \text{ m s}^{-2}}{1.4 \text{ m s}^{-1}} = 14 \text{ s}^{-1}.$$

The function $v(t)$ satisfying the conditions enumerated in the task is the following:

$$v(t) = 1.4\frac{e^{14t} - 1}{e^{14t} + 1}.$$

(ii) During the stage in which the fall occurs at constant velocity v_L, from the principle of inertia (Newton's first law), the sum of the forces acting on the baking cup must be zero.

$$F_f = W \Longrightarrow kv_L^2 = mg \Longrightarrow v_L = \sqrt{\frac{mg}{k}}$$

whence

$$\alpha = v_L = \sqrt{\frac{mg}{k}}.$$

From the condition that initial acceleration must be equal to g, one has:

$$\beta = \frac{2g}{\alpha} = \frac{2g}{V_L} = 2g\sqrt{\frac{k}{mg}} = 2\sqrt{\frac{kg}{m}}.$$

Therefore, the equation of $v(t)$ can be written as

$$v(t) = \sqrt{\frac{mg}{k}}\frac{e^{2t\sqrt{kg/m}} - 1}{e^{2t\sqrt{kg/m}} + 1}.$$

(iii) We have set

$$z = \sqrt{k} \quad \text{and} \quad b = 2t\sqrt{\frac{g}{m}}.$$

Let us calculate the limit for z approaching zero of the given function. By De l'Hopital's theorem, one has:

$$v = \sqrt{mg}\lim_{z\to 0}\frac{e^{bz} - 1}{z(e^{bz} + 1)} = \sqrt{mg}\lim_{z\to 0}\frac{be^{bz}}{e^{bz} + 1 + zbe^{bz}} = \frac{\sqrt{mg}}{2}b.$$

From position $b = 2t\sqrt{g/m}$, it follows that

$$v = \frac{\sqrt{mg}}{2} \cdot 2t\sqrt{\frac{g}{m}} = gt.$$

The significance of this result can be interpreted in the following way: in absence of friction, the baking cup falls with constant acceleration equal to g.

(iv) In the expression for $v(t)$, the fraction turns out to be adimensional. Therefore, we need the 1.4 factor to be homogeneous with velocity, and, thus, to be expressed in metres per second.
The derivative of the function $F(t)$ is

$$F'(t) = A\frac{2}{e^{14t}+1}\frac{14e^{14t}\cdot 2}{4} + B$$

which, after some algebraic manipulations, can be rewritten in the form

$$F'(t) = \frac{(14A + B)e^{14t} + B}{e^{14t}+1}.$$

Hence, the function

$$F'(t) = 0.2\ln\left(\frac{e^{14t}+1}{2}\right) - 1.4t$$

is a primitive of the function $v(t)$.

The function $v(t)$ is continuous in the time interval (in seconds) [0, 1]. The average value in the time interval (in seconds) [0, 1] is given by

$$v_{\text{avg}} = \frac{\int_0^1 v(t)\,\mathrm{d}t}{1-0} = F(1) - F(0) = 0.2\ln\left(\frac{e^{14}+1}{2}\right) - 1.4 = 1.26\ \mathrm{m\ s^{-1}}.$$

The integral $\int_0^1 v(t)\,\mathrm{d}t$ is the area of the surface subtended by the graph of $v(t)$ in the time interval (in seconds) [0, 1]. This area corresponds to the distance travelled by the baking cup in one second. Therefore, the value of v_{avg} represents the average velocity of the baking cup during the first second of its fall.

Solution: Exercise 2.11
Mainly BOOKWORK.

Solution: Exercise 2.12

(i) Taking the initial horizontal velocity in the x-direction and the vertical to be the z-direction, with the origin of coordinates at the child's feet, we can apply the general result

$$r(t) = r_0 + ut + \frac{1}{2}at^2$$

with initial displacement

$$r_0 = h\hat{k}$$

initial velocity

$$u = u\cos\alpha\hat{i} + u\sin\alpha\hat{k}$$

and force

$$F = -mg\hat{k} \Longrightarrow a = -g\hat{k},$$

and therefore

$$r(t) = u\cos(\alpha)t\hat{i} + \left[+u\sin(\alpha)t - \frac{gt^2}{2}\right]\hat{k}.$$

Therefore, the horizontal displacement is

$$x(t) = u\cos(\alpha)t$$

and the vertical displacement is

$$z(t) = h + u\sin(\alpha)t - \frac{1}{2}gt^2.$$

(ii) Substituting

$$t = \frac{x}{u\cos(\alpha)},$$

we obtain

$$z = h + x\tan(\alpha) - \frac{gx^2}{2u^2\cos^2(\alpha)}.$$

Using the result

$$\sec^2(\alpha) = 1 + \tan^2(\alpha)$$

gives

$$z = h + x\tan(\alpha) - \frac{gx^2}{2u^2}[1 + \tan^2(\alpha)],$$

or

$$\frac{gx^2}{2u^2}\tan^2(\alpha) - x\tan(\alpha) + \left(z - h + \frac{gx^2}{2u^2}\right) = 0.$$

(iii) We need the coconut to be at a height $z = h_{\text{target}} = 0.3$ m when it has moved through a horizontal distance $x = d = 2$ m. Thus,

$$\frac{gd^2}{2u^2}\tan^2(\alpha) - d\tan(\alpha) + h_{\text{target}} - h + \frac{gd^2}{2u^2} = 0.$$

This is a quadratic equation in $\tan(\alpha)$ of the form

$$a\tan^2(\alpha) + b\tan(\alpha) + c = 0$$

with

$$a = \frac{gd^2}{2u^2} = 1.23 \text{ m}; \quad b = -d = -2 \text{ m}; \quad c = h_{\text{target}} - h + \frac{gd^2}{2u^2} = 0.53 \text{ m}.$$

Hence,

$$\tan(\alpha) = \frac{-b \pm \sqrt{b^2 - 4ac}}{2a} = 1.301 \quad \text{or } 0.330.$$

It follows that (taking solutions in the quadrant $0 \leqslant \alpha \leqslant \pi/2$)

$$\alpha = \tan^{-1}(1.301) = 0.916 \text{ rad} \quad \text{or } 52.5°,$$

or

$$\alpha = \tan^{-1}(0.330) = 0.319 \text{ rad} \quad \text{or } 18.3°.$$

Solution: Exercise 2.13

(i) Initial velocity

$$\boldsymbol{u} = u\cos\alpha\hat{\boldsymbol{i}} + u\sin\alpha\hat{\boldsymbol{j}},$$

where the vertical (upwards) direction is $\hat{\boldsymbol{j}}$ and the horizontal (towards the stage) direction is $\hat{\boldsymbol{i}}$.

Subsequent velocity at time t is

$$\boldsymbol{v} = \boldsymbol{u} - gt\hat{\boldsymbol{j}}$$

and displacement is

$$\boldsymbol{s} = \boldsymbol{u}t - \frac{1}{2}gt^2\hat{\boldsymbol{j}}.$$

At the maximum height, $\hat{\boldsymbol{j}}$ component of $\boldsymbol{v}$ is zero and thus

$$u\sin\alpha - gt = 0$$

and

$$t_{\max} = \frac{u\sin\alpha}{g}.$$

Vertical displacement, i.e. peak height at time $t_{\max}$ is

$$y_{\max} = u\sin(\alpha)t_{\max} - \frac{1}{2}gt_{\max}^2$$
$$= u\sin(\alpha)\left(\frac{u\sin(\alpha)}{g}\right) - \frac{g}{2}\left(\frac{u\sin(\alpha)}{g}\right)^2 = \frac{u^2\sin^2(\alpha)}{2g},$$

as required.

(ii) Horizontal displacement in this time

$$x_m = u\cos\alpha\left(\frac{u\sin\alpha}{g}\right).$$

The vertical distance from the peak height to the stage is

$$y_{\max} - h.$$

The time taken to fall through this distance t_s satisfies

$$y_{\max} - h = \frac{1}{2}gt_s^2$$

since the initial vertical velocity at the peak height is zero, i.e.

$$t_s = \sqrt{\frac{2y_{\max} - 2h}{g}} = \sqrt{\frac{u^2\sin^2(\alpha)}{g^2} - \frac{2h}{g}} = \frac{1}{g}\sqrt{u^2\sin^2(\alpha) - 2gh}$$

and the horizontal distance is

$$t_s u\cos\alpha.$$

Therefore, the total distance from launch point is

$$\frac{u\cos\alpha}{g}\left(u\sin\alpha + \sqrt{u^2\sin^2(\alpha) - 2gh}\right)$$

and the distance on stage is

$$\frac{u\cos\alpha}{g}\left(u\sin\alpha + \sqrt{u^2\sin^2(\alpha) - 2gh}\right) - d.$$

There will be variants here on how to get to the answer.

Solution: Exercise 2.14
No solution provided.

Solution: Exercise 2.15
We have a projectile which starts from $x = -X$, $y = H$ with $\dot{x} = U$ and $\dot{y} = 0$, and we want it to hit the origin. Its x-position (no horizontal acceleration) is

$$x = -X + Ut,$$

and its vertical position is

$$y = H - \frac{1}{2}gt^2.$$

It hits the ground when

$$\frac{1}{2}gt^2 = H,$$

i.e. at

$$t = \sqrt{\frac{2H}{g}},$$

and there we have

$$x = -X + U\sqrt{\frac{2H}{g}}.$$

We need this to be zero, therefore

$$X = U\sqrt{\frac{2H}{g}}.$$

Solution: Exercise 2.16: Safety zone
If the position of the aeroplane is (X, Y), then we would like to find condition(s) that the shell cannot reach the point (X, Y) no matter what angle α is used. Thus, if $X \neq 0$, from the Cartesian equation for the path of the particle, we want (X, Y) such that at (X, Y):

$$Y = X\tan(\alpha) - \frac{gX^2}{2U^2}(1 + \tan^2(\alpha))$$

has no real solutions in α, or, equivalently, $\tan(\alpha)$.

We thus rewrite our equation as a quadratic in $\tan(\alpha)$ and we get

$$\frac{gX^2}{2U^2}\tan^2(\alpha) - X\tan(\alpha) + \left(Y + \frac{gX^2}{2U^2}\right) = 0.$$

This has no real solutions in tan(α) if

$$X^2 < 4\left(\frac{gX^2}{2U^2}\right)\left(Y + \frac{gX^2}{2U^2}\right).$$

Hence, the plane is safe if

$$Y > \frac{U^2}{2g} - \frac{gX^2}{2U^2}$$

when $X \neq 0$. This formula also gives the right solution when $X = 0$. Thus, the aeroplane is safe if it flies above the parabola

$$Y = \frac{U^2}{2g} - \frac{gX^2}{2U^2}.$$

Solution: Exercise 2.17
Initially,

$$u_x = u\cos\alpha = u\cos 45° = \frac{u}{\sqrt{2}}$$
$$u_y = u\sin\alpha = u\sin 45° = \frac{u}{\sqrt{2}}.$$

When the ball hits the wall,

$$y = x\tan(\alpha) - \frac{1}{2}\frac{gx^2}{u^2\cos^2(\alpha)}.$$

Using $y = H$, $x = d$, and $\alpha = 45°$,

$$H = d\left(1 - \frac{gd}{u^2}\right).$$

If the collision of the ball with the wall is perfectly elastic, then at the point of impact, the horizontal component of the velocity u_x' will be reversed, the magnitude remaining constant, while both the direction and magnitude of the vertical component v_y' are unaltered. Let t be the time taken for the ball to bounce back and R be the range.

$$y = v_y't - \frac{1}{2}gt^2.$$

Using

$$t = \frac{R}{u\cos 45°} = \sqrt{2}\frac{R}{u}$$

and

$$y = -(H + h)$$

and

$$v_y' t = u \sin 45^\circ - g\frac{d}{u \cos 45^\circ} = \frac{u}{\sqrt{2}} - \sqrt{2}\frac{gd}{u}.$$

Combining all the equations, we get a quadratic equation in R, which has the acceptable solution

$$R = \frac{u^2}{2g} + \sqrt{\frac{u^2}{4g^2} + H + h}\,.$$

Solution: Exercise 2.18

Take the reference x-axis in the direction of plane.

The component of initial velocity parallel and perpendicular to the plane are equal to

$$u_\perp = u \cos\theta$$
$$u_\parallel = u \sin\theta.$$

The component of g along and perpendicular to the plane

$$a_\parallel = -\,g \sin\alpha$$
$$a_\perp = g \cos\alpha.$$

Time of flight is

$$T = \frac{2u \sin\theta}{g \cos\alpha}.$$

Maximum height on an inclined plane is

$$H = \frac{u^2 \sin^2(\theta)}{2g \cos\alpha}.$$

For one-dimensional motion,

$$s = ut + \frac{1}{2}at^2.$$

On an inclined plane, the horizontal range is given by

$$R = u_\parallel T + \frac{1}{2}a_\parallel T^2$$
$$R = u \cos(\theta)T - \frac{1}{2}g \sin(\alpha)T^2$$
$$R = u \cos\theta\left(\frac{2u \sin\theta}{g \cos\alpha}\right) - \frac{1}{2}g \sin\alpha\left(\frac{2u \sin\theta}{g \cos\alpha}\right)^2.$$

By resolving,

$$R = \frac{2u^2}{g}\frac{\sin\theta\cos(\theta+\alpha)}{\cos^2(\alpha)}.$$

Solution: Exercise 2.19
Vector equation of motion is:

$$\boldsymbol{F} = \boldsymbol{F}_D - mg\hat{\boldsymbol{k}}$$

$$\Longrightarrow m\frac{\mathrm{d}\boldsymbol{v}}{\mathrm{d}t} = -m\kappa\boldsymbol{v} - mg\hat{\boldsymbol{k}}.$$

Initial velocity at $t = 0$

$$\boldsymbol{v} = u(\cos\alpha\hat{\boldsymbol{i}} + \sin\alpha\hat{\boldsymbol{j}}).$$

From the vector equation of motion, one obtains two scalar equations of motions. For the $\hat{\boldsymbol{i}}$-component of velocity

$$\frac{\mathrm{d}v_x}{\mathrm{d}t} + \kappa v_x = 0$$

whereas for the $\hat{\boldsymbol{j}}$-component

$$\frac{\mathrm{d}v_z}{\mathrm{d}t} + \kappa v_z = -g$$

and the aforementioned initial conditions. Separating variables for velocity components and applying initial conditions leads to

$$v_x = (u\cos\alpha)e^{-\kappa t}, \quad v_z = (u\sin\alpha)e^{-\kappa t} - \frac{g}{\kappa}(1 - e^{-\kappa t}).$$

Integrating again to find the components of position vector, and applying initial condition $x = 0$, $z = 0$ at $t = 0$, gives

$$x = \frac{u\cos\alpha}{\kappa}(1 + e^{-\kappa t}), \quad z = \frac{\kappa u\sin\alpha + g}{\kappa^2}(1 - e^{-\kappa t}) - \frac{g}{\kappa}t.$$

The trajectory is not a parabola any more, but rather complex to derive analytically.

Solution: Exercise 2.20
BOOKWORK.

Solution: Exercise 2.21

(i) The impulse is given by the integral of the force with respect to time (over the duration of the interaction):

$$\boldsymbol{I} = \int_0^2 \boldsymbol{F}(t)\,\mathrm{d}t.$$

With the form of the force given, we find

$$\boldsymbol{I} = \int_0^2 [(2t - t^2)\hat{\boldsymbol{i}} + 4\hat{\boldsymbol{j}}]\,\mathrm{d}t = \left[\left(t^2 - \frac{t^3}{3}\right)\hat{\boldsymbol{i}} + 4t\hat{\boldsymbol{j}}\right]_{t=0}^{t=2} = \left(\frac{4}{3}\hat{\boldsymbol{i}} + 8\hat{\boldsymbol{j}}\right)\,\mathrm{N\,s}.$$

The change in the momentum $\Delta\boldsymbol{p}$ of the object is given by the total impulse. The body is initially at rest; therefore, the final velocity is

$$\boldsymbol{v}_{\text{final}} = \boldsymbol{v}_{\text{initial}} + \Delta\boldsymbol{v} = 0 + \frac{\Delta\boldsymbol{p}}{m} = \left(\frac{2}{3}\hat{\boldsymbol{i}} + 4\hat{\boldsymbol{j}}\right)\,\mathrm{m\,s^{-1}}.$$

(ii) To find the position of the object, we must integrate again. The velocity after a general time $t < 2$ s will be

$$\boldsymbol{v}(t) = \frac{1}{2}\int_0^t [(2\tilde{t} - \tilde{t}^2)\hat{\boldsymbol{i}} + 4\hat{\boldsymbol{j}}]\,\mathrm{d}\tilde{t} = \frac{1}{2}\left[\left(t^2 - \frac{t^3}{3}\right)\hat{\boldsymbol{i}} + 4t\hat{\boldsymbol{j}}\right].$$

Therefore, the displacement up to time t will be

$$\Delta\boldsymbol{r} = \int_0^t \boldsymbol{v}(\tilde{t})\,\mathrm{d}\tilde{t} = \int_0^t \left[\left(\frac{\tilde{t}^2}{2} - \frac{\tilde{t}^3}{6}\right)\hat{\boldsymbol{i}} + 2\tilde{t}\hat{\boldsymbol{j}}\right]\mathrm{d}\tilde{t} = \left[\left(\frac{t^3}{6} - \frac{t^4}{24}\right)\hat{\boldsymbol{i}} + t^2\hat{\boldsymbol{j}}\right]\mathrm{m}.$$

Putting $t = 2$ gives

$$\Delta\boldsymbol{r} = \left(\frac{2}{3}\hat{\boldsymbol{i}} + 4\hat{\boldsymbol{j}}\right)\mathrm{m}.$$

The magnitude of the displacement is therefore

$$|\Delta\boldsymbol{r}| = \sqrt{\left(\frac{2}{3}\right)^2 + 4^2}\ \mathrm{m} \approx 4.06\ \mathrm{m},$$

as required; the direction will be in the direction of the vector $\Delta\boldsymbol{r}$, i.e. in the xy-plane making an angle of

$$\tan^{-1}\left(4 \times \frac{3}{2}\right) = \tan^{-1}(6) = 1.41\ \mathrm{rad} = 80.5°$$

to the x-axis.

(iii) After the force stops acting, the particle will continue to move in a straight line with the same velocity vector (Newton's first law).
Therefore, after a further three seconds, the total displacement will be

$$\Delta \boldsymbol{r}(t = 5\ \mathrm{s}) = \Delta \boldsymbol{r}(t = 2\ \mathrm{s}) + (3\ \mathrm{s})\boldsymbol{v}_{\mathrm{final}} = \left[\frac{2}{3}\hat{\boldsymbol{i}} + 4\hat{\boldsymbol{j}} + 3\left(\frac{2}{3}\hat{\boldsymbol{i}} + 4\hat{\boldsymbol{j}}\right)\right]\ \mathrm{m}$$

$$= \left(\frac{8}{3}\hat{\boldsymbol{i}} + 16\hat{\boldsymbol{j}}\right)\ \mathrm{m}.$$

Solution: Exercise 2.22
No solution will be provided.

IOP Publishing

Classical Mechanics
A professor-student collaboration
Mario Campanelli

Chapter 13

Selected solutions to Chapter 3: Kinematic relations

Solution: Exercise 3.5.

(i) Using conservation of mechanical energy, we have that

$$K_i + V_i = K_f + V_f$$
$$\frac{1}{2}mv_0^2 + 0 = 0 + mgh_{\max}.$$

Note that initially the potential energy is zero, because the height is zero, while at the maximum height of the trajectory the speed is zero.

From this equation, we can find the maximum height:

$$h_{\max} = \frac{v_0^2}{2g} \approx \frac{(60)^2}{2 \times 9.81}\ \text{m} \approx 183\ \text{m}.$$

To find the total time it takes to fall back into the lake, consider the fact that the only force acting on the particle is the gravitational force. Then, the speed is given by $v = v_0 - gt$, where $v_0 = 60\ \text{ms}^{-1}$. So, when the particle is at $h_{\max}$, and thus $v = 0$, a time $t_{\max} = v_0/g$ has passed. The total time is just the double of $t_{\max}$, i.e.

$$t_{\text{tot}} = 2t_{\max} = \frac{2v_0}{g} \approx \frac{2 \times 60}{9.81}\ \text{s} \approx 12\ \text{s}.$$

(ii) A force of 0.001 ms^{-1} on 1 g corresponds to an acceleration of 1ms^{-1}. So, the horizontal displacement over 12 s is

$$s = \frac{1}{2}at^2 = 72\ \text{m}.$$

doi:10.1088/978-0-7503-2690-2ch13

(iii) No, it's not a parabola, because the horizontal velocity increases instead of being constant as in the case of the ballistic motion.

Solution: Exercise 3.7. Let's evaluate the integral $\int \boldsymbol{F} \cdot \mathrm{d}\boldsymbol{r}$ along the path given.

- First the line from (0, 0, 0) to (1, 0, 0). Along here $y = z = 0$ and the displacement is purely in the x-direction (i.e. $\mathrm{d}\boldsymbol{r} = \mathrm{d}x\hat{\boldsymbol{x}}$), so the contribution to the integral is

$$\int_{x=0}^{x=1} x^2 \, \mathrm{d}x = \left[\frac{x^3}{3}\right]_{x=0}^{x=1} = \frac{1}{3}.$$

- Next take the line from (1, 0, 0) to (1, 1, 0). Along this line $x = 1$ and $z = 0$, and the displacement is in the y-direction, i.e. $\mathrm{d}\boldsymbol{r} = \mathrm{d}y\hat{\boldsymbol{y}}$, so the contribution is

$$\int_{y=0}^{y=1} y \, \mathrm{d}y = \left[\frac{y^2}{2}\right]_{y=0}^{y=1} = \frac{1}{2}.$$

- Finally consider the line from (1, 1, 0) to (1, 1, 1). Along this line $x = y = 1$ and the displacement is in the z-direction, so $\mathrm{d}\boldsymbol{r} = \mathrm{d}z\hat{\boldsymbol{z}}$. We therefore get

$$\int_{z=0}^{z=1} 1 \, \mathrm{d}z = [z]_{z=0}^{z=1} = 1.$$

The total integral is then $\frac{1}{3} + \frac{1}{2} + 1 = \frac{11}{6}$, which is different from the value of 1 we found earlier.

Solution: Exercise 3.8. The potential energy function $V(\boldsymbol{r})$ is given by

$$V(\boldsymbol{r}) = -\int_{r_0}^{r} \boldsymbol{F}(\tilde{\boldsymbol{r}}) \, \mathrm{d}\tilde{\boldsymbol{r}} = -\int_{0}^{r} \boldsymbol{F}(\tilde{\boldsymbol{r}}) \, \mathrm{d}\tilde{\boldsymbol{r}},$$

where $\tilde{\boldsymbol{r}}$ is a dummy variable.

Given the potential V, $\boldsymbol{F}$ can be found by taking the gradient of the potential, i.e.

$$\boldsymbol{F} = -\nabla V.$$

The force is found as just mentioned above. That is,

$$\boldsymbol{F}(x, y) = -\nabla V(x, y) = -\frac{\partial V}{\partial x}\hat{\boldsymbol{x}} - \frac{\partial V}{\partial y}\hat{\boldsymbol{y}} = -3Kx^2y^2\hat{\boldsymbol{x}} - 2Kx^3y\hat{\boldsymbol{y}}.$$

Let's now calculate the values of the potential at the two end points, i.e. (0, 0) and (2, 4). We have

$$V(0, 0) = 0 \text{ J}$$
$$V(2, 4) = 128K \text{ J}.$$

The work is then given by

$$W = -\Delta V = V(0, 0) - V(2, 4) = -128K \text{ J}.$$

Solution: Exercise 3.9. To find the force, we need to take the gradient of the potential, i.e.

$$\boldsymbol{F}(x, y, z) = -\nabla V = -\begin{pmatrix} \partial_x V \\ \partial_y V \\ \partial_z V \end{pmatrix} = -\begin{pmatrix} 4xy^2 \\ 3y^2 + 4x^2y \\ 2z \end{pmatrix}.$$

To find the change in kinetic energy in going from (0, 0, 0) to (1, 1, 2), let's calculate the potential energy at those points. We have

$$V(0, 0, 0) = 0$$
$$V(1, 1, 2) = (4 + 1 + 2) \text{ J} = 7 \text{ J}.$$

Therefore, the change in potential energy is $\Delta V = 7$ J. Since we have seen that $W = -\Delta V$ but also $W = \Delta K$, we have that $\Delta K = -\Delta V$. Thus, the change in kinetic energy is $\Delta K = -7$ J.

Finally, we are given the force acting on a second particle and we need to find the potential energy. To do so, we need to go through a procedure that can be a bit confusing. Let's divide it into steps.

Let's start by calculating the integral of F_x and F_y. That is,

$$V(x, y) = -\int_0^x F_{\tilde{x}} \, \mathrm{d}\tilde{x} = -\int_0^x (-2\tilde{x}y^3) \, \mathrm{d}\tilde{x} = x^2y^3 + g(y)$$
$$V(x, y) = -\int_0^y F_{\tilde{y}} \, \mathrm{d}\tilde{y} = -\int_0^y (-3x^2\tilde{y}^2) \, \mathrm{d}\tilde{y} = x^2y^3 + h(x),$$

where $\tilde{x}$ and $\tilde{y}$ are dummy variables.

As you can see, in the solution of the integral some functions $g(y)$ and $h(x)$ have been considered. This has been done because, when a partial derivative is taken, the other variables are considered constant so a term depending only on another variable would be discarded. Therefore, to find the complete expression of the potential energy we need to add these functions. To find an expression for these two functions $g(y)$ and $h(x)$, we now equal the integrals of the components, i.e.

$$x^2y^3 + g(y) = x^2y^3 + h(x).$$

We see that the two sides are the same if $g(y) = h(x) = c$, where c is a constant. Therefore, the potential is given by $V(x, y) = x^2y^3 + c$. However, we are told that $V(0, 0) = 0$, thus $c = 0$ and our final result is

$$V(x, y) = x^2y^3.$$

It is not required, but we can check that this is the right result by taking the negative gradient of the potential which has to give us the force acting on the particle. Let's do this:

$$\boldsymbol{F}(x, y) = -\nabla V = -\nabla(x^2y^3) = -2xy^3\hat{\boldsymbol{x}} - 3x^2y^2\hat{\boldsymbol{y}},$$

which is correct.

Solution: Exercise 3.10. We'll have to calculate some integrals.

(i)

$$W_1 = \int_{r_1}^{r_2} \boldsymbol{F}_1 \cdot \mathrm{d}\boldsymbol{r} = 3\int_1^2 \mathrm{d}x + 2\int_1^2 \mathrm{d}y + \int_3^5 \mathrm{d}z = 3 + 2 + (5 - 3) = 7 \text{ J}$$

(ii)

$$W_2 = \int_{r_1}^{r_2} \boldsymbol{F}_2 \cdot \mathrm{d}\boldsymbol{r} = 2y^3 \int_1^2 x \,\mathrm{d}x + 3x^2 \int_1^2 y^2 \,\mathrm{d}y$$
$$= 2y^3\left[\frac{x^2}{2}\right]_1^2 + 3x^2\left[\frac{y^3}{3}\right]_1^2 = 3y^3 + 7x^2.$$

To check whether the forces $\boldsymbol{F}_1$ and $\boldsymbol{F}_2$ are conservative, we need to calculate their curls. That is,

(i)

$$\nabla \times \boldsymbol{F}_1 = \begin{pmatrix} \partial_x \\ \partial_y \\ \partial_z \end{pmatrix} \times \begin{pmatrix} 3 \\ 2 \\ 1 \end{pmatrix} = \boldsymbol{0}$$

(ii)

$$\nabla \times \boldsymbol{F}_2 = \begin{pmatrix} \partial_x \\ \partial_y \\ \partial_z \end{pmatrix} \times \begin{pmatrix} 2xy^3 \\ 3x^2y^2 \\ 0 \end{pmatrix} = \begin{pmatrix} 0 - 0 \\ 0 - 0 \\ 6xy^2 - 6xy^2 \end{pmatrix} = \boldsymbol{0}.$$

Therefore, both forces are conservative.

Alternatively, this could also be proved by finding a potential V such that $\boldsymbol{F} = -\nabla V$. In this case, we have

(i)

$$V_1 = -(3x + 2y + z)$$

(ii)

$$V_2 = -x^2y^3.$$

Solution: Exercise 3.12. To find the work done by the force on the object, given the potential energy, we need to use the formula $W = -\Delta V$.

The potential energy at (0, 0) and at (2, 0) is given by

$$V(0, 0) = 0 \text{ J}$$
$$V(2, 0) = 8 \times 4 = 32 \text{ J}.$$

Thus, since the object goes from (2, 0) to (0, 0),

$$\Delta V = V(0, 0) - V(2, 0) = 0 - 32 = -32 \text{ J}$$

and $W = 32$ J. Let's now find a general expression for the force associated with the given potential. We have

$$\boldsymbol{F} = -\nabla V = -\begin{pmatrix} 16x \\ 4y + 4\alpha y^3 \\ 0 \end{pmatrix}.$$

Finally, since there exists a potential V, the force F is conservative and, thus, mechanical energy is conserved.

IOP Publishing

Classical Mechanics
A professor-student collaboration
Mario Campanelli

Chapter 14

Selected solutions to Chapter 4: Oscillatory motion

Solution: Exercise 4.2. The simple pendulum

The first observation to make is that since the string is not extendable, the tension $\boldsymbol{T}$ will exactly compensate the projection of the gravitational force along the length of the rope $F_{||}$, i.e.

$$|T| = mg\cos\theta.$$

The perpendicular component $F_{\perp} = -mg\sin\theta$, seen in red in the figure below, must therefore be the only contribution to the acceleration of the pendulum.

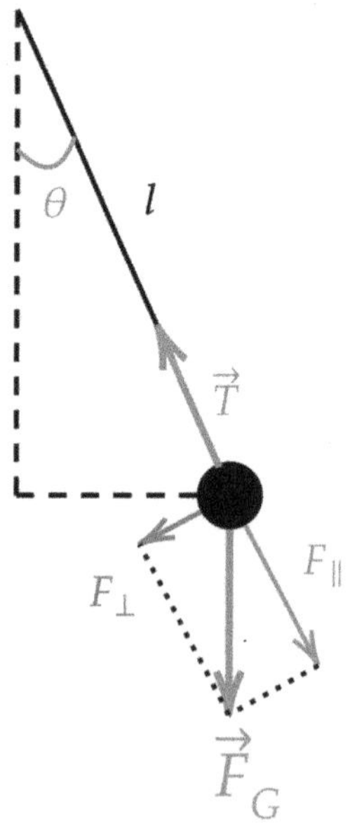

We therefore have that:

$$F_{\perp} = ma = -mg \sin\theta.$$

Considering **small oscillations** we can look at an arc distance which the pendulum will move.

$$\widehat{AB} = l\theta.$$

Therefore, the velocity and acceleration will be given by

$$\dot{\widehat{AB}} = l\dot{\theta}$$
$$\ddot{\widehat{AB}} = l\ddot{\theta}.$$

We can therefore state that:

$$m\boldsymbol{a} = -mg \sin\theta = ml\ddot{\theta}.$$

Rearranging and considering the small angle approximation $\sin\theta = \theta$ we obtain

$$\ddot{\theta} + \frac{g}{l}\theta = 0.$$

We can recognise that this *linear homogeneous second order differential equation* is that of a **simple harmonic oscillator** where the angular velocity ω can be easily determined.

$$\ddot{\theta} + \omega^2\theta = 0$$
$$\therefore \quad \omega \equiv \sqrt{\frac{g}{l}}.$$

We know that the frequency f is related to the angular velocity by the relation $\omega = 2\pi f$. Therefore, the frequency of the pendulum's oscillations is given by:

$$f = \frac{1}{2\pi}\sqrt{\frac{g}{l}}.$$

Solution: Exercise 4.3. As in exercise 4.2, the string is inextensible, so the best way to start approaching the problem is by looking at the forces acting on the spring.

Since the body is moving with uniform circular motion we know that the angular velocity ω will remain constant. We also know that there will be a centripetal force acting inwards on the body, which will be equal to:

$$F_C = m\omega^2 r.$$

Firstly we will apply Newton's second law, dividing the forces acting on the mass into the planes perpendicular and parallel to the circular motion, as seen in the figure below.

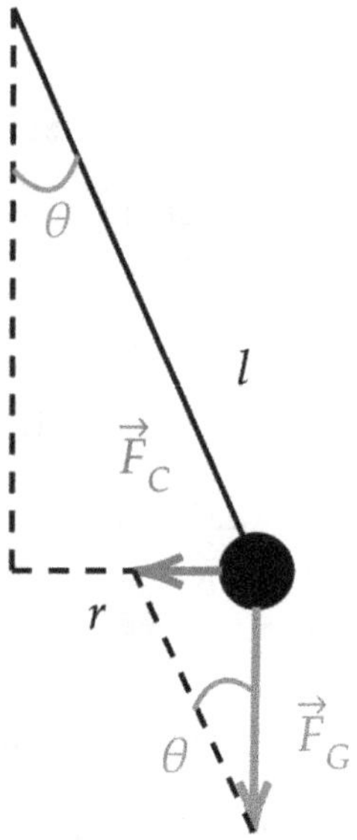

The other force acting in the $-\hat{k}$ is the gravitational force, $F_G = mg$. These two forces are balanced by the string tension T.

If we look more closely at the triangle formed by the three balanced forces, we can derive that

$$\tan\theta = \frac{F_C}{F_G},$$

by rearranging and substituting the equations of the forces we obtain:

$$mg\frac{\sin\theta}{\cos\theta} = m\omega^2 r.$$

By observing the above figure we can find the trigonometric relation between r and l. This is given by:

$$\sin\theta = \frac{r}{l}.$$

Substituting this into the relation above and cancelling out we get,

$$g\frac{1}{\cos\theta} = \omega^2 l,$$

then rearranging the equation above in terms of ω we obtain the desired solution of:

$$\omega = \sqrt{\frac{g}{l\cos\theta}}.$$

Solution: Exercise 4.4. Let's start by defining what the form of the gravitational force, acting on the ball will be.

$$F_g = -K\frac{M_i}{r^2}$$

where $K = mG$ and M_i is the mass of the Earth producing the gravitational force.

The gravitational force pulling the ball towards the center of the Earth will only be produced by the fraction of the Earth's mass between the ball and the central point. Hence to calculate the force at a given instant in time we must express the internal mass M_i as a fraction of the total mass of the Earth M_T, when the ball is in the radial position $r < R$. Recall that the mass of a sphere of radius r and density ρ is given by $\frac{4}{3}\pi r^3\rho$.

$$M_i = \frac{M_r}{M_T}M_T$$

$$M_i = \frac{4/3\pi r^3\rho}{4/3\pi R^3\rho}M_T$$

$$M_i = \frac{r^3}{R^3}M_T.$$

Now let's look at what the instantaneous gravitational force will be.

$$F_g = -K\frac{M_i}{r^2} = -K\frac{M_T}{r^2}\frac{r^3}{R^3}.$$

We can replace the equation for the total mass of the Sun to obtain:

$$F_g = -K\frac{4/3\pi R^3\rho}{r^2}\frac{r^3}{R^3}.$$

Simplifying we find the instantaneous gravitational force acting on the ball to be:

$$F_g = -K\frac{4}{3}\pi\rho r.$$

Now we can apply Newton's second law to obtain the equation of motion of the ball.

$$m\ddot{r} = -K\frac{4}{3}\pi\rho r$$

$$\ddot{r} + \frac{K}{m}\frac{4}{3}\pi\rho r = 0$$

where we can easily determine the angular frequency ω as:

$$\ddot{r} + \omega^2 r = 0$$

$$\therefore \quad \omega \equiv \sqrt{\frac{K}{m}\frac{4}{3}\pi\rho}.$$

We now need to prove that $r(t) = R\cos\omega t$ is a solution to the equation of motion seen above. This can be simply done by inserting the solution in the equation above and verifying that the identity holds.

$$\ddot{r}(t) = -R\omega^2\cos\omega t.$$

We can substitute this into the equation of motion obtained and verify the identity.

$$\begin{aligned}\ddot{r}(t) + \omega^2 r(t) &= 0\\ -R\omega^2\cos\omega t + \omega^2 R\cos\omega t &= 0\\ \therefore\ 0 &= 0.\end{aligned}$$

Since the identity holds, we have proven that a solution in the form $r(t) = R\cos\omega t$ is acceptable.

Solution: Exercise 4.5. In this exercise we need to determine the constants given the boundary conditions. Let's first figure out what the boundary conditions tell us mathematically. Firstly at $t = 0$ the position $x(t)$ will have to be equal to x_0, secondly since the body will be at rest its velocity must also be zero at $t = 0$. This gives us the two conditions below:

$$\begin{cases} x(t=0) = (A + Bt)e^{\frac{-\lambda t}{2m}} = x_0 \\ \dot{x}(t=0) = 0 \end{cases}.$$

Let's look at the first condition:

$$\begin{aligned}x(0) &= (A + B\times 0)e^0 = x_0\\ e^0 &\equiv 1\\ \therefore\ A &= x_0.\end{aligned}$$

Now that we have determined the first constant, we can look at the second boundary condition. Firstly we need to calculate the derivative, *it is important to expand the brackets making sure to apply the product rule for derivatives.*

$$\begin{aligned}x(t) &= x_0 e^{\frac{-\lambda t}{2m}} + Bte^{\frac{-\lambda t}{2m}}\\ \dot{x}(t) &= \frac{-x_0\lambda}{2m}e^{\frac{-\lambda t}{2m}} + Be^{\frac{-\lambda t}{2m}} + \frac{-Bt\lambda}{2m}e^{\frac{-\lambda t}{2m}}\\ \dot{x}(0) &= \frac{-x_0\lambda}{2m}\times 1 + B\times 1 + \frac{-B\lambda}{2m}\times 0\times 1\end{aligned}$$

$$\frac{-x_0\lambda}{2m} + B = 0$$

$$\therefore\ B = \frac{x_0\lambda}{2m}.$$

So our final solution, determined by the boundary conditions is given by:

$$x(t) = x_0(1 + \frac{t\lambda}{2\,m})e^{\frac{-\lambda t}{2m}}.$$

Solution: Exercise 4.11.

(i) To determine the equilibrium position we must determine the point in which the *first derivative* of the potential goes to **zero**. Since the potential can be decomposed into the two directions, the equilibrium position can be found analytically as seen below.

$$\begin{aligned}\frac{\mathrm{d}V}{\mathrm{d}x} &= \frac{1}{2}k_x \times 2(x - x_0)^2 \\ \frac{\mathrm{d}V}{\mathrm{d}x} &= 0 \\ \therefore\ (x - x_0) &= 0 \\ x &= x_0.\end{aligned}$$

The analogous calculation can be made for the y direction, hence obtaining the equilibrium position $P\ (x_0, y_0)$. The force can be obtained from the identity:

$$\begin{aligned}\boldsymbol{F} &= -\nabla V \\ \therefore\ F_x &= -\frac{\mathrm{d}V}{\mathrm{d}x}.\end{aligned}$$

Since we have already done the calculations above for the two dimensions, we can write the component of the force in each of the two directions as:

$$\begin{cases}F_x = -k_x(x - x_0) \\ F_y = -k_y(y - y_0)\end{cases}.$$

(ii) The force induced by the potential is the only one acting on the object of mass m. We can apply Newton's second law and hence obtain the equation of motion in the two separate x, y components.

$$\begin{cases}F_x = -k_x(x - x_0) = m\ddot{x} \\ F_y = -k_y(y - y_0) = m\ddot{y}\end{cases}.$$

Rearranging we obtain the two equations of motion, which are second order inhomogeneous differential equations.

$$\begin{cases}m\ddot{x} + k_x x = k_x x_0 \\ m\ddot{y} + k_y y = k_y y_0\end{cases}.$$

(iii) First let's interpret the boundary conditions we are given. At a time $t = 0$ the position of the particle will be (a, b),

$$\begin{cases} x(t = 0) = a \\ y(t = 0) = b. \end{cases}$$

At $t = 0$ the particle will also be at rest:

$$\begin{cases} \dot{x}(t = 0) = 0 \\ \dot{y}(t = 0) = 0. \end{cases}$$

Now, the first step in solving the second order inhomogeneous differential equation is to solve its *corresponding homogeneous version*. The equations are identical in both directions, for simplicity we will only be solving the x component, remember that the y component will have an identical solution.

Given that there is an initial zero velocity we will guess a solution in the form:

$$x(t) = A\cos(\omega_x t + \phi_x)$$
$$\therefore\ \ddot{x}(t) = -A\omega_x^2\cos(\omega_x t + \phi_x).$$

Let's make an *ansatz*, and see if our solution is valid for the homogeneous differential equation.

$$m\ddot{x} + k_x x = 0$$
$$-Am\omega_x^2\cos(\omega_x t + \phi_x) + k_x A\cos(\omega_x t + \phi_x) = 0$$
$$-m\omega_x^2 + k_x = 0$$
$$\therefore\ \omega_x = \sqrt{\frac{k_x}{m}}.$$

This is a correct solution, and also helps us determine the value of the angular frequency of the motion. Now to complete the full solution we must determine the particular solution to the *inhomogeneous differential equation*, and add this onto the one shown above.

To do this we will guess a polynomial function of order 0, coherent with the order of $k_x x_0$. *If you have any doubts go back to section 1.4.4 in the text.*

$$x_p(t) = \alpha$$
$$\therefore\ \ddot{x}_p(t) = 0.$$

We can substitute this into the differential equation and find the value of the constant α.

$$m0 + k_x\alpha = k_x x_0$$
$$\therefore \alpha \equiv x_0.$$

Our solution for motion in the x direction is seen below, from which we can obtain that for the y direction.

$$\begin{cases} x(t) = A\cos(\omega_x t + \phi_x) + x_0 \\ y(t) = B\cos(\omega_y t + \phi_y) + y_0 \end{cases}.$$

We can now apply the boundary conditions and obtain a set of simultaneous equations necessary to obtain the values of the constants A and B and also to define the phase angles ϕ_x and ϕ_y.

$$\begin{cases} a = A\cos(0 + \phi_x) + x_0 \\ b = B\cos(0 + \phi_y) + y_0 \\ 0 = -A\sin(0 + \phi_x) \\ 0 = -B\sin(0 + \phi_y) \end{cases}.$$

We can immediately deduce that for the last two identities to hold $\phi_x = \phi_y \equiv 0$. Hence the second two will become:

$$\begin{cases} a = A\cos(0) + x_0 \\ b = B\cos(0) + y_0. \end{cases}$$

Therefore,

$$\begin{cases} A = a - x_0 \\ B = b - y_0. \end{cases}$$

So our final solutions to the equation of motion in the two directions are:

$$\begin{cases} x(t) = (a - x_0)\cos(\omega_x t + \phi_x) + x_0 \\ y(t) = (b - y_0)\cos(\omega_y t + \phi_y) + y_0 \end{cases}.$$

IOP Publishing

Chapter 15

Selected solutions to Chapter 5: Angular momentum and central forces

Solution: Exercise 5.1. Particle on a plane in circular motion

The first part of the exercise asks us to sketch the directions of the unit vectors $\hat{r}$ and $\hat{\theta}$. We are told that the r is the distance between the point and the origin. Hence $\hat{r}$ would be in this direction increasing as the distance increases.

We are also told that θ is the angle between the fixed axis and the position vector. The direction of $\hat{\theta}$ is therefore in the direction of increasing θ.

The second part of the exercise requires us to write expressions for the velocity and the acceleration in this coordinate system.

We can see in equation (1.3) that differentiating the position vector $\boldsymbol{r}$ we get the velocity vector.

$$v = \frac{\mathrm{d}\boldsymbol{r}}{\mathrm{d}t} = \frac{\mathrm{d}}{\mathrm{d}t}(r\hat{\boldsymbol{r}})$$
$$v = \frac{\mathrm{d}r}{\mathrm{d}t}\hat{\boldsymbol{r}} + r\frac{\mathrm{d}\hat{\boldsymbol{r}}}{\mathrm{d}t}$$
$$v = \dot{r}\hat{\boldsymbol{r}} + r\dot{\hat{\boldsymbol{r}}}.$$

This gives us the velocity vector in this coordinate system. To obtain the acceleration vector, we have differentiated the velocity vector.

$$
\begin{aligned}
\boldsymbol{a} &= \frac{\mathrm{d}\boldsymbol{v}}{\mathrm{d}t} \\
\boldsymbol{a} &= \frac{\mathrm{d}}{\mathrm{d}t}\left(\frac{\mathrm{d}r}{\mathrm{d}t}\hat{\boldsymbol{r}} + r\frac{\mathrm{d}\theta}{\mathrm{d}t}\hat{\boldsymbol{\theta}}\right) \\
\boldsymbol{a} &= \frac{\mathrm{d}^2r}{\mathrm{d}t^2}\hat{\boldsymbol{r}} + \frac{\mathrm{d}r}{\mathrm{d}t}\frac{\mathrm{d}\hat{\boldsymbol{r}}}{\mathrm{d}t} + \frac{\mathrm{d}r}{\mathrm{d}t}\frac{\mathrm{d}\theta}{\mathrm{d}t}\hat{\boldsymbol{\theta}} + r\frac{\mathrm{d}^2\theta}{\mathrm{d}t^2}\hat{\boldsymbol{\theta}} + \frac{\mathrm{d}\theta}{\mathrm{d}t}\frac{\mathrm{d}\hat{\boldsymbol{\theta}}}{\mathrm{d}t} \\
\boldsymbol{a} &= \ddot{r}\hat{\boldsymbol{r}} + \dot{r}\dot{\hat{\boldsymbol{r}}} + \dot{r}\dot{\theta}\hat{\boldsymbol{\theta}} + r\ddot{\theta}\hat{\boldsymbol{\theta}} + r\dot{\theta}\dot{\hat{\boldsymbol{\theta}}} \\
\boldsymbol{a} &= \ddot{r}\hat{\boldsymbol{r}} + \dot{r}\dot{\hat{\boldsymbol{r}}} + \dot{r}\dot{\theta}\hat{\boldsymbol{\theta}} + r\ddot{\theta}\hat{\boldsymbol{\theta}} + r\dot{\theta}\dot{\hat{\boldsymbol{\theta}}} \\
\boldsymbol{a} &= \ddot{r}\hat{\boldsymbol{r}} + \dot{r}\dot{\theta}\hat{\boldsymbol{\theta}} + \dot{r}\dot{\theta}\hat{\boldsymbol{\theta}} + r\ddot{\theta}\hat{\boldsymbol{\theta}} - r\dot{\theta}\dot{\theta}\hat{\boldsymbol{r}} \\
\boldsymbol{a} &= (\ddot{r} - r\dot{\theta}^2)\hat{\boldsymbol{r}} + (2\dot{r}\dot{\theta} + r\ddot{\theta})\hat{\boldsymbol{\theta}}.
\end{aligned}
$$

The third part of the exercise requires us to determine the force required to maintain a mass m in uniform circular motion. We know the formula of force from Newton's second law is $\boldsymbol{F} = m\boldsymbol{a}$. The mass of the particle is known so we need to find the acceleration. We know that in uniform circular motion the angular velocity $\dot{\boldsymbol{\theta}}$ remains constant. This tells us that the angular acceleration $\ddot{\boldsymbol{\theta}}$ is zero. We can also deduce that because the circular motion is uniform, the radius of motion is constant. With the information we are given we can calculate the acceleration vector:

$$
\begin{aligned}
\boldsymbol{a} &= (\ddot{r} - r\dot{\theta}^2)\hat{\boldsymbol{r}} + (2\dot{r}\dot{\theta} + r\ddot{\theta})\hat{\boldsymbol{\theta}} \\
\boldsymbol{a} &= -(r\dot{\theta}^2)\hat{\boldsymbol{r}}.
\end{aligned}
$$

The angular velocity is ω and the radius is r. We can substitute these to find the acceleration for the particle in uniform circular motion.

$$\boldsymbol{a} = -(r\omega^2)\hat{\boldsymbol{r}}.$$

Now that we have the acceleration we can just substitute everything into the formula for force to find the solution.

$$
\begin{aligned}
\boldsymbol{F} &= m\boldsymbol{a} \\
\boldsymbol{F} &= -m(r\omega^2)\hat{\boldsymbol{r}}.
\end{aligned}
$$

Solution: Exercise 5.3. Angular momentum and torque of particle about origin

We are given the position of the particle with respect to the origin along with the mass, velocity and force. The position of the particle with respect to the origin doesn't change. This means that the motion of the particle is similar to circular motion. To find the angular momentum of the particle, we just have to apply the formula and input the values given.

$$\boldsymbol{L} = m\boldsymbol{r} \times \boldsymbol{v}$$
$$\boldsymbol{L} = m\boldsymbol{r} \times (\dot{r}\hat{\boldsymbol{r}} + r\dot{\theta}\hat{\boldsymbol{\theta}})$$
$$\boldsymbol{L} = mr^2\dot{\theta}\hat{\boldsymbol{\theta}}.$$

To find the torque we need to find the derivative of the angular momentum.

$$\boldsymbol{\tau} = \frac{\mathrm{d}\boldsymbol{L}}{\mathrm{d}t}$$
$$\boldsymbol{\tau} = 2mr\dot{\theta} + mr^2\ddot{\theta}$$
$$\boldsymbol{\tau} = mr(2\dot{\theta} + r\ddot{\theta}).$$

Solution: Exercise 5.4. Angular momentum subject to a central force
Angular momentum is defined as the cross product between the position vector and the momentum of a particle.

$$\boldsymbol{L} = \boldsymbol{r} \times \boldsymbol{p}$$
$$\boldsymbol{L} = m\boldsymbol{r} \times \boldsymbol{v}.$$

We can use the definition of angular momentum and velocity to simplify this equation for Circular motion.

$$\boldsymbol{L} = m\boldsymbol{r} \times \boldsymbol{v}$$
$$\boldsymbol{L} = m\boldsymbol{r} \times (\dot{r}\hat{\boldsymbol{r}} + r\dot{\theta}\hat{\boldsymbol{\theta}})$$
$$\boldsymbol{L} = mr^2\dot{\theta}\hat{\boldsymbol{\theta}}.$$

The direction of angular momentum is defined by $\hat{\boldsymbol{\theta}}$.

The motion defined is circular motion because only a central force is acting on the object under discussion. This means that force is in the radial $\hat{\boldsymbol{r}}$ direction only. We know from that the formula for a force is $\boldsymbol{F} = m\boldsymbol{a}$ This means that the acceleration only has a radial component. We know that acceleration can be written as:

$$\boldsymbol{a} = (\ddot{r} - r\dot{\theta}^2)\hat{\boldsymbol{r}} + (2\dot{r}\dot{\theta} + r\ddot{\theta})\hat{\boldsymbol{\theta}}.$$

Since the transverse acceleration is zero, we can thus say:

$$2\dot{r}\dot{\theta} + r\ddot{\theta} = 0.$$

Solution: Exercise 5.5. Central forces

In the first part of the exercise, we are asked to define central forces. Central forces are defined as forces which only act in the radial $\hat{r}$ direction. This means that they are directed either away from or towards the centre of motion. The second part of the exercise requires us to prove that angular momentum remains constant under a central force. Angular momentum of an object under a central force is already known to be:

$$\boldsymbol{L} = mr^2\dot{\theta}.$$

We will take the time derivative of the angular momentum to check if it remains constant.

$$\frac{\mathrm{d}L}{\mathrm{d}t} = \frac{\mathrm{d}}{\mathrm{d}t}(mr^2\dot{\theta})$$
$$\frac{\mathrm{d}L}{\mathrm{d}t} = m(2r\dot{r}\dot{\theta} + r^2\ddot{\theta})$$
$$\frac{\mathrm{d}L}{\mathrm{d}t} = mr(2\dot{r}\dot{\theta} + r\ddot{\theta}).$$

The term inside the bracket in the last step is the acceleration in the transverse direction. As we know, this is zero when a central force is being applied. Thus we can say that:

$$\frac{\mathrm{d}L}{\mathrm{d}t} = 0.$$

This proves that the angular momentum remains constant when a central force is being applied.

Solution: Exercise 5.6. Object under the influence of a potential

In the first part of the question we are told that the object is under the influence of a potential and we are told to find the force associated with this potential.

Solution: Exercise 5.7. Particle attached to a light in extensible string

We are told that the particle is a distance r_0 from the hole and that it moves in circular motion with a constant angular velocity ω_0. The first part of the exercise requires us to find the angular momentum of the particle. We have all the required variables so we can just input them into the formula to find the angular momentum.

$$\boldsymbol{L} = mr^2\dot{\theta}$$
$$\boldsymbol{L} = m{r_0}^2\omega_0.$$

In the second part of the exercise, we are told that the string starts being pulled down with a constant speed resulting in a decrease in the radius of circular motion. This can be stated as $r(t) = (r_0 - Vt)$. The decrease in the radius doesn't affect the angular momentum and it remains constant.

We are then asked to determine the angular velocity ω and the velocity v at time t.

Since we know that the angular momentum remains constant, we can use it to find the angular velocity ω in terms of ω_0.

$$m{r_0}^2\omega_0 = m(r_0 - Vt)\omega$$
$$\omega = \frac{{r_0}^2\omega_0}{r_0 - Vt}.$$

The tension in the string is radial in direction. This means that it can be written as $\boldsymbol{T} = m(\ddot{r} - r\dot{\theta}^2)$. We know that the speed $\dot{r} = v$ of the particle remains constant. This means that the acceleration $\ddot{r} = 0$. Hence we can write the tension as:

$$\boldsymbol{T} = -\,m(r_0 - Vt)\omega^2$$
$$\boldsymbol{T} = -\,m(r_0 - Vt)\frac{{r_0}^4{\omega_0}^2}{(r_0 - Vt)^4}$$
$$\boldsymbol{T} = \frac{-m{r_0}^4{\omega_0}^2}{(r_0 - Vt)^3}.$$

Solution: Exercise 5.8. Angular momentum of a particle with variable angular frequency

In this exercise we are given that a particle is rotating with a time dependent angular frequency $\dot{\theta}$ and a radial velocity that is decreasing at a constant rate. We are asked to show that the angular momentum is conserved and find the equation for the angular velocity $\dot{\theta}$ at a general time t. We are given that the rate of change of the radius r is $\dot{r} = -\alpha$ and that the particle is at a position R_0 at time $t = 0$. The general position of the particle is thus $r = R_0 - \alpha t$. At time $t = 0$ the angular momentum of the particle is $\boldsymbol{L} = m{R_0}^2\alpha$. In circular motion, angular momentum is conserved so we can equate:

$$m(R_0 - \alpha t)^2\dot{\theta} = m{R_0}^2\alpha$$
$$\dot{\theta} = \frac{{R_0}^2\alpha}{(R_0 - \alpha t)^2}.$$

In the next part of the exercise we are asked to show that the angular component of the acceleration is zero. The formula for angular acceleration is:

$$\boldsymbol{a}\dot{\boldsymbol{\theta}} = 2\dot{r}\dot{\theta} + r\ddot{\theta}.$$

To compute the angular component of acceleration, we first need to calculate the angular acceleration $\ddot{\theta}$. This will be done by differentiating the angular velocity that we calculated in the previous part of the exercise:

$$\ddot{\theta} = \frac{2{R_0}^2\alpha^2}{(R_0 - \alpha t)^3}.$$

Thus the angular component of acceleration can be calculated as:

$$\begin{aligned}
\boldsymbol{a}\hat{\boldsymbol{\theta}} &= 2\dot{r}\dot{\theta} + r\ddot{\theta} \\
\boldsymbol{a}\hat{\boldsymbol{\theta}} &= 2\alpha\frac{R_0{}^2\alpha^2}{(R_0 - \alpha t)^2} + (R_0 - \alpha t)\frac{2R_0{}^2\alpha^2}{(R_0 - \alpha t)^3} \\
\boldsymbol{a}\hat{\boldsymbol{\theta}} &= \frac{-R_0{}^2\alpha^2}{(R_0 - \alpha t)^2} + \frac{-R_0{}^2\alpha^2}{(R_0 - \alpha t)^2} \\
\boldsymbol{a}\hat{\boldsymbol{\theta}} &= 0.
\end{aligned}$$

Solution: Exercise 5.9. Ladybird on a stick

We are told that a ladybird is at rest on the midpoint of a stick. The stick then starts rotating about one end of the stick and the angular velocity is $\dot{\theta} = \omega_0 t$ and the ladybird starts to walk away from the midpoint. The exercise requires us to calculate the angular momentum of the ladybird. We can use the equation $\boldsymbol{L} = mr^2\dot{\theta}$. The radius of the ladybird from the axis of rotation is increasing and can be written as $r = R_0 + vt$. The angular momentum can therefore be calculated as:

$$\begin{aligned}
\boldsymbol{L} &= mr^2\dot{\theta} \\
\boldsymbol{L} &= m(R_0 + vt)^2\omega_0 t.
\end{aligned}$$

Solution: Exercise 5.10. Beads on a hoop

The mass of the hoop is M and it has a radius R. The force equation for the bead is:

$$\begin{aligned}
mg\cos\theta + N &= \frac{mv^2}{r} \\
N &= \frac{mv^2}{r} - mg\cos\theta.
\end{aligned}$$

In the initial position the bead is at rest and so only had potential energy. When it starts sliding down, it has a combination of potential and kinetic energy. Using conservation of energy, we can write the equation:

$$\begin{aligned}
mgR &= \frac{1}{2}mv^2 + mgR\cos\theta \\
2mgR(1 - \cos\theta) &= mv^2 \\
v^2 &= 2gR(1 - \cos\theta).
\end{aligned}$$

This allows us to write the normal force as:

$$\begin{aligned}
N &= 2mg(1 - \cos\theta) - mg\cos\theta \\
N &= mg(2 - 3\cos\theta).
\end{aligned}$$

From this equation we can see that N changes value when $\cos\theta > \frac{2}{3}$. If we consider the whole system with the hoop, there are two normal forces from the two beads acting upwards. Hence we can balance the vertical forces as:

$$
\begin{aligned}
2Ncos\theta &= Mg \\
2(mg(2 - 3\cos\theta)\cos\theta &= Mg \\
4mg\cos\theta - 6mg\cos^2\theta &= Mg \\
6mg\cos^2\theta - 4mg\cos\theta + Mg &= 0.
\end{aligned}
$$

This is a quadratic equation. Using the quadratic formula, we see that the value of $\cos\theta$ is:

$$
\cos\theta = \frac{1}{3}\left(1 \pm \sqrt{1 - \frac{3M}{2m}}\right).
$$

A solution for $\cos\theta$ exists when the discriminant is greater than 0:

$$
\begin{aligned}
\sqrt{1 - \frac{3M}{2m}}\Bigg) &> 0 \\
1 - \frac{3M}{2m} &> 0 \\
\frac{3M}{2m} &< 1 \\
\frac{M}{m} &< \frac{2}{3} \\
\frac{m}{M} &> \frac{3}{2}.
\end{aligned}
$$

Hence the hoop will rise if the ratio of masses of the hoop with the bead is $\frac{m}{M} > \frac{3}{2}$.

Chapter 16

Solutions to Chapter 6: Centre of mass and collisions

Solution: Exercise 6.1.

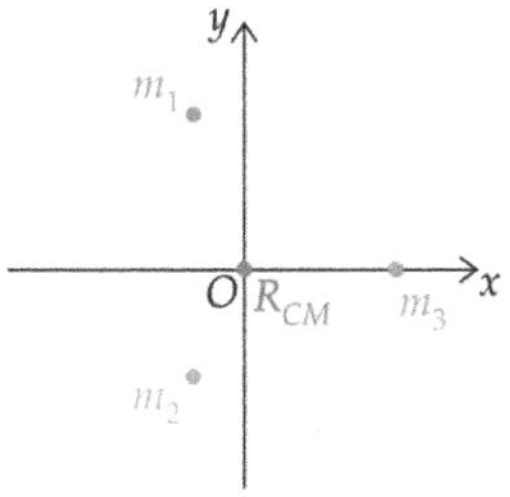

We start by noticing that to answer the first question we need to find the value of three variables, meaning that we probably need to solve a system of three equations. The first equation is easy to find because it is given already,

$$m_1 + m_2 = 4. \tag{16.1}$$

The only other information we have is that the position of the centre of mass is at the origin of the axis. Let's write this explicitly

$$\begin{aligned}\boldsymbol{R}_{\text{CM}} &= \frac{m_1\boldsymbol{r}_1 + m_2\boldsymbol{r}_2 + m_3\boldsymbol{r}_3}{m_1 + m_2 + m_3} \\ &= \frac{-m_1\hat{\boldsymbol{i}} + 3m_1\hat{\boldsymbol{j}} - m_2\hat{\boldsymbol{i}} - 2m_2\hat{\boldsymbol{j}} + 3m_3\hat{\boldsymbol{i}}}{4 + m_3} \\ &= \frac{(-m_1 - m_2 + 3m_3)\hat{\boldsymbol{i}}}{4 + m_3} + \frac{(3m_1 - 2m_2)\hat{\boldsymbol{j}}}{4 + m_3} \\ &= 0\hat{\boldsymbol{i}} + 0\hat{\boldsymbol{j}}\,.\end{aligned}$$

doi:10.1088/978-0-7503-2690-2ch16

By comparison of the last two expressions we obtain equations (16.2) and (16.3):

$$-m_1 - m_2 + 3m_3 = 0 \tag{16.2}$$

$$3m_1 - 2m_2 = 0. \tag{16.3}$$

The system of equations we need to solve now looks like this

$$\begin{cases} m_1 + m_2 = 4 \\ -m_1 - m_2 + 3m_3 = 0. \\ 3m_1 - 2m_2 = 0 \end{cases}$$

Using equation (16.1), equation (16.2) can be rewritten as

$$3m_3 = m_1 + m_2$$
$$m_3 = 4/3.$$

Now rewrite equation (16.3) as

$$m_1 = \frac{2}{3}m_2$$

and sub this in equation (16.1) to obtain an expression for m_2

$$\frac{2}{3}m_2 + m_2 = 4$$
$$m_2 = \frac{12}{5}.$$

If follows then that

$$m_1 = \frac{8}{5}.$$

We have now solved the three simultaneous equations and found the value of the three masses:

$$m_1 = \frac{8}{5}$$
$$m_2 = \frac{12}{5}$$
$$m_3 = \frac{4}{3}.$$

Solution: Exercise 6.2.

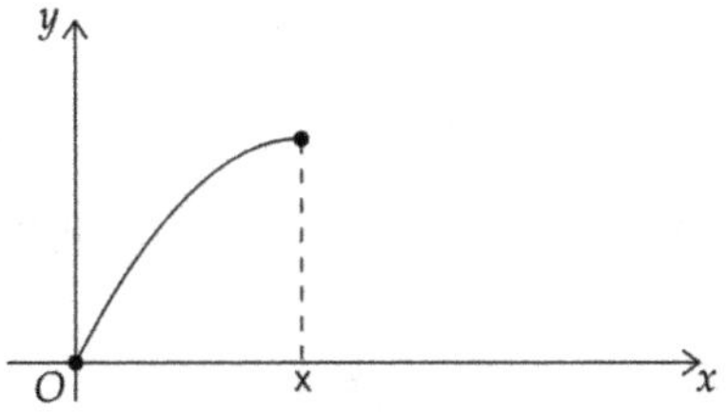

This problem is an emblematic example of how sometimes symmetries in the centre of mass frame are very useful in order to find a solution.

Take a look at the situation before the collision in the centre of mass frame. At this point we still have only one body in the system, with velocity $\boldsymbol{u} = u_x\hat{\boldsymbol{i}} + u_y\hat{\boldsymbol{j}}$. This coincides with the centre of mass velocity V_{CM} of the system. We also know that V_{CM} will not change after the collision since **no external forces** are involved. It is therefore easy to predict where the centre of mass will be after the collision as this is simply a projectile motion without air friction. To picture this just imagine the projectile doesn't break up in two parts.

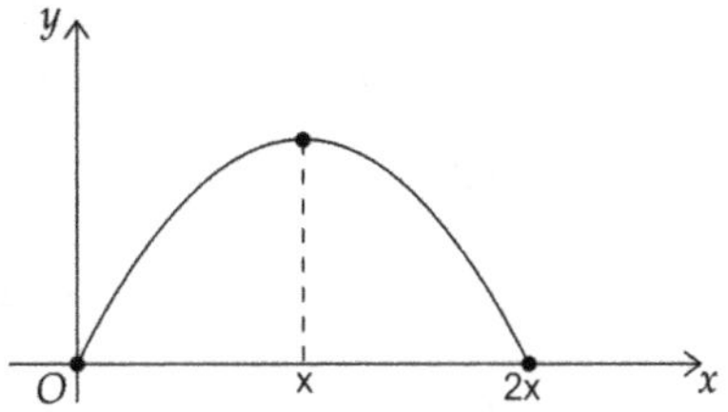

We now know that the centre of mass final position, i.e. when both parts of the original projectile reach the ground, is at a distance $2x$ from the launching point. Since the two parts have the same mass and we know that the first piece lands at the launching point, at the origin of the axis, it is trivial to predict that the second piece must land at a distance $4x$ from the origin. Just for the sake of clarity, if the original projectile has mass M:

$$R_{CM} = \frac{\frac{M}{2}r_1 + \frac{M}{2}r_2}{M} = 2\text{x}$$

$$\frac{M}{2}r_2 = M2\text{x}$$

$$\Rightarrow r_2 = 4\text{x}$$

Since $r_1 = 0$.

Solution: Exercise 6.3. The two equations we need to solve are the equation of conservation of momentum and the equation of conservation of energy.

We start by calculating the total momentum before the collision,

$$p_{tot} = m_1 u_1 + m_2 u_2 = 6,$$

we know then that total momentum after the collision must remain 6. This is our first equation:

$$4v_1 + v_2 = 6. \quad (16.4)$$

We follow the same procedure to find the conservation of energy equation, we only need to consider the kinetic energy of the two objects since there is no energy loss and no other forms of energy.

$$K_{tot} = \frac{1}{2} m_1 u_1^2 + \frac{1}{2} m_2 u_2^2 = 36.$$

After the collision we then require

$$2v_1^2 + \frac{1}{2} v_2^2 = 36. \quad (16.5)$$

The system of simultaneous equations we need to solve is therefore:

$$\begin{cases} 4v_1 + v_2 = 6 \\ 2v_1^2 + \frac{1}{2} v_2^2 = 36 \end{cases}.$$

Sub $v_2 = 6 - 4v_1$ from equation (16.4) into equation (16.5)

$$2v_1^2 + 18 - 24v_1 + 8v_1^2 = 36$$
$$5v_1^2 - 12v_1 - 9 = 0.$$

We have obtained a quadratic equation, this will give us two solutions for v_1.

$$v_1 = \frac{6 \pm \sqrt{36 + 45}}{5}$$
$$= \frac{6 \pm 9}{5}$$
$$\Rightarrow v_1 = 3 \quad \vee \quad \frac{-3}{5}.$$

To decide which solution we should accept we check which one satisfies the relative velocity relation before and after an elastic collision in a two-body system, i.e.

$$v = -u. \quad (16.6)$$

Using the velocities before the collision we find u:

$$u = u_1 - u_2 = 3 + 6 = 9$$
$$v = -u = -9.$$

For $v_1 = 3$, we see that $v_2 = -6$, this does not satisfy the above relation

$$v = v_1 - v_2 = -3 + 6 \neq -9.$$

For $v_1 = \frac{-3}{5}$ instead we have $v_2 = \frac{42}{5}$ which indeed satisfies the relation

$$v = v_1 - v_2 = \frac{-3}{5} - \frac{42}{5} = \frac{-45}{5} = -9.$$

These two solutions for v_1 and v_2 coincide with the results obtained in example 6.3. It should be clear that in this case working in the centre of mass frame requires fewer calculations and therefore gives fewer chances of making mistakes. Not making mistakes is particularly useful in exams.

Solution: Exercise 6.4.
BOOKWORK

Solution: Exercise 6.5.
BOOKWORK

Solution: Exercise 6.6. We are told that no external forces are involved in the collision, which means that if we describe the problem in the centre of mass frame the centre of mass velocity of the system must remain constant before and after the collision.

This information is already enough to answer the first question. If V_{CM} is constant then also the centre of mass kinetic energy must remain constant since $K_{CM} = \frac{1}{2}MV_{CM}^2$.

However, the collision described is inelastic, meaning that there must be some loss of kinetic energy, i.e. $\Delta K \neq 0$. We have seen in section 6.1.3 that the total kinetic energy K can be written as the sum of the relative kinetic energy, $K_{rel} = \frac{1}{2}\sum_i^n m_i u'_i$, and the centre of mass kinetic energy, $K_{CM} = \frac{1}{2}MV_{CM}^2$, i.e.

$$K = K_{rel} + K_{CM}.$$

Since we know that K_{CM} remains unchanged after the collision there must be a difference in relative kinetic energy before and after the collision.

In the case of two bodies, $n = 2$, the relative kinetic energy can be written as

$$K_{\text{rel}} = \frac{1}{2}\sum_{i=1}^{2} m_i u'_i{}^2 = \frac{1}{2}\mu u^2 \tag{16.7}$$

where u is the relative velocity between the two bodies before the collision. We are given also the coefficient of restitution e, which relates the relative velocities before and after an inelastic collision,

$$v = eu$$

such that $v < u$. We can use this relation to write the final relative kinetic energy in terms of the initial relative velocity,

$$K_{\text{rel}} = \frac{1}{2}\mu v^2 = \frac{1}{2}\mu(eu)^2, \tag{16.8}$$

if we now compare the initial relative kinetic energy, equation (16.7), with the final relative kinetic energy, equation (16.8),

$$\frac{1}{2}\mu u^2 < \frac{1}{2}\mu e^2 u^2$$

it is clear that the relative kinetic energy has been reduced by a factor of e^2.

Solution: Exercise 6.7. Since the chicken is immovable from the support, its velocity will be constant and remain equal to zero, $u_2 = v_2 = 0$, and the same for its kinetic energy. This means that all the energy transferred to the chicken will be in the form of heat. Our goal is to find the necessary initial velocity such that the kinetic energy loss of the bowling ball is equal to -500 kg, i.e.

$$\begin{aligned}\Delta K &= \frac{1}{2}mv_1^2 - \frac{1}{2}mu_1^2 \\ &= -\ 500.\end{aligned}$$

By looking at the problem in the laboratory frame of reference, the coefficient of restitution determines the factor by which the initial velocity of the bowling ball is reduced after the collision. We can then investigate the change of kinetic energy of the bowling ball.

$$\begin{aligned}v &= eu \\ \Delta K &= \frac{1}{2}m(eu_1)^2 - \frac{1}{2}mu_1^2 \\ &= \frac{1}{2}m_1(e^2 - 1)u^2 \\ &= -\ 500.\end{aligned}$$

With a little bit of rearranging we can determine u_1 and get the following expression

$$
\begin{aligned}
u_1 &= \sqrt{\frac{\Delta K \times 2}{m(e^2 - 1)}} \\
&= \sqrt{\frac{-500\,000 \times 2}{7.27(0.01^2 - 1)}} \\
&= 371 \text{ m s}^{-1}.
\end{aligned}
$$

In conclusion, you would need to break the sound barrier in order to cook a chicken with a single throw of a bowling ball. Moreover, the ball would bounce back at a speed $v = eu = 3.71$ m s^{-1} which would definitely be a risk to take into account when throwing the ball.

IOP Publishing

Chapter 17

Solutions to Chapter 7: Orbits

Solution: Exercise 7.1.

$$\frac{\mathrm{d}}{\mathrm{d}t}\left(r^2\dot{\theta}\right) = 2r\dot{r}\dot{\theta} + r^2\ddot{\theta} = r\left(r\dot{r}\dot{\theta} + r\ddot{\theta}\right) = 0$$

Solution: Exercise 7.2. Note that the shape of the effective potential does not change. You could actually show all four orbit shapes in the same graph. However, we will draw them separately for clarity.

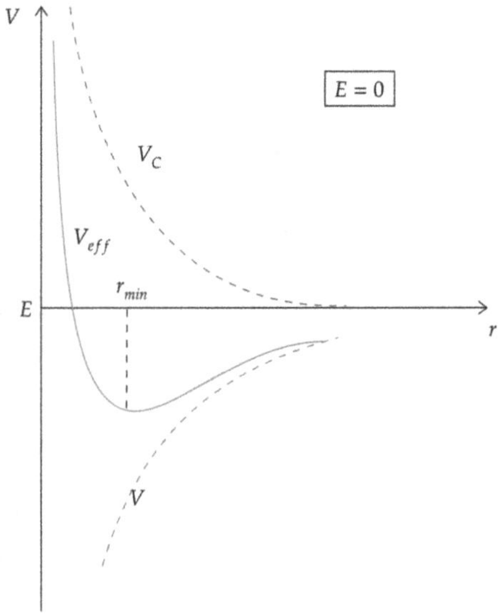

doi:10.1088/978-0-7503-2690-2ch17

Secondly,

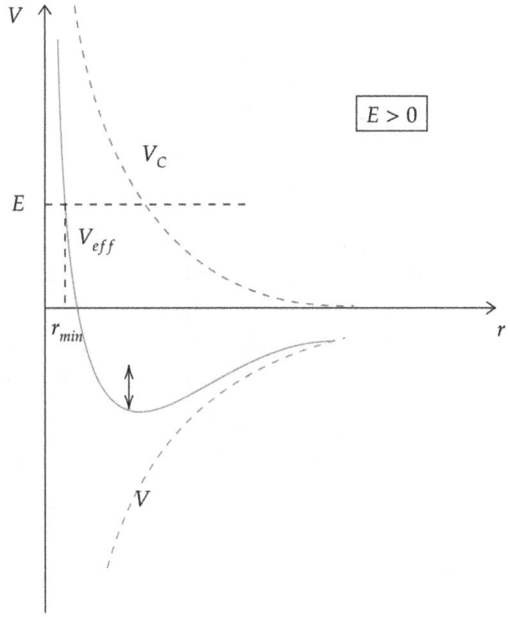

Solution: Exercise 7.3.

BOOKWORK

Solution: Exercise 7.4. We find the reduced mass of the Earth and Sun, and the consequent change in the period of the Earth's orbit from what we would have calculated assuming the Sun was fixed.

The reduced mass is

$$\mu = \frac{M_\odot M_{\text{Earth}}}{M_\odot + M_{\text{Earth}}} = M_{\text{Earth}} \frac{1}{1 + M_{\text{Earth}}/M_\odot}.$$

We have $M_\odot = 1.99 \times 10^{30}$ kg and $M_{\text{Earth}} = 5.97 \times 10^{24}$ kg, so

$$M_{\text{Earth}}/M_\odot = 3 \times 10^{-6}$$

and

$$\mu = 0.999\,997 M_{\text{Earth}}.$$

The period was

$$T = 2\pi \sqrt{\frac{ma^3}{|K|}}.$$

The constant $K = GM_\odot M_{\text{Earth}}$ is determined by the law of gravitation and is unchanged. If a is kept the same, we have $T \propto \sqrt{m}$ so the period is reduced by a factor $\sqrt{\mu/M_{\text{Earth}}}$. This corresponds to a fractional change

$$\frac{\Delta T}{T} = \frac{1}{2}\left(1 - \frac{\mu}{M_{\text{Earth}}}\right) = 1.5 \times 10^{-6}.$$

In a year this corresponds to $\Delta T = 34$ s, which is small but measurable.

Solution: Exercise 7.5. The orbital radius R can be found by equating the centripetal force required to keep the stars moving around one another with the gravitational force between them (or equivalently, by requiring the gravitational force to cancel the apparent centrifugal force in their rotating frame of reference). Let $\boldsymbol{r} = \boldsymbol{r}_1 - \boldsymbol{r}_2$ be the position vector of star 1 relative to star 2. The gravitational force on star 1 is

$$\boldsymbol{F} = -\frac{GM_1M_2}{R^2}\hat{\boldsymbol{r}}.$$

On the other hand if the stars move around one another in circular orbits of radius R and angular frequency ω, the required centripetal acceleration

$$\ddot{\boldsymbol{r}} = -\omega^2 R\hat{\boldsymbol{r}}.$$

From general results of two-body motion we know that the relative coordinate $\boldsymbol{r}$ behaves like the position vector of a single body with a mass equal to the reduced mass

$$\mu = \frac{M_1M_2}{M_1 + M_2}$$

so

$$\boldsymbol{F}_{grav} = \mu\ddot{\boldsymbol{r}}$$

$$\Longrightarrow \frac{GM_1M_2}{R^2} = \mu\omega^2 R$$

$$\Longrightarrow R^3 = \frac{GM_1M_2}{\mu\omega^2} = \frac{T^2G(M_1 + M_2)}{4\pi^2}$$

since $T = 2\pi/\omega$. Hence

$$R = \left(\frac{T^2G(M_1 + M_2)}{4\pi^2}\right)^{1/3}.$$

Solution: Exercise 7.6.

(i) The force acting on the feather are

$$\boldsymbol{F} = -(m_f g - v\beta\sin\alpha)\hat{j} - v\cos\alpha\hat{i}$$

while on the droppings it is only $\boldsymbol{F} = -mg\hat{j}$.

(ii) The position at which the duck loses material (assuming the duck was at rest at the origin) is

$$x_d = v\cos\alpha t$$
$$y_d = v\sin\alpha t.$$

The terminal velocity of the feather will be

$$v_T = \frac{m_f g}{\beta}$$

so the time taken to fall in the water is

$$t_f = \frac{v\sin\theta t\beta}{m_f g}$$

and the distance is obtained adding to the position of the droppings the distance travelled during this time.

(iii) If the initial motion of the feather cannot be neglected, its motion will follow that of a mass under air resistance, separately on the x- and y-axis. On x:

$$\frac{\mathrm{d}v_x}{\mathrm{d}t} = -\frac{\beta v_x}{m}$$
$$\frac{\mathrm{d}v_x}{v_x} = -\frac{\beta \mathrm{d}t}{m}$$
$$v_x(t) = -v\cos\alpha e^{-\beta t/m}.$$

On the y-axis:

$$\frac{\mathrm{d}v_y}{\mathrm{d}t} = -mg - \frac{\beta v_y}{m}$$
$$-\frac{\mathrm{d}v_y}{g + \frac{\beta v_y}{m}} = \mathrm{d}t$$
$$\frac{\mathrm{d}v_y}{v_T + v_y} = -\frac{\beta \mathrm{d}t}{m}$$

and imposing the condition that velocity is v_T for $t \to \infty$ and $v_y = v_0 \sin\alpha$ at $t = 0$, the velocity is

$$v_y = (v_0 \sin\alpha + v_T)e^{-\beta t/m} - v_T.$$

(iv) The dropping will go higher since it's not stopped by air friction.

Chapter 18

Selected solutions to Chapter 8: Rigid bodies

Solution: Exercise 8.9.

Let's first calculate the moment of inertia of the solid cylinder about its central axis. Since the cylinder has uniform density, an infinitesimal element of mass is given by

$$\mathrm{d}m = \rho\ \mathrm{d}V = \rho\ r\ \mathrm{d}r\ \mathrm{d}\phi\ \mathrm{d}z = \frac{M}{\pi R^2 L} r\ \mathrm{d}r\ \mathrm{d}\phi\ \mathrm{d}z,$$

where we have used cylindrical coordinates.

The moment of inertia of the solid cylinder is thus given by

$$I = \int r^2\ \mathrm{d}m = \frac{M}{\pi R^2 L}\int_0^R r^3\ \mathrm{d}r \int_0^{2\pi} \mathrm{d}\phi \int_0^L \mathrm{d}z = \frac{M}{\pi R^2 L} \times \frac{R^4}{4} \times 2\pi \times L = \frac{MR^2}{2}.$$

Now let's find the moment of inertia of the hollow cylinder about its central axis. Since the hollow cylinder has negligible thickness, it is only formed by a curved surface. Its density is uniform and given by

$$\sigma = \frac{M}{2\pi RL}.$$

An infinitesimal element of mass is thus given by

$$\mathrm{d}m = \sigma\ \mathrm{d}S = \frac{M}{2\pi RL} R\ \mathrm{d}\phi\ \mathrm{d}z = \frac{M}{2\pi L}\mathrm{d}\phi\ \mathrm{d}z,$$

where we are again using cylindrical coordinates.

Since in this case the radius is constant, the moment of inertia is given by

$$I = R^2 \int \mathrm{d}m = \frac{MR^2}{2\pi L}\int_0^{2\pi} \mathrm{d}\phi \int_0^L \mathrm{d}z = \frac{MR^2}{2\pi L} \times 2\pi \times L = MR^2.$$

doi:10.1088/978-0-7503-2690-2ch18

Thus, the moment of inertia of the hollow cylinder is higher than the one of the solid cylinder. To understand this, we need to think about what the moment of inertia actually represents, i.e. the resistance to a change in the rotation of the body. Since the hollow cylinder has its mass concentrated far away from the axis of rotation, it will have a higher moment of inertia with respect to the solid cylinder, whose mass is on average closer to the axis.

Let's now move on to the second part of the exercise. To understand which cylinder is going to arrive first, we need to find the acceleration acting on both of them.

Let's start by considering that the cylinders are undergoing a pure roll motion. This means that

$$v_{\mathrm{CM}} = v_{\mathrm{ROT}} = \omega R,$$

and thus

$$a_{\mathrm{CM}} = \dot{\omega} R = \alpha R, \tag{18.1}$$

where α is the angular acceleration.

There are two forces acting on the cylinder, the gravitational force and the frictional force. As usual, we can decompose the gravitational force into a parallel and a perpendicular component. We have that

$$F_{\parallel} = Mg \sin\theta$$
$$F_{\perp} = Mg \cos\theta.$$

The normal force of the surface, $\boldsymbol{N}$, is equal and opposite to $F_{\perp}$, so the vector sum of the perpendicular forces is zero. Instead, the frictional force, $\boldsymbol{F}_f$, acts along the surface in the direction opposite to the motion. The only force that does not pass through the axis of rotation is the frictional force, so this is the only force that matters to calculate the torque about the axis of rotation. This is given by

$$\tau = RF_f = I\alpha = I\frac{a_{\mathrm{CM}}}{R}, \tag{18.2}$$

where I is the moment of inertia of the cylinder about the central axis, and we have used equation (18.1).

Since we do not know both F_f and a_{CM}, we need another equation. We can use the good old Newton's second law. In the direction parallel to the surface we have that

$$Ma_{\mathrm{CM}} = Mg \sin\theta - F_f. \tag{18.3}$$

Combining equations (18.2) and (18.3), we obtain

$$Ma_{\mathrm{CM}} = Mg \sin\theta - I\frac{a_{\mathrm{CM}}}{R^2}$$

$$a_{\mathrm{CM}}\left(M + \frac{I}{R^2}\right) = Mg \sin\theta$$

$$a_{\mathrm{CM}} = \frac{MR^2 g \sin\theta}{MR^2 + I}.$$

For a solid cylinder, we have seen that $I = \frac{MR^2}{2}$. Therefore, the acceleration of a solid cylinder is

$$a_{CM} = \frac{2}{3} g \sin\theta.$$

We can see that the acceleration does not depend on the radius or the mass of the cylinder! Thus, it doesn't matter if one solid cylinder is much heavier or bigger than the other. The two cylinders will arrive at the same time at the end of the slope!

Finally, let's solve the last part of the problem. We just need to find the acceleration of a hollow cylinder under the same conditions as the previous part of the exercise. Since for a thin hollow cylinder the moment of inertia is $I = MR^2$, we have that

$$a_{CM} = \frac{1}{2} g \sin\theta.$$

We observe that the acceleration of a solid cylinder down a slope is greater than the acceleration of a thin hollow cylinder. Therefore, the solid cylinder will always arrive first, irrespective of the mass or radius of the two cylinders.

Solution: Exercise 8.10.

(i) If the wheel is rolling, without skidding or slipping, then it said to be undergoing a pure roll motion. Under these conditions, we have that

$$v_{LIN} = v_{ROT} = \omega R.$$

(ii) The kinetic energy will have a linear and a rotational component. The linear component is given by

$$K_{LIN} = \frac{1}{2} M v^2,$$

where $v \equiv v_{LIN}$, while the rotational component is given by

$$K_{ROT} = \frac{1}{2} I_0 \omega^2 = \frac{I_0 v^2}{2R^2}.$$

Thus, the total kinetic energy is given by

$$K = K_{LIN} + K_{ROT} = \frac{Mv^2}{2} + \frac{I_0 v^2}{2R^2} = \frac{v^2}{2R^2}\left(MR^2 + I_0\right).$$

(iii) The forces acting on the wheel are the gravitational force, the normal force of the surface, and the frictional force. The normal force is equal and opposite to the perpendicular component of the gravitational force. Thus, we are left with the parallel component of the gravitational force and the frictional force, which have opposite directions.

(iv) Using conservation of mechanical energy, we must have that $\Delta K = -\Delta V$. Since the wheel is released from rest, the initial kinetic energy is zero. Therefore, when the centre of mass has travelled a distance ℓ, we have that

$$\frac{v^2}{2R^2}\left(MR^2 + I_0\right) = Mg\ell \sin\theta,$$

where we have used the fact that $h = \ell \sin\theta$. Rearranging, we obtain

$$\begin{aligned} v^2(MR^2 + I_0) &= 2MR^2 g\ell \sin\theta \\ v^2 &= \frac{2MR^2 g\ell \sin\theta}{MR^2 + I_0} \\ v &= \sqrt{\frac{2MR^2 g\ell \sin\theta}{MR^2 + I_0}}. \end{aligned}$$

However, we know that the moment of inertia of a wheel (approximated to a disk) is $I_0 = \frac{MR^2}{2}$. Thus, the above equation simplifies to

$$v = 2\sqrt{\frac{g\ell \sin\theta}{3}}.$$

Solution: Exercise 8.11: Cats!

(i) Given the moment of inertia I_{CM} calculated around an axis passing through the centre of mass of a rigid body of mass M, the theorem of parallel axis states that the moment of inertia of the same body around an axis passing through a point at a distance d from the centre of mass, and parallel to the first axis, is given by

$$I = I_{\mathrm{CM}} + Md^2.$$

This theorem is valid for any arbitrary non-planar object, but only holds if the first axis passes through the centre of mass (the first axis cannot pass through any other point in the body for this theorem to work).

(ii) Let's consider separately the contributions of the cylinder and of the tail to the total moment of inertia of the body.

Assuming uniform density, an infinitesimal element of mass of the cylinder is given by

$$\mathrm{d}m = \rho\, \mathrm{d}V = \frac{m}{\pi R^2 L} r\, \mathrm{d}r\, \mathrm{d}\phi\, \mathrm{d}z,$$

where we are, of course, using cylindrical coordinates. Calling L the length of the cylinder, its moment of inertia is then given by

$$I_c = \int r^2\, \mathrm{d}m = \frac{m}{\pi R^2 L}\int_0^R r^3\, \mathrm{d}r \int_0^{2\pi} \mathrm{d}\phi \int_0^L \mathrm{d}z = \frac{m}{\pi R^2 L} \times \frac{R^4}{4} \times 2\pi \times L = \frac{MR^2}{2}.$$

Let's insert the values $m = 3$ kg and $R = 0.1$ m:

$$I_c = \frac{3 \times (0.1)^2}{2} = 1.5 \times 10^{-2}\ \text{kg m}^2.$$

Let's now consider the tail. The moment of inertia of the uniform stick around its centre of mass is $I = \frac{1}{12} m\ell^2$. The distance between the centre of mass and the axis of the cylinder is $d = r + \ell/2$. Thus, we can apply the theorem of parallel axis, and the moment of inertia of the stick around the axis of the cylinder is given by

$$\begin{aligned} I_t &= \frac{1}{12} m_t \ell^2 + m_t\left(r + \frac{\ell}{2}\right)^2 \\ &= \frac{1}{12} \times 0.2 \times (0.2)^2 + 0.2 \times (0.1 + 0.1)^2 \approx 8.7 \times 10^{-3}\ \text{kg m}^2 \end{aligned}$$

(iii) Since angular momentum is conserved, we have

$$0 = I_t\omega_t - I_c\omega_c$$
$$I_t\omega_t = I_c\omega_c.$$

Thus, we can find ω_c:

$$\omega_c = \frac{I_t}{I_c}\omega_t \approx \frac{0.87}{1.5}\omega_t = 0.58\omega_t.$$

Solution: Exercise 8.12.

(i) Assuming constant density, the moment of inertia of the stick around an axis passing through its centre is given by

$$I = \int x^2\, \mathrm{d}m = \lambda \int_{-\ell/2}^{\ell/2} x^2\, \mathrm{d}x = \frac{m}{l}\left[\frac{x^3}{3}\right]_{-\ell/2}^{\ell/2} = \frac{m}{l} \times 2 \times \frac{\ell^3}{24} = \frac{1}{12} m\ell^2.$$

However, we are asked to find the moment of inertia of the stick around an axis passing through the pivot P. Thus, we need to use the theorem of parallel axis. Since the pivot and the centre of the stick are at a distance d from each other, we obtain

$$I = \frac{1}{12} m\ell^2 + md^2.$$

(ii) Since the gravitational force is acting on the centre of mass, at a distance d from the pivot and forming an angle of θ with the stick, the torque is given by

$$\tau = |\boldsymbol{r} \times \boldsymbol{F}| = d\ m\ g\ \sin\theta \approx m\ g\ d\ \theta.$$

Note that, as we are told θ is small, we have used the small-angle approximation, i.e. $\sin\theta \approx \theta$.

(iii) The differential equation of the system is

$$I\ddot{\theta} = \tau$$

$$\left(\frac{m\ell^2}{12} + md^2\right)\ddot{\theta} = -\,m\ g\ d\ \theta$$

$$\ddot{\theta} + \frac{g\ d}{\ell^2/12 + d^2}\theta = 0.$$

This differential equation is in the same form as the differential equation for simple harmonic motion, i.e. $\ddot{x} + \omega^2 x = 0$, with

$$\omega = \sqrt{\frac{d\ g}{\ell^2/12 + d^2}}.$$

Thus, the frequency of oscillation is given by

$$f = \frac{\omega}{2\pi} = \frac{1}{2\pi}\sqrt{\frac{d\ g}{\ell^2/12 + d^2}}.$$

Solution: Exercise 8.13.

(i) We consider one corner of the triangle at the origin, and the second at $(1, 0)\ell$. The third point will be at $\left(\frac{1}{2}, \frac{\sqrt{3}}{2}\right)\ell$. The centre of the triangle will be the centre of mass, so

$$r_{\mathrm{CM}} = \left(\frac{1}{2}, \frac{1}{3} \times \frac{\sqrt{3}}{2}\right)\ell = \left(\frac{1}{2}, \frac{\sqrt{3}}{6}\right)\ell.$$

To calculate the distance of a disk from the centre of the triangle, we can calculate the distance of the centre of mass from the origin:

$$d = \ell\sqrt{\left(\frac{1}{2}\right)^2 + \left(\frac{\sqrt{3}}{6}\right)^2} = \ell\sqrt{\frac{1}{4} + \frac{1}{12}} = \frac{\sqrt{3}}{3}\ell.$$

Since this is a discrete distribution, we need to use the formula

$$I = \sum_i m_i\, r_i^2 = \sum_i m_i\left(x_i^2 + y_i^2\right).$$

(a) If the discs are considered point-like objects, the total moment of inertia is three times the moment of inertia of a single disc with respect to a distance d:

$$I = 3\,md^2 = m\ell^2.$$

(b) If the discs are instead assumed to have radius $\ell/4$ and a uniform mass distribution, the moment of inertia of each of them around its axis will be

$$I_{\text{disc}} = \frac{mr^2}{2} = \frac{m\ell^2}{32}.$$

These three moments of inertia need to be added to the point-like case to obtain

$$I_{\text{tot}} = \frac{35}{32}m\ell^2.$$

(ii) A speed v at a distance $d = \frac{\sqrt{3}}{3}\ell$ means that the system will have angular momentum

$$L = 3\,mv\frac{\sqrt{3}}{3}\ell = \sqrt{3}\,mv\ell.$$

Note that there are three masses moving, so the total angular momentum is three times the angular momentum of a single mass. To find the angular speed, we need to divide the angular momentum by the moment of inertia:

$$\omega = \frac{L}{I} = \frac{\sqrt{3}\,mv\ell}{m\ell^2} = \frac{\sqrt{3}\,v}{\ell}.$$

(iii) We decide to remove the top disc, so the centre of mass will move to $\left(\frac{1}{2}, 0\right)\ell$. The moment of inertia with respect to the geometrical centre is

$$I_{\text{centre}} = 2\,md^2 = \frac{2}{3}m\ell^2,$$

while with respect to the new centre of mass it is the moment of inertia of two points separated from the centre of mass by a distance $\frac{\ell}{2}$, i.e.

$$I_{\text{CM}} = 2 \times \frac{1}{4}m\ell^2 = \frac{1}{2}m\ell^2.$$

(iv) After the rotation is applied again, the system will rotate around the centre, even if it is not anymore the centre of mass. So its angular speed will be again the ratio between the new angular momentum (which only assumes two rotating discs) and the new moment of inertia about the centre. The new angular momentum is given by

$$L_2 = 2\,mv\frac{\sqrt{3}}{3}\ell,$$

therefore the angular speed is

$$\omega_2 = \frac{L_2}{I_{\text{centre}}} = \frac{2\sqrt{3}mv\ell/3}{m\ell^2/2} = \frac{\sqrt{3}v}{\ell},$$

exactly as before.

On the other hand, the fact that the centre of mass does not coincide with the centre of rotation will generate a force on the rotational centre. We can assume that all the mass of the system ($2m$) is concentrated in the centre of mass, at a distance $\frac{\sqrt{3}}{2}\ell$ from the geometrical centre. So the centrifugal force will be given by

$$F = 2\,m\omega^2\left(\frac{\sqrt{3}}{2}\ell\right) = 2\,m \times \left(\frac{\sqrt{3}v}{\ell}\right)^2 \times \left(\frac{\sqrt{3}}{2}\ell\right) = \frac{3\sqrt{3}mv^2}{\ell}.$$

Solution: Exercise 8.14.

(i) The moment of inertia of a uniform stick of length ℓ about a perpendicular axis passing through one of its ends is $I_1 = \frac{m\ell^2}{3}$.

The total moment of inertia of the windmill is six times the moment of inertia of a stick, i.e.

$$I_w = 6\,I_1 = 6 \times \frac{m\ell^2}{3} = 2m\ell^2.$$

(ii) Let's set the origin of the coordinate system on the ground, at the base of the windmill. Thus, the boy is sitting at position $(-d,\,0)$ from the origin. Since the stone is thrown with initial velocity v, which forms an angle α with the horizontal, we have that

$$v_x = v\cos\alpha$$
$$v_y = v\sin\alpha - gt.$$

When the stone reaches the maximum height, the y-component of the velocity is zero. Thus, we can find the time $t_{\max}$ taken by the stone to reach the maximum height:

$$0 = v \sin \alpha - g t_{\max}$$

$$t_{\max} = \frac{v \sin \alpha}{g}.$$

At any instant t, the position of the stone is given by

$$x = v \cos \alpha t - d$$

$$y = v \sin \alpha t - \frac{1}{2} g t^2.$$

At $t = t_{\max}$, $x = 0$. Therefore,

$$0 = v \cos \alpha \cdot \frac{v \sin \alpha}{g} - d$$

$$d = \frac{v^2 \cos \alpha \sin \alpha}{g} = \frac{v^2 \sin 2\alpha}{2g}.$$

Analogously, at $t = t_{\max}$, $y = h$. Thus,

$$h = v \sin \alpha \cdot \frac{v \sin \alpha}{g} - \frac{1}{2} g \frac{v^2 \sin^2 \alpha}{g^2} = \frac{v^2 \sin^2 \alpha}{g} - \frac{1}{2} \frac{v^2 \sin^2 \alpha}{g} = \frac{1}{2} \frac{v^2 \sin^2 \alpha}{g}.$$

(iii) The stone hits the windmill with a horizontal speed $v_x = v \cos \alpha$. Its angular momentum with respect to the pivot is

$$L = |\boldsymbol{r} \times \boldsymbol{p}| = \ell\, p = m_s\, v \cos \alpha\, \ell.$$

The moment of inertia of the stone with respect to the pivot is $I_s = m_s\, \ell^2$. Thus, after the collision, the total moment of inertia of the system (windmill + stone) will be

$$I_f = I_w + I_s = 2\, m\ell^2 + m_s \ell^2 = \left(2\, m + m_s\right)\ell^2.$$

Conservation of angular momentum gives us

$$m_s\, v \cos \alpha\, \ell = I_f\, \omega$$

$$\omega = \frac{m_s\, v \cos \alpha\, \ell}{(2\, m + m_s)\ell^2} = \frac{m_s\, v \cos \alpha}{(2\, m + m_s)\ell},$$

so we have found the angular velocity after the collision. Note that since this is a totally inelastic collision, angular momentum is conserved (if we ignore air resistance), but kinetic energy is not.

(iv) The kinetic energy before the collision is

$$K_b = \frac{1}{2} m_s \, v^2 \cos^2 \alpha,$$

while the kinetic energy after the collision is given by

$$K_a = \frac{1}{2} I_f \, \omega^2 = \frac{1}{2}\left(2\,m + m_s\right)\ell^2\left(\frac{m_s \, v \, \cos\alpha}{(2\,m + m_s)\ell}\right)^2 = \frac{1}{2}\frac{m_s^2 \, v^2 \cos^2\alpha}{2\,m + m_s}.$$

The difference in potential energy between the case of the stone at the bottom and at the top of the windmill is $\Delta V = 2m_s g\ell$, so, using conservation of mechanical energy, the condition for the stone to reach the highest point is

$$K_a \geqslant 2m_s g\ell.$$

Chapter 19

Selected solutions to Chapter 9: Accelerating frames of reference

Solution: Exercise 9.3. We know that the Earth's orbit is a stable uniform circular motion, hence the forces acting on the Earth must be in balance. The only real force acting on the Earth is the gravitational force F_G, pulling it radially inwards towards the Sun.

$$F_G = G\frac{mM}{r^2}$$

where $G = 6.67 \times 10^{-11}\, m^3\, kg^{-1}\, s^{-2}$ *is the gravitational constant,* M *and* m *the mass of the Sun and Earth, respectively, and* r *is the radius of the Earth's orbit around the Sun.*

For the Earth to continue in the uniform motion we must introduce the fictitious centrifugal force F_C in its frame of reference, to balance the gravitational force F_G. This will take the form:

$$F_C = m\omega^2 r.$$

The two forces must be balanced, hence

$$G\frac{mM}{r^2} = m\omega^2 r.$$

Rearranging for r we have that:

$$r^3 = \frac{MG}{\omega^2}.$$

We now need to determine the angular frequency ω of the Earth's rotation around the Sun. We know that $\omega = 2\pi f$ and that $f = 1/T$, where T is the period of rotation

doi:10.1088/978-0-7503-2690-2ch19

in seconds. We can find T knowing that the duration in days of the Earth's orbit is approximately 365.

$$T = 365 \times 24 \times 60 \times 60 = 31\,536\,000 \text{ s}.$$

Therefore,

$$\omega = \frac{2\pi}{31\,536\,000}.$$

This gives us an equation for the radius of the Earth's orbit given by:

$$r = \sqrt[3]{\frac{MGT^2}{4\pi^2}}.$$

To check that this equation is correct, it is always good practice to do a ***dimension check*** *and be sure that the result is equal to a length. This is done as a quick example below, by considering the dimensions of the components of the equation above.*

$$r = \sqrt[3]{\frac{\text{kg m}^3\,\text{kg}^{-1}\,\text{s}^{-2}\,\text{s}^2}{1}}$$

$$\therefore\; r = \sqrt[3]{\text{m}^3} = \text{m}.$$

Now that we have checked that the equation is correct we can continue by substituting in the numerical values we obtained.

$$r = \sqrt[3]{\frac{1.99 \times 10^{30} \times 6.67 \times 10^{-11} \times 31\,536\,000^2}{4\pi^2}}.$$

This gives us a radius of

$$r = 1.495\,35 \times 10^{11}\,\text{m} = 1.4954 \times 10^{8}\,\text{km}.$$

Comparing our estimate to the real 149.60×10^6 km of the Earth's average orbit radius around the Sun, we see that are calculations come very close to the accepted astronomical measurement.

Solution: Exercise 9.5. The Coriolis force is given by:

$$\boldsymbol{F_C} = -\,2m\boldsymbol{\omega_E} \times \boldsymbol{v}$$
$$F_C = -\,2m\omega_E v \sin(90 - 41.84)$$

where $\boldsymbol{\omega_E} = \frac{2\pi}{T} \equiv \frac{2\pi}{60 \times 60 \times 24}$ is the angular velocity of the Earth (counterclockwise direction), while the angle is given by the angle between the angular velocity and the velocity of the electron. Since we already have ω_E, the angle between $\boldsymbol{\omega_E}$ and $\boldsymbol{v}$ and the mass of the electron, the only missing piece of information to solve the problem is the magnitude of the velocity of the electron.

To find the magnitude of the velocity we need to inspect the Lorentz force generated by the magnetic field, acting on the electron. The full equation for the Lorentz force reads:

$$\boldsymbol{F_L} = q\boldsymbol{v} \times \boldsymbol{B}.$$

Since the orbit of the electron lies on the r, θ plane we know $\boldsymbol{v}$ must always lie on this plane as well. $\boldsymbol{B}$ instead is along the $\hat{\boldsymbol{k}}$ direction, i.e. is perpendicular to the r, θ plane. The Lorentz force must therefore always point towards the centre of the orbit, its magnitude equation then reads:

$$F_L = qvB$$

Note that since the electron is in a circular orbital motion, the inwards Lorentz force must be equal in magnitude to the outwards centrifugal force experienced by the electron.

$$\begin{aligned} F_L &= F_{\text{centrifugal}} \\ qvB &= mv^2/R_0 \\ \Rightarrow v &= \frac{R_0 qB}{m}. \end{aligned}$$

We can now obtain an expression for the magnitude of the Coriolis force:

$$\begin{aligned} F_C &= -\,2m\omega_E v \sin{(90 - 41.84)} \\ &= -\,2m\frac{2\pi}{60 \times 60 \times 24}\frac{R_0 qB}{m}\sin{(48.16)}. \end{aligned}$$

Solution: Exercise 9.7. The first step is to draw an image so we have a clear picture of how the stone is being dropped.

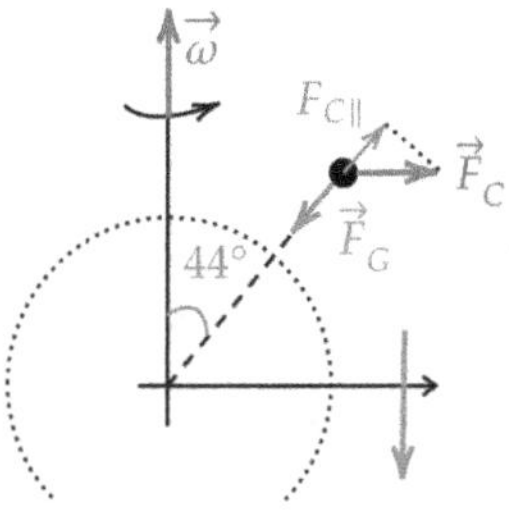

A note on latitude conventions: *In physics when discussing the latitude angle, it is always measured from the North pole, as seen in the figure above.*

The first step is to determine the magnitude and direction of the Coriolis force F_C acting on the coin. We know the Coriolis force is given by:

$$\boldsymbol{F}_{\text{Coriolis}} = -2m\boldsymbol{\omega} \times \boldsymbol{v}.$$

We must now use the *right-hand rule*, to determine the direction of the Coriolis force. The normal cross product between the two vectors seen in the figure above can be seen in the left part of the figure below. However, if the equation for the Coriolis force is preceded by a *minus sign*, this will result in an **inversion of the direction**. Hence the real direction of the Coriolis force is as seen in the right part of the figure below, i.e. pushing the falling coin east.

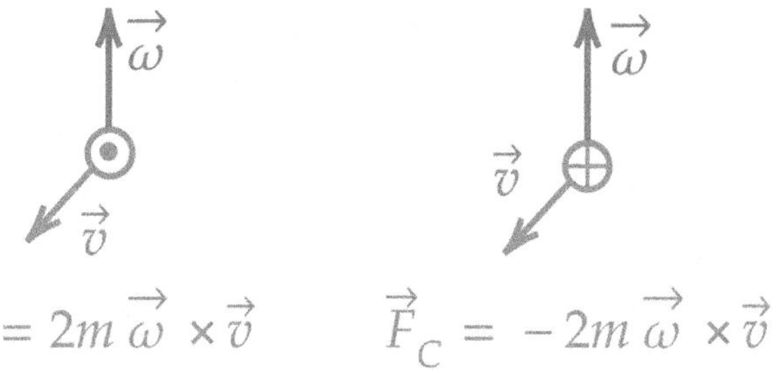

It is a very common mistake to forget the minus sign causing the inversion of direction of the Coriolis force, so make sure you always take this into account when solving exercises of this kind.

We can now find the magnitude of the Coriolis force by applying the cross product definition,

$$F_{\text{Coriolis}} = -2m\omega v \sin(44).$$

The exercise requires us to determine the magnitude when the coin is about to touch the ground. This means we must **calculate the velocity** v the coin will have after having travelled 1.7 m in uniform accelerated motion under the effects of the gravitational force F_G. However, we must not forget that the falling motion of the coin will also be influenced by the centrifugal force given by the Earth's rotation. The centrifugal force F_{Cent} points perpendicularly outwards from the axis of rotation, with modulus of:

$$F_{\text{Cent}} = m\omega^2 r.$$

To find the total force causing the accelerated motion of the coin we must find the component of the centrifugal force which is parallel to F_G and then subtract the two.

$$F_{\text{Cent}\|} = m\omega^2 r\cos(46)$$
$$\therefore\ F_{\|} = mg - m\omega^2 r\cos(46).$$

Using Newton's second law of motion we can find the acceleration of the body subject to such a force.

$$F_{\|} = ma = mg - m\omega^2 r \cos(46)$$
$$\therefore\ a = g - \omega^2 r \cos(46).$$

Once we have the acceleration we can use Suvat's equation to determine the final velocity once the coin will have travelled 1.7 m.

$$v^2 = 2as$$

Substituting in the known values:

$$v = \sqrt{2\,(g - \omega^2 r \cos(46)) \times 1.7}$$

where r is the radius of the Earth and ω the angular frequency of rotation of the Earth. Hence we can substitute all this to get the modulus of the Coriolis force when the coin has fallen 1.7 m.

$$F_{\text{Coriolis}} = -2m\omega \sin(44)\sqrt{(2g - \omega^2 r \cos(46)) \times 1.7}.$$

Solution: Exercise 9.9.

(i) In the rotating frame of reference we use the existence of the fictitious forces. Therefore we know the Coriolis force acting on the arrow is given by:

$$\boldsymbol{F}_{\text{Coriolis}} = -2m\boldsymbol{\omega} \times \boldsymbol{v}.$$

Since the two vectors are exactly perpendicular the modulus will simplify to:

$$F_{\text{Coriolis}} = -2m\omega v.$$

Where v is the velocity of the arrow and ω the angular frequency of rotation of the Earth, which can be expressed as $\omega = \frac{2\pi}{T} \equiv \frac{2\pi}{60 \times 60 \times 24}$.

We can substitute the above values in, and also apply Newton's second law to derive the acceleration of the arrow, caused by the Coriolis force.

$$F_{\text{Coriolis}} = -2m\frac{2\pi}{60 \times 60 \times 24} \times 10 = ma_C$$

Hence the constant acceleration will be given by:

$$a_C = \frac{4\pi}{8640}\ \text{m s}^{-2}$$

We can use the equation for uniform accelerated motion to find the displacement of the arrow caused by the Coriolis force, this will be of:

$$s = \frac{1}{2}a_c t^2 = 50 \times \frac{4\pi}{8640}$$

where the time t of the arrow's flight is determined by the horizontal velocity of 10 m s^{-1} *and the fact that the arrow must travel* 100 m, *hence it will have a* 10 s *flight.*

$$s \approx \frac{63}{863} = 0.072\ \text{m}.$$

Since the bear's head is 0.14 m wide, and the arrow was aimed at the center of the bear's head, the arrow will miss the bear by 0.002 m.

(ii) Now let's calculate the displacement as seen from an external observer. As mentioned above, the arrow will be in flight for 10 s, during which time the Earth will have rotated by a small angle $\Delta\theta$.

$$\Delta\theta = \omega t = \frac{2\pi}{86\,400} \times 10$$

where we have used the angular velocity calculated in part i of this exercise.

Whilst the arrow will travel straight down from the pole, the bear standing on the surface 100 m south will have moved with the rotating surface of the Earth. Hence by the time the arrow reaches the bear, it will be shifted a distance Δs from the flight direction of the arrow.
We can apply the identity $\mathrm{d}s = \mathrm{d}r\mathrm{d}\theta$, to obtain the shift of the bear on the surface.

$$\Delta s = \Delta\theta \times \Delta r$$
$$\therefore \Delta s = \frac{2\pi}{8640} \times 100 = 0.072 \text{ m}.$$

This, as we would expect agrees with the solution found for the previous approach.

Solution: Exercise 9.10.

(i) As stated in the question, a geostationary satellite has the same angular velocity as the Earth, i.e. $\omega = \frac{2\pi}{T}$ where T is the period of 24 h. To be in a geostationary orbit around the Earth the radial forces acting on the satellite must be balanced. The two forces involved are indeed gravity and the centrifugal force:

$$\boldsymbol{F}_{\text{centr}} = \boldsymbol{F}_G$$
$$M_S\,\omega^2 r_0\,\hat{\boldsymbol{r}} = G\frac{M_E M_S}{r_0^2}\hat{\boldsymbol{r}}$$

Where M_E and M_S are, respectively, the masses of the Earth and the satellite. From the last expression with a bit of rearranging for r_0 we obtain the answer for part (i)

$$r_0^3 = \frac{GM_E}{\omega^2}.$$

(ii) In the second part of the question the satellite breaks up in two parts of masses m and $2m$. In these cases when the system changes it is in generally useful to study the conservation of momentum and the conservation of energy of the two systems. The problem asks for the radial velocities of the two parts, notice that the radial velocity of the satellite before the explosion was zero, and hence its radial component of the momentum. From the conservation of momentum we therefore have

$$P_{\text{tot}} = 0 = mv_1 + 2mv_2$$
$$\Rightarrow v_1 = 2v_2.$$

For the conservation of energy it has to be taken into account that there is an increase E of the total kinetic energy of the system. This increase is due to the radial velocities of the two parts of the satellite.

$$E = \frac{1}{2}mv_1^2 + \frac{1}{2}2mv_1^2$$
$$= 3mv_2^2$$

where we subbed in the relation between velocities $v_1 = 2v_2$. We can now write the two velocities in terms of E

$$v_2 = \sqrt{E/(3m)} \tag{19.1}$$

$$v_1 = 2\sqrt{E/(3m)}. \tag{19.2}$$

(iii) We are given that the initial angular momentum is $L = m\omega r_o^2$ and the centrifugal force is $F_C = \frac{L^2}{mr^3}$. It is then suggested to start with a first order Taylor expansion of the centrifugal force as well as the gravitational force. We will then get an expression for the resulting force and compare this to the general equation of a harmonic oscillator, $m\ddot{x} = -kx$
Remember the Taylor expansion of $f(r)$ around r_0 is

$$f(r) \simeq \sum_{n=0}^{\infty} \frac{1}{n} \frac{\mathrm{d}^n f}{\mathrm{d}r^n}\bigg|_{r=r_0} (r - r_o)^n. \tag{19.3}$$

First order expansion of F_C with respect to r_0 is

$$F_C \simeq \frac{L^2}{mr_0^3} - 3\frac{L^2}{mr_0^4}(r - r_0).$$

We then expand the gravitational force as well to the first order

$$F_G \simeq \frac{GM_E m}{r_0^2} - 2\frac{GM_E m}{r_0^3}(r - r_0).$$

If we now calculate the resulting force, $\boldsymbol{F}_{\text{tot}} = \boldsymbol{F}_C + \boldsymbol{F}_G$

$$F_{\text{tot}} = \frac{L^2}{mr_0^3} - 3\frac{L^2}{mr_0^4}(r - r_0) - \left(\frac{GM_E m}{r_0^2} - 2\frac{GM_E m}{r_0^3}(r - r_0)\right).$$

Notice how we used here the magnitudes instead of their vector forms, hence the minus sign. Vectors $\boldsymbol{F}_C$ and $\boldsymbol{F}_G$ are in fact opposite directions. Now using the fact that to be in equilibrium, as we looked in part (i),

$$L^2/(mr_0^3) = GM_E m/r_0^2$$

the resulting force is then

$$F_{\text{tot}} = -\frac{3L^2}{mr_0^4}(r - r_0) + \frac{2GM_E m}{r_0^3}(r - r_0)$$

which if compared to the equation of a harmonic oscillator, in its general form

$$m\ddot{x} = -kx$$

it corresponds to a harmonic oscillator with elastic constant

$$k = \frac{3L^2}{mr_0^4} - \frac{2GM_E m}{r_0^3}$$

using again now the fact that at equilibrium $L^2/(mr_0^3) = GM_E m/r_0^2$

$$\begin{aligned} k &= \frac{3}{r_0}\frac{L^2}{mr_0^3} - \frac{2GM_E m}{r_0^3} \\ &= \frac{3}{r_0}\frac{GM_E m}{r_0^2} - \frac{2GM_E m}{r_0^3} \\ \Rightarrow k &= \frac{GM_E m}{r_0^3} = \omega^2 m. \end{aligned}$$

Solution: Exercise 9.11.

(i) The first part is easier to solve using the external point of view, i.e. using the fixed Cartesian reference frame shown in the figure. The ball is travelling at constant velocity v_B along the y axis, hence it will reach the edge of the carousel in time $t = R/v_B$. To calculate the angle α we look at the carriage velocity v_c and the distance d covered in time t.

$$\begin{aligned} v_c &= \omega R \\ d &= v_c t \\ &= \omega R^2/v_B. \end{aligned}$$

We can now find the angle α by diving the distance by the radius:

$$\begin{aligned} \alpha &= d/R \\ \Rightarrow \alpha &= \omega R/v_B. \end{aligned}$$

(ii) The Coriolis force is

$$\boldsymbol{F}_C = -2m\boldsymbol{\omega} \times \boldsymbol{v}$$

where in our solution $\boldsymbol{v} \equiv \boldsymbol{v}_B$. We know the carousel is rotating in an anticlockwise direction, hence using the right-hand rule we know the vector $\boldsymbol{\omega}$ must be along the positive $\hat{\boldsymbol{k}}$ direction with respect to the external frame of reference. Notice this direction is still upwards from the point of view of an internal observer rotating with the carousel. Since the velocity of the ball is parallel to the plane of the carousel, this must be perpendicular to the angular velocity of the attraction. Thus the Coriolis force magnitude can simply be written as

$$|\boldsymbol{F}_C| = 2m\omega|\boldsymbol{v}_B|.$$

The direction of the force instead, given by the right-hand rule, is always perpendicular to the velocity of the ball due to the cross product. The force is therefore perpendicular to the infinitesimal displacement of the ball and the work on the ball must be zero due to the dot product.

$$W = \boldsymbol{F} \cdot \mathrm{d}\boldsymbol{S} = 0.$$

(iii) A change in angular velocity induces, in the rotating non-inertial frame, an azimuthal force.

$$F_{\mathrm{azm}} = -m\frac{\mathrm{d}\boldsymbol{\omega}}{\mathrm{d}t} \times \boldsymbol{r}.$$

Notice that the infinitesimal term $\frac{\mathrm{d}\omega}{\mathrm{d}t}$ will be less than zero since the carousel is stopping. Therefore, knowing the direction of $\boldsymbol{\omega}$ which is perpendicular to the plane upwards and the direction of $\boldsymbol{r}$ is towards the centre of the carousel, the azimuthal force must point along the $\hat{\boldsymbol{x}}'$ direction in our rotating frame of reference (see figure below).

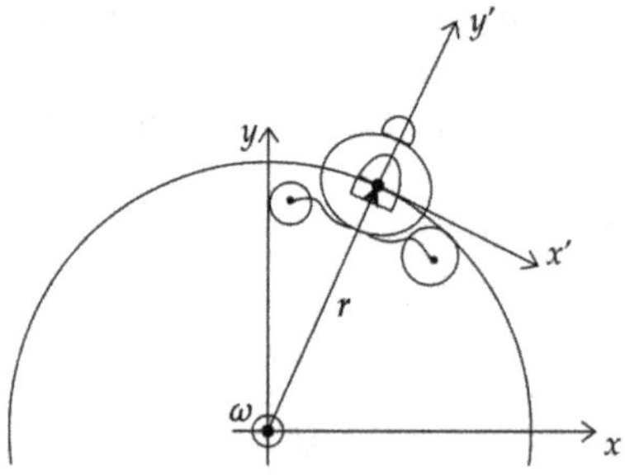

IOP Publishing

Classical Mechanics
A professor–student collaboration
Mario Campanelli

Chapter 20

Solutions to Chapter 10: Fluid mechanics

Solution: Exercise 10.1.

(i) Recall $dm = \rho dA_1 dv_1 dt$.

(ii) BOOKWORK

Solution: Exercise 10.2.

(i) BOOKWORK

(ii) BOOKWORK

(iii) Start by looking at the mass of fluid

$$dm = \rho dA_1 v dt,$$

where v is the volume in the hosepipe.

This must be equal to the volume coming out

$$\rho dA_1 v dt = \rho dA_2 v_0 dt,$$

where v is the volume coming out of the hosepipe.

The speed inside the hosepipe is therefore:

$$v = v_0 \frac{dA_2}{dA_1} = v_0 \frac{A_2}{A_1}.$$

So the excess pressure is given by

$$p + \frac{1}{2}\rho v^2 + \rho gh = p_0 + \frac{1}{2}\rho v_0^2 + \rho gh,$$

the ρgh terms cancel out giving:

$$p - p_0 = +\frac{1}{2}\rho v_0^2 - \frac{1}{2}\rho v^2.$$

Subbing in the relation found earlier we get:

$$p - p_0 = +\frac{1}{2}\rho v_0^2 - \frac{1}{2}\rho v_0^2\left(\frac{A_2}{A_1}\right)^2$$
$$= +\frac{1}{2}\rho v_0^2\left[1 - \left(\frac{A_2}{A_1}\right)^2\right].$$

(iv) BOOKWORK

Solution: Exercise 10.3.

(i) Apply Bernoulli's equation together with the continuity equation to solve this problem.
Continuity states that:

$$v_0 \mathrm{d}A_0 = v\mathrm{d}A.$$

Bernoulli requires that:

$$p_1 + \frac{1}{2}\rho v_0^2 + \rho g R = p_2 + \frac{1}{2}\rho v^2 + \rho g r.$$

By rearranging, we get

$$\Delta p + \rho g(R - r) = \frac{1}{2}\rho\left(v^2 - v_0^2\right),$$

by substituting in $v = v_0\frac{A_0}{A}$, we get:

$$\Delta p + \rho g(R - r) = \frac{1}{2}\rho v_0^2\left(\frac{A_0^2}{A^2} - 1\right).$$

Hence, the speed of flow v_0 is

$$v_0 = \left[\frac{\Delta p + \rho g(R - r)}{\frac{1}{2}\rho\left(\frac{A_0^2}{A^2} - 1\right)}\right]^{\frac{1}{2}}.$$

(ii) BOOKWORK

Solution: Exercise 10.4.

(i) This question is simply a disguised application of Newton's first law. In fact, assuming that the spinning ball is not acted upon by external forces, and its weight balances the jet, then our scenario is possible!

(ii) It might be helpful to note/confirm that we are indeed in the case of a **rough** ball. If the ball were to be smooth, it would not drag air round with it. One may want to refer to the example which precedes this exercise and see that at one point the 'roughness' of the ball drags the air around it and increases the speed at some other point. So we'll have two different velocities at those points, which will imply inversely proportional respective pressures and will produce a resultant force from the point at higher pressure to the one at lower pressure. Therefore, the trajectory will be deflected.

Appendix A

Index notation

> Remark A.1. *This chapter goes beyond the scope of this introductory course in classical mechanics and is presented here for your own interest and pleasure.*

A.0 Introduction

The idea behind index notation is to simplify expressions and calculations by making them very short and succinct. It may be a very conceptually challenging topic for some students, but we provide here some information for the more curious ones.

In standard vector notation we write the vector $\boldsymbol{a}$ as

$$\begin{aligned}\boldsymbol{a} &= a_1\hat{\boldsymbol{i}} + a_2\hat{\boldsymbol{j}} + a_3\hat{\boldsymbol{k}} \\ &= a_1\hat{\boldsymbol{e}}_1 + a_2\hat{\boldsymbol{e}}_2 + a_3\hat{\boldsymbol{e}}_3 \\ &= \sum_{i=1}^{3} a_i\hat{\boldsymbol{e}}_i.\end{aligned} \tag{A.1}$$

Notice that in the expression within the summation, the index i is repeated. Repeated indices are always contained within summations, or phrased differently *a repeated index implies a summation*. Therefore, the summation symbol is typically dropped, and we can express $\boldsymbol{a}$ as

$$\boldsymbol{a} = a_i\hat{\boldsymbol{e}}_i. \tag{A.2}$$

This repeated index notation is known as Einstein's convention. Any repeated index is called a *dummy index*. Since a repeated index implies a summation over all possible values of the index, it doesn't matter what index we choose, i.e.

$$
\begin{aligned}
\boldsymbol{a} &= \sum_{i=1}^{3} a_i \hat{\boldsymbol{e}}_i = a_i \hat{\boldsymbol{e}}_i \\
&= \sum_{j=1}^{3} a_j \hat{\boldsymbol{e}}_j = a_j \hat{\boldsymbol{e}}_j \\
&= \sum_{k=1}^{3} a_k \hat{\boldsymbol{e}}_k = a_k \hat{\boldsymbol{e}}_k.
\end{aligned}
$$

A.1 Kronecker delta

Now, let's introduce the **Kronecker delta**, δ_{ij}, which has the following properties

$$
\delta_{ij} = \begin{cases} 0 & \text{if } i \neq j \\ 1 & \text{if } i = j \end{cases}. \tag{A.3}
$$

Clearly,

$$
\hat{\boldsymbol{e}}_i \cdot \hat{\boldsymbol{e}}_j = \delta_{ij} \quad i, j = 1, 2, 3.
$$

A.2 Scalar product

Now, it's easy to show that

$$
\boldsymbol{a} \cdot \boldsymbol{b} = (a_i \hat{\boldsymbol{e}}_i) \cdot (b_j \hat{\boldsymbol{e}}_j) = a_i b_j \hat{\boldsymbol{e}}_i \cdot \hat{\boldsymbol{e}}_i = a_i b_j \delta_{ij} = a_i b_i. \tag{A.4}
$$

Remark A.2. Minkowski space–time

In the first chapter of the course, we have seen how the geometry of a 3D space is encoded in the distance (or length) function

$$
|\boldsymbol{a}|^2 = a_1^2 + a_2^2 + a_3^2,
$$

which we employed to define the scalar product

$$
\boldsymbol{a} \cdot \boldsymbol{b} = a_1 b_1 + a_2 b_2 + a_3 b_3.
$$

In Newtonian physics, different observers might represent points in space differently according to different coordinate systems, but distances between points and angles between line segments will be independent of such a choice of reference frame. Einstein showed that observers moving relatively to each other will not generally agree on distances between points or even time intervals between events. However, they will agree on the speed of light in vacuum, regardless of the speed of its source. Hermann

Minkowski, who was one of Einstein's professors of mathematics, realised that observers would agree on the value of

$$c^2t^2 - x^2 - y^2 - z^2, \tag{A.5}$$

where c is the speed of light, rather than the separate values of t and $(x^2 + y^2 + z^2)$. The quantity (A.5) can be perceived as a generalisation of the square of the magnitude of a vector, but it may have any sign. This *distance* gives rise to a type of scalar product encoding the geometry of what is now know as Minkowski space-time.

A.3 Levi-Civita symbol

$$\varepsilon_{ijk} = \begin{cases} -1 & \text{if } (i, j, k) \text{ is an odd permutation of } (1, 2, 3) \\ 0 & \text{if two or more subscripts are the same} \\ 1 & \text{if } (i, j, k) \text{ is an even permutation of } (1, 2, 3) \end{cases}. \tag{A.6}$$

This means

$$\begin{aligned} \varepsilon_{123} = \varepsilon_{312} = \varepsilon_{231} &= 1 \\ \varepsilon_{213} = \varepsilon_{321} = \varepsilon_{132} &= -1 \\ \varepsilon_{111} = \varepsilon_{122} = \varepsilon_{133} &= \ldots = 0. \end{aligned}$$

A.4 Vector product

It can be shown that

$$\hat{\boldsymbol{e}}_i \times \hat{\boldsymbol{e}}_j = \varepsilon_{ijk}\hat{\boldsymbol{e}}_k \tag{A.7}$$

and

$$\boldsymbol{a} \times \boldsymbol{b} = \varepsilon_{ijk} a_i b_j \hat{\boldsymbol{e}}_k. \tag{A.8}$$

A.5 Important identities

Three important identities which are often employed in calculations are:

$$\varepsilon_{ijk}\varepsilon_{imn} = \delta_{jm}\delta_{kn} - \delta_{jn}\delta_{km}, \tag{A.9}$$

$$\varepsilon_{imn}\varepsilon_{jmn} = 2\delta_{ij}, \tag{A.10}$$

$$\varepsilon_{ijk}\varepsilon_{ijk} = 6. \tag{A.11}$$

A.6 Triple vector product

Let $\boldsymbol{a} = a_i \boldsymbol{e}_i$, $\boldsymbol{b} = b_j \boldsymbol{e}_j$, and $\boldsymbol{c} = c_k \boldsymbol{e}_k$. From equation (A.7), one may write

$$\boldsymbol{a} \times \boldsymbol{b} = d_k \boldsymbol{e}_k,$$

where $d = \varepsilon_{ijk} a_i b_j$. Thus,

$$(\boldsymbol{a} \times \boldsymbol{b}) \times \boldsymbol{c} = \varepsilon_{\ell mn} d_\ell c_m \boldsymbol{e}_n = \varepsilon_{\ell mn} \varepsilon_{ij\ell} a_i b_j c_m \boldsymbol{e}_n = \varepsilon_{\ell mn} \varepsilon_{\ell ij} a_i b_j c_m \boldsymbol{e}_n.$$

Now, using identity (A.9), one has

$$\begin{aligned}(\boldsymbol{a} \times \boldsymbol{b}) \times \boldsymbol{c} &= (\delta_{mi}\delta_{nj} - \delta_{mj}\delta_{ni}) a_i b_j c_m \boldsymbol{e}_n = \delta_{mi}\delta_{nj} a_i b_j c_m \boldsymbol{e}_n - \delta_{mj}\delta_{ni} a_i b_j c_m \boldsymbol{e}_n \\ &= a_m c_m b_n \boldsymbol{e}_n - b_m c_m a_n \boldsymbol{e}_n = (\boldsymbol{a} \cdot \boldsymbol{c})\boldsymbol{b} - (\boldsymbol{b} \cdot \boldsymbol{c})\boldsymbol{a}.\end{aligned}$$

A.7 The scalar triple product

The expression $(\boldsymbol{a} \times \boldsymbol{b}) \cdot \boldsymbol{c}$ is called the *scalar triple product* and is denoted by $[\boldsymbol{a}, \boldsymbol{b}, \boldsymbol{c}]$. Using index notation, one has

$$[\boldsymbol{a}, \boldsymbol{b}, \boldsymbol{c}] = (\boldsymbol{a} \times \boldsymbol{b}) \cdot \boldsymbol{c} = \varepsilon_{ijk} a_i b_j c_k.$$

A.8 Examples

Let $\boldsymbol{a} = a_i \hat{e}_i$ and $\boldsymbol{b} = b_j \hat{e}_j$. We want to try and express each of the following in terms of $\boldsymbol{a} \cdot \boldsymbol{b}$ or $\boldsymbol{a} \times \boldsymbol{b}$.

Example A.1. $\delta_{ij}\delta_{jk}\delta_{ki}\delta_{mn} a_m b_n$
First consider the *action* of δ_{mn}. We know that if $m = n$ then $\delta_{mn} = 1$, otherwise it is zero. So the index of b, i.e. n must be equal to m, that is

$$\delta_{ij}\delta_{jk}\delta_{ki}\delta_{mn} a_m b_n = \delta_{ij}\delta_{jk}\delta_{ki} a_m b_m.$$

We now look at δ_{ki}, clearly we must have $k = i$ for this to make sense, so we replace the other k present in the expression with an i,

$$\delta_{ij}\delta_{jk}\delta_{ki} a_m b_m = \delta_{ij}\delta_{ji} a_m b_m.$$

We do the same as before with δ_{ji}, i.e. j must be equal to i so we are left with

$$\delta_{ij}\delta_{ji} a_m b_m = \delta_{ii} a_m b_m.$$

You have proved in lectures (or maybe we proved it together) that

$$\delta_{ii} = 3.$$

Hence,

$$\delta_{ii}a_m b_m = 3a_m b_m = 3\boldsymbol{a} \cdot \boldsymbol{b},$$

since m is a dummy variable and we saw before that $a_i b_i$ is a scalar product.

Example A.2. $\delta_{ij}\varepsilon_{ijk}\varepsilon_{k\ell m}a_\ell b_m$
Now, consider δ_{ij}, it implies that i must be equal to j, otherwise it is zero. So

$$\delta_{ij}\varepsilon_{ijk}\varepsilon_{k\ell m}a_\ell b_m = \varepsilon_{jjk}\varepsilon_{k\ell m}a_\ell b_m.$$

We notice that we have ε_{jjk}. When we defined the Levi-Civita symbol, we defined it to be zero if two or more subscripts are the same. Hence, everything is zero because $\varepsilon_{jjk} = 0$.

Example A.3. $\varepsilon_{ijk}\delta_{i\ell}a_\ell b_k \hat{\mathbf{e}}_j$
First let's look at $\delta_{i\ell}$. It implies $\ell = i$, so

$$\varepsilon_{ijk}\delta_{i\ell}a_\ell b_k \hat{\boldsymbol{e}}_j = \varepsilon_{ijk}a_i b_k \hat{\boldsymbol{e}}_j.$$

This expression resembles the vector product, however, in the vector product we have the following order $\boldsymbol{a} \times \boldsymbol{b} = \varepsilon_{ijk}a_i b_j \hat{\boldsymbol{e}}_k$. So we want our ε_{ijk} to be in the form where the last subscript represents the unit vector and the second represents the second vector. In order to accomplish that, we must go from ε_{ijk} to ε_{ikj} but in doing so we are permuting in an odd fashion, hence, we will need a minus sign, i.e.

$$\varepsilon_{ijk} = -\varepsilon_{ikj}.$$

Therefore,

$$\varepsilon_{ijk}a_i b_k \hat{\boldsymbol{e}}_j = -\varepsilon_{ikj}a_i b_k \hat{\boldsymbol{e}}_j = -\boldsymbol{a} \times \boldsymbol{b}.$$

Alternatively, if instead we go from ε_{ijk} to ε_{kij}, i.e. an even permutation which doesn't affect the sign, we then want to have

$$\varepsilon_{kij}b_k a_i \hat{\boldsymbol{e}}_j = \boldsymbol{b} \times \boldsymbol{a} = -\boldsymbol{a} \times \boldsymbol{b}.$$

It may help to think of the subscripts of ε_{ijk} as the order in which the terms with the relative subscript appear …

Appendix B

Solving differential equations

> **Remark B.1.** *This chapter goes beyond the scope of this introductory course in classical mechanics and is presented here for your own interest and pleasure.*

Mathematically, Newton's second law can be expressed as a **differential equation**: assuming the force on a body is determined by the position (and perhaps velocity) of that body, the motion of the particle (or its velocity) can be inferred. It is a **second order** differential equation for the position vector, and therefore its solution will contain two arbitrary constants, that could be fixed knowing for instance the initial position and initial velocity of the particle.

In general, to solve the motion of a particle we need two ingredients:

- **Newton's second law:** $F = ma$, where F is the net force acting on the particle;
- **a specific force law:** the dependence of the force F on the variables x, v and t.

Hence, in general, we have to solve a second order ordinary differential equation of the form

$$m\ddot{x} = F(t, x, v)$$

whose general solution $x(t)$ can be quite complex, depending on the nature of F and its dependences. The solution will depend on two integration constants, which can be determined by applying the **initial conditions**, for instance the particle's initial position $x_0 = x(0)$ and its initial velocity $v_0 = v(0)$. Since as we said the general solution can be quite complex, we'll consider the following specific cases, all of them referring to particles with constant mass moving in one dimension (even if the extension to multiple dimensions is trivial):

doi:10.1088/978-0-7503-2690-2ch22

(i) $F = 0$, free particle, will move at constant speed.

(ii) $F =$ **constant**.

If $F =$ constant, then the acceleration is a constant: $a(t) = a = F/m$. We then, integrate once to get:

$$v(t) = v_0 + \int_0^t a(\tilde{t})\,\mathrm{d}\tilde{t} = at + v_0.$$

We then integrate once again to get

$$x(t) = x_0 + \int_0^t v(\tilde{t})\,\mathrm{d}t = \frac{1}{2}at^2 + v_0 t + x_0.$$

In this simple case, the two constants of integration are the initial position and velocity x_0 and v_0. No further algebra is needed to express the solution in terms of the initial conditions (usually things are not so simple!). For example, falling body in absence of air resistance, friction between solid surfaces, or motion in uniform electric field.

(iii) ***F* is a function of *t* only, i.e.** $F \equiv F(t)$.

From Newton's second law, we have

$$F \equiv F(t) = m\frac{\mathrm{d}v}{\mathrm{d}t}.$$

Separating variables and integrating both sides:

$$\int_{t_0}^{t} F(\tilde{t})\,\mathrm{d}\tilde{t} = m\int_{v(t_0)}^{v(t)} \mathrm{d}\tilde{v}.$$

We have put tildes on the integration variables so that we dont confuse them with the limits of integration.

Since the integral of $\mathrm{d}\tilde{v}$ is just $\tilde{v}$, then the previous equation yields v as a function of t, i.e. $v(t)$. Separating variables in $v(t) = \frac{\mathrm{d}x}{\mathrm{d}t}$ and integrating yields

$$\int_{x(t_0)}^{x(t)} \mathrm{d}\tilde{x} = \int_{t_0}^{t} v(\tilde{t})\,\mathrm{d}\tilde{t}.$$

This yields x as a function of t, i.e. $x(t)$. It is a seemingly *fiddly* way of integrating twice, but this technique is more useful in the following.

(iv) ***F* is a function of *x* only,** i.e. $F \equiv F(x)$. This is the most important case because many fundamental forces of physics are position-dependent. We can use this example, for a generic formulation of F, to introduce the key concept of **energy**, that will be further developed later.

We introduce the indefinite integral of $F(x)$, namely the potential energy:

$$V(x) = -\int F(x)\,\mathrm{d}x,$$

where the minus sign is inserted for convenience. Any value can be chosen for the constant of integration as long as we have

$$F(x) = -\frac{\mathrm{d}V}{\mathrm{d}x}.$$

Now, Newton's second law is

$$m\frac{\mathrm{d}^2x}{\mathrm{d}t^2} = F(x).$$

Multiplying both sides by $\mathrm{d}x/\mathrm{d}t$ (formulating the operation this way is not really mathematically correct, but it works), we get

$$m\frac{\mathrm{d}^2x}{\mathrm{d}t^2}\frac{\mathrm{d}x}{\mathrm{d}t} = F(x)\frac{\mathrm{d}x}{\mathrm{d}t},$$

and notice (by the chain rule) that both sides are the derivative w.r.t. time (i.e. $\mathrm{d}/\mathrm{d}t$) of something, namely

$$m\frac{\mathrm{d}^2x}{\mathrm{d}t^2}\frac{\mathrm{d}x}{\mathrm{d}t} = \frac{\mathrm{d}}{\mathrm{d}t}\left[\frac{1}{2}m\left(\frac{\mathrm{d}x}{\mathrm{d}t}\right)^2\right]$$

where the quantity K. E. $=\frac{1}{2}mv^2$ is the kinetic energy, and

$$F(x)\frac{\mathrm{d}x}{\mathrm{d}t} = -\frac{\mathrm{d}V}{\mathrm{d}x}\frac{\mathrm{d}x}{\mathrm{d}t} = \frac{\mathrm{d}}{\mathrm{d}t}[-V(x)].$$

Bringing everything to the left-hand side, we see that the Newtonian differential equation can therefore be written as

$$\frac{\mathrm{d}}{\mathrm{d}t}\left[\frac{1}{2}m\left(\frac{\mathrm{d}x}{\mathrm{d}t}\right)^2 + V(x)\right] = 0,$$

And this equation has an easy first integral, namely

$$\frac{1}{2}m\left(\frac{\mathrm{d}x}{\mathrm{d}t}\right)^2 + V(x) = \text{constant} \equiv E, \quad \text{total energy.}$$

which is the law of conservation of energy: the sum of kinetic and potential energy is a constant, regardless of the expression of v (therefore of F).

We can then solve this for $\mathrm{d}x/\mathrm{d}t$:

$$\frac{\mathrm{d}x}{\mathrm{d}t} = \pm\sqrt{\frac{2[E - V(x)]}{m}}.$$

This is now a separable first order differential equation for the unknown function $x(t)$; it can be solved by writing

$$dt = dx\sqrt{\frac{m}{2[E - V(x)]}}$$

and integrating both sides. This process gives you t as a function of x; you have to algebraically invert this to get the desired x as a function of t. Note that there will appear a constant of integration. By evaluating both sides of the equation at $t = 0$, you can solve for this constant of integration in terms of the initial position $x_0 = x(0)$, and then re-express everything in terms of x_0.

The law of conservation of energy is one of the most important concepts in all of physics, as we shall see.

Note that the kinetic energy K. E. $= \frac{1}{2}mv^2$ is always non-negative. Therefore, any motion with total energy E is restricted to the region of space $\{x: V(x) \leqslant E\}$. In physics, *conservation of X*, or *X is conserved*, means that X is constant in time, i.e. $dX/dt = 0$.

(v) ***F* is a function of v only,** i.e. $F \equiv F(v)$.

Write Newton's second law as

$$F \equiv F(v) = m\frac{dv}{dt}.$$

This is a separable first order differential equation for the unknown function $v(t)$.

It can be solved by writing

$$dt = \frac{m}{F(v)}dv$$

and integrating both sides

$$m\int_{v(t_0)}^{v(t)} \frac{d\tilde{v}}{F(\tilde{v})} = \int_{t_0}^{t} d\tilde{t}.$$

This process gives you t as a function of v; you have to algebraically invert this to get the desired v as a function of t. (This inversion is not always doable in terms of elementary functions.) Note that there will appear a constant of integration. By evaluating both sides of the equation at $t = 0$, you can solve for this constant of integration in terms of the initial velocity $v_0 = v(0)$, and then re-express everything in terms of v_0.

Finally, integrate once more to obtain $x(t)$; the second initial condition $x_0 = x(0)$ will come in as a second constant of integration.

IOP Publishing

Classical Mechanics
A professor-student collaboration
Mario Campanelli

Appendix C

The Lagrangian method

Remark C.1. *This chapter goes beyond the material needed for this introductory course in classical mechanics and is presented here just for your interest.*

C.0 Introduction

Operating with vectors is a powerful tool but it can also be inconvenient when dealing with a large number of particles. There is an alternative way of solving classical mechanics problems that comes from conservation of energy. In chapter 3, we have seen that there exists a quantity E—the mechanical energy—which is conserved when energy is not dissipated, i.e. when only conservative forces are acting on the system. We write $E = K + V$, where K is the kinetic energy and V is the potential energy. If mechanical energy is conserved, we can also write:

$$\frac{\mathrm{d}E}{\mathrm{d}t} = \frac{\mathrm{d}K}{\mathrm{d}t} + \frac{\mathrm{d}V}{\mathrm{d}t} = 0. \tag{C.1}$$

Remark C.2. From now on, a dot on top of a variable will be used to represent its derivative with respect to time.

As always, let's first look at a one-dimensional problem with one particle.

C.1 One dimension

Since $K = \frac{1}{2}mv^2$ and $v = \dot{x}$, it follows that

$$\dot{K} = \frac{\mathrm{d}}{\mathrm{d}t}\left(\frac{1}{2}m\dot{x}^2\right) = \frac{1}{2}m \times 2\dot{x}\ddot{x} = m\dot{x}\ddot{x} \tag{C.2}$$

doi:10.1088/978-0-7503-2690-2ch23

where the chain rule has been used to differentiate $\dot{x}^2$ with respect to time.

Let's do the same for the potential energy. Since V depends only on position, we can use the chain rule to write

$$\dot{V} = \frac{\mathrm{d}V}{\mathrm{d}x}\dot{x}. \tag{C.3}$$

By substituting equations (C.2) and (C.3) into equation (C.1), we finally obtain

$$m\dot{x}\ddot{x} + \frac{\mathrm{d}V}{\mathrm{d}x}\dot{x} = \left(m\ddot{x} + \frac{\mathrm{d}V}{\mathrm{d}x}\right)\dot{x} = 0.$$

It follows that

$$m\ddot{x} = -\frac{\mathrm{d}V}{\mathrm{d}x}. \tag{C.4}$$

But we know that, for a conservative force, $F = -\frac{\mathrm{d}V}{\mathrm{d}x}$. Therefore, this is simply $F = m\ddot{x}$, i.e. Newton's second law in 1D.

Example C.1. Simple harmonic oscillator

Consider a spring system with no friction and no driving force. This is a simple harmonic oscillator and will be discussed in the next chapter. Given a displacement x from equilibrium and a spring constant k, the kinetic energy of the spring is $K = \frac{1}{2}m\dot{x}^2$ while its potential energy is $V = \frac{1}{2}kx^2$. We have that

$$\frac{\mathrm{d}V}{\mathrm{d}x} = kx.$$

Therefore, substituting in equation (C.4), we obtain

$$m\ddot{x} = -kx$$

which is just Hooke's law.

This approach hasn't saved us time here, but in many cases it can be much simpler to deal with energies (scalar quantities) rather than vectors (which can get very messy).

♦

We will now generalise these results to three dimensions.

C.2 Three dimensions

In the previous part of the chapter, the symbol x represented a specific coordinate of a specific particle. If there is more than one particle, then each one will have its own specific coordinates. For instance, $x_1, x_2, \ldots, x_N$ for N particles in 1D.

In 3D, each particle has three coordinates. For simplicity, let's consider the case of a single particle in three dimensions. The kinetic and the potential energy of the particle are now given, respectively, by $K = \frac{1}{2}m(\dot{x}^2 + \dot{y}^2 + \dot{z}^2)$ and $V \equiv V(x, y, z)$.

As when solving problems with Newton's second law, we want to obtain separated equations for each coordinate (and, if there were more than one particle, also for each one of them). We can write

$$F_x = -\frac{\partial V(x, y, z)}{\partial x}; \quad F_y = -\frac{\partial V(x, y, z)}{\partial y}; \quad F_z = -\frac{\partial V(x, y, z)}{\partial z}.$$

To obtain separate equations of motion for each dimension, we also need to consider the kinetic energy. We have shown in equation (C.2) that in 1D its time derivative is given by $\dot{K} = m\dot{x}\ddot{x}$. Since we want to write an equation of motion, we need to find a way to express $m\ddot{x}$. By just taking the partial derivative of $\dot{K}$ with respect to $\dot{x}$, we have

$$\frac{\partial}{\partial \dot{x}}\left(\frac{\mathrm{d}K}{\mathrm{d}t}\right) = \frac{\partial}{\partial \dot{x}}(m\dot{x}\ddot{x}) = m\ddot{x}.$$

We can write this in an equivalent way as

$$\frac{\mathrm{d}}{\mathrm{d}t}\left(\frac{\partial K}{\partial \dot{x}}\right) = m\ddot{x}. \tag{C.5}$$

Following an analogous reasoning, we can find also $m\dot{y}$ and $m\dot{z}$.

Therefore, for a general coordinate, we can combine equations (C.4) and (C.5) to obtain

$$\frac{\mathrm{d}}{\mathrm{d}t}\left(\frac{\partial K}{\partial \dot{x}_i}\right) = -\frac{\partial V}{\partial x_i}.$$

Let's try to find a single scalar function $\mathcal{L}$ that unifies K and V, so that we can rewrite this last equation in a better way. *This not only an aesthetic requirement, we want to have a differential equation of a single function.* We want to obtain

$$\frac{\mathrm{d}}{\mathrm{d}t}\left(\frac{\partial \mathcal{L}}{\partial \dot{x}_i}\right) = \frac{\partial \mathcal{L}}{\partial x_i}.$$

If you look at this equation a bit, you will convince yourself that $\mathcal{L} = K - V$ does the job. In fact, remember that $V \equiv V(x)$, i.e. the potential does not depend on the speed $\dot{x}_i$, therefore

$$\frac{\partial \mathcal{L}}{\partial \dot{x}_i} = \frac{\partial K}{\partial \dot{x}_i} - \frac{\partial V}{\partial \dot{x}_i} = \frac{\partial K}{\partial \dot{x}_i}.$$

It follows that

$$\frac{\mathrm{d}}{\mathrm{d}t}\left(\frac{\partial \mathcal{L}}{\partial \dot{x}_i}\right) = \frac{\mathrm{d}}{\mathrm{d}t}\left(\frac{\partial K}{\partial \dot{x}_i}\right).$$

Similarly, since the kinetic energy does not depend on the position x_i,

$$\frac{\partial \mathcal{L}}{\partial x_i} = \frac{\partial K}{\partial x_i} - \frac{\partial V}{\partial x_i} = -\frac{\partial V}{\partial x_i}.$$

Remark C.3. Note that these are partial derivatives so you derive with respect to the variable shown as if it were independent from the others and as if it didn't depend on the time t.

The quantity $\mathcal{L}$ is called the Lagrangian.

Definition C.2.1: Lagrangian
The Lagrangian is defined as

$$\mathcal{L} = K - V. \tag{C.6}$$

Since it is a difference of energies, it will also have units of energy.

Definition C.2.2: Euler–Lagrange equation
The differential equation

$$\frac{\mathrm{d}}{\mathrm{d}t}\left(\frac{\partial \mathcal{L}}{\partial \dot{x}_i}\right) = \frac{\partial \mathcal{L}}{\partial x_i} \tag{C.7}$$

is called the Euler–Lagrange equation.
This differential equation is coordinate independent, meaning that it has the same form in all coordinate systems.

The Lagrangian is a very powerful tool for solving systems of many particles. In most cases, it offers a simpler and more elegant alternative to the vectorial approach. For instance, it is used in the Standard Model of particle physics. All we know about particles is summed up in a huge Lagrangian[1].

C.3 The principle of stationary action

The Euler–Lagrange equation comes from one of the most important principles in physics, namely the principle of stationary action. Before discussing it, let's first define what an action is.

[1] There is a good explanatory article: [3]. Search it online, it's free to access.

Definition C.3.1: Action
An action S is defined as the quantity

$$S = \int_{t_1}^{t_2} \mathcal{L}(x, \dot{x}, t)\, \mathrm{d}t. \tag{C.8}$$

It has units of (energy) × (time).

Such an integral is also called a *functional.*

Theorem C.3.1. If the function $x_0(t)$, with fixed endpoints $x(t_1) = x_1$ and $x(t_2) = x_2$, yields a stationary value (that is, a local minimum, maximum, or saddle point) of S, then

$$\frac{\mathrm{d}}{\mathrm{d}t}\left(\frac{\partial \mathcal{L}}{\partial \dot{x}_0}\right) = \frac{\partial \mathcal{L}}{\partial x_0}.$$

Law C.3.1: The principle of stationary action
The path taken by a particle is the one that yields a stationary value of the action.

Remark C.4. Why is this principle so important?
This law is very simple, written in just one line, yet it has been proven that **the whole classical—and also quantum!—mechanics can be derived from it**. For instance, Einstein's field equation can be obtained from the action of Einstein–Hilbert.

Remark C.5. This is often called *the principle of least action* and stated saying that the path taken by a particle is the one that minimises the action. This is, however, incorrect, since we have seen that the action could have also a local maximum or a saddle point.

For a good explanation of the quantum mechanical reasons from which this principle originates, see [2 section 6.2, page 225] and [1 volume II, chapter 19].

References

[1] Feynman R 2010 *The Feynman Lectures on Physics* (New York: Basic Books)
[2] Morin D 2016 *Introduction to Classical Mechanics With Problems and Solutions* (Cambridge: Cambridge University Press)
[3] Woithe J, Wiener G J and der Veken F F V 2017 Let's have a coffee with the Standard Model of particle physics! *Phys. Educ.* **52**

Further reading

There are a lot of books covering the topics of this textbook in a similar or more advanced fashion. The following non-exhaustive list should be considered as a guidance to explore topics on different resources.

In order to refresh your memory or broaden your mathematical knowledge, we recommend referring to

[1] Riley K F, Hobson M P and Bence S J 2006 *Mathematical Methods for Physics and Engineering* 3rd edn (Cambridge: Cambridge University Press)

This book contains all the mathematical knowledge you need for this course (and much more).

An inspiring reading which is generally considered to be a fundamental element for the background knowledge of a physicist is

[2] Feynman R 2010 *The Feynman Lectures on Physics* (New York: Basic Books)

whose online version can be found at https://www.feynmanlectures.caltech.edu/

However, note that in general this book is more helpful to expand the understanding of topics previously learnt rather than to learn those concepts for the first time. The following books are more suitable to aid you in the first steps of the learning process.

Two slightly less advanced books that can help you grasp the theory better are

[3] Halliday D, Resnick R and Walker J 2015 *Fundamentals of Physics* (New York: Wiley)

[4] Serway A R and Jewett J W 2019 *Physics for Scientists and Engineers with Modern Physics* 10th edn (Boston, MA: Cengage)

They both contain a great number of examples and exercises, about 80–100 exercises per chapter. Although the exercises are not necessarily as difficult as those in the following textbooks in this list, the reader may find them very useful to reinforce their understanding of the covered topics.

The former official textbook for the course at UCL, which provides students with a variety of solved problems at a similar as well as more advanced level, is

[5] Morin D 2016 *Introduction to Classical Mechanics with Problems and Solutions* (Cambridge: Cambridge University Press)

This book is highly recommended if you are looking for a more rigorous introduction to the subject and/or harder problems. It also contains more advanced exercises which involve the use of differential equations and Lagrangian mechanics, introduced in chapter 6 of the same textbook.

For a more thorough introduction to Lagrangian mechanics it is suggested to refer to the first chapter of:

[6] Goldstein H, Poole C P and Safko J L 2002 *Classical Mechanics* (Reading, MA: Addison Wesley)

which contains a clear and satisfactory explanation of phase space, constraints and D'Alembert's Principle.

More advanced textbooks, at a level beyond the one of an introductory course, are

[7] Douglas G R 2006 *Classical Mechanics* (Cambridge: Cambridge University Press)

This book has a rigorous mathematical approach and it also covers Hamiltonian and Lagrangian mechanics. It is generally recommended in Departments of Mathematics for their Newtonian Mechanics courses.

[8] Kleppner D and Kolenkow R 2012 *An Introduction to Mechanics* (Cambridge: Cambridge University Press)

[9] Forshaw J and Smith G 2009 *Dynamics and Relativity* (New York: Wiley)

[10] Kibble T W B and Berkshire F H 2004 *Classical Mechanics* 5th edn (London: Imperial College Press)

[11] de Lange O L and Pierrus J 2010 *Solved Problems in Classical Mechanics* (Oxford: Oxford University Press)

This book contains a wide variety of problems at a slightly higher level than the one expected from an introductory course in Classical Mechanics. All the solutions are presented with additional comments and questions to broaden the understanding of each topic both from a physical and a mathematical point of view.